PRINCIPLES OF CELLULAR ENGINEERING: UNDERSTANDING THE BIOMOLECULAR INTERFACE

Whenever the body of an animal is subdivided to its ultimate parts, one always finally arrives at a limited number of chemical atoms . . . . One draws the conclusion in harmony with this observation, that all forms of activity arising in the animal body must be a result of the simple attractions and repulsions which would be observed in the coming together of those elementary objects.

Carl Friedrich Wilhelm Ludwig, 1858

■ CONTENTS

CONTRIBUTORS

Numbers in parentheses indicate the pages on which the authors' contributions begin.

Anne-Marie Benoliel (143) INSERM UMR_S 600, CNRS FRE2059, Laboratoire d'Immunologie, Hôpital de Sainte-Marguerite, Marseille, France

Craig D. Blanchette (195) Biophysics Graduate Group, University of California, Davis, California

David Boettiger (51) Department of Microbiology, University of Pennsylvania, Philadelphia, Pennsylvania

Pierre Bongrand (143) INSERM U600, CNRS FRE2059, Laboratoire d'Immunologie, Hôpital de Sainte-Marguerite, Marseille, France

Thomas R. Gaborski (123) Department of Biomedical Engineering, University of Rochester, Rochester, New York

Douglas J. Goetz (213) Department of Chemical Engineering, Ohio University, Athens, Ohio

Alan H. Goldstein (173) School of Engineering, New York State College of Ceramics at Alfred University, Alfred, New York

Daniel A. Hammer (3) Department of Bioengineering, University of Pennsylvania, Philadelphia, Pennsylvania

Brian P. Helmke (25) Department of Biomedical Engineering, University of Virginia, Charlottesville, Virginia

Donald E. Ingber (81) Departments of Pathology and Surgery, Children's Hospital and Harvard Medical School, Boston, Massachusetts

Michael R. King (255, 267) Department of Biomedical Engineering, University of Rochester, Rochester, New York

Dooyoung Lee (255) Department of Chemical Engineering, University of Rochester, Rochester, New York

Laurent Limozin (143) INSERM U600, CNRS FRE2059, Laboratoire d'Immunologie, Hôpital de Sainte-Marguerite, Marseille, France

Elena Lomakina (105) Department of Pharmacology and Physiology, University of Rochester, Rochester, New York

Marjorie L. Longo (195) Department of Chemical Engineering and Materials Science, University of California, Davis, California

James L. McGrath (123) Department of Biomedical Engineering, University of Rochester, Rochester, New York

Nipa A. Mody (267) Department of Chemical Engineering, University of Rochester, Rochester, New York

Rosalind E. Mott (25) Department of Biomedical Engineering, University of Virginia, Charlottesville, Virginia

Gerard B. Nash (235) Department of Physiology, Medical School, University of Birmingham, Birmingham, United Kingdom

Anne Pierres (143) INSERM U600, CNRS FRE2059, Laboratoire d'Immunologie, Hôpital de Sainte-Marguerite, Marseille, France

G. Ed Rainger (235) Center for Cardiovascular Sciences, Medical School, University of Birmingham, Birmingham, United Kingdom

Timothy V. Ratto (195) Biophysical and Interfacial Science Group, Chemistry and Materials Science, Lawrence Livermore National Laboratory, Livermore, California

Cynthia A. Reinhart-King (3) Department of Bioengineering, University of Pennsylvania, Philadelphia, Pennsylvania

Philippe Robert (143) INSERM U600, CNRS FRE2059, Laboratoire d'Immunologie, Hôpital de Sainte-Marguerite, Marseille, France

Dimitrije Stamenović (81) Department of Biomedical Engineering, Boston University, Boston, Massachusetts

Prithu Sundd (213) Department of Physics and Astronomy, Ohio University, Athens, Ohio

David F.J. Tees (213) Department of Physics and Astronomy, Ohio University, Athens, Ohio

Ning Wang (81) Physiology Program, Harvard School of Public Health, Boston, Massachusetts

Richard E. Waugh (105) Department of Biomedical Engineering, University of Rochester, Rochester, New York

■ PREFACE

Cellular engineering can be simply defined as the application of engineering tools and concepts to the study and manipulation of living cells. Cellular engineering, and bioengineering in general, is necessarily interdisciplinary, combining tools and ideas from the fields of biology, engineering, applied mathematics, physics, and chemistry. Research engineers and scientists converge on cellular engineering from several different directions: cell and molecular biologists looking for new ways to quantify and model their results, engineers from traditional disciplines (chemical, electrical, mechanical) with an interest in tackling biomedically important problems, and, of course, bioengineers specializing in cell and tissue engineering applications. To gain a mechanistic rather than "black box" understanding of cellular behavior, it is necessary to concern oneself with the phenomena that occur one length scale below that of single cells — the domain of atoms, molecules, and supramolecular assemblies. Further, to motivate the study of cellular behavior and connect it to clinical outcomes, it is also necessary to concern oneself with the next higher length scale, that of cell aggregates and tissues. Thus, cellular engineering is inherently multiscale, but focuses on the behavior of single cells. In this book we further refine the list of topics to a specific, but very important aspect of cellular engineering, the cell surface, where the fundamental biological processes of adhesion and signal transduction occur.

The term *biomolecular interface* has recently been used to describe the intersection of biological science and materials science [A. L. Plant, et al., *Langmuir* 19: 1449–1450 (2003)].

This subdiscipline of cellular engineering – involving the study of soft organic matter, the chemistry and physics of self-assembly, and the application of such knowledge toward new medical therapies and research – is by its nature highly interdisciplinary. Recent work on biological interfaces has revealed important new applications based on spatially controlled molecular layers, determination of the forces involved in biomolecular assemblies that provide the structural integrity and motility of cells, and new ways to assay for effects of protein conformation on outside-in and inside-out biochemical signaling. Advances in microfabrication technologies allow us to exploit the physical milieu of microscale flows to encapsulate and transport single cells within immiscible droplets, or carry out enzymatic reactions at pinned fluid interfaces. Polymer-supported lipid bilayer membranes preserve the lateral mobility of transmembrane proteins, enabling electrophoretic fractionation of protein mixtures or precise modeling of the biological effects of receptor clustering. Many of these studies have been carried out in basic science laboratories isolated from researchers and practitioners in engineering; here we work to bring together this knowledge and find new ways to integrate recent findings and approaches into the growing field of cellular engineering.

This book aims to bring together in one place the important results, concepts, and opinions of how cells interact with biomolecular surfaces. Of particular interest are the interaction and adhesion of cells with each other and with surfaces in complex mechanical environments such as the cardiovascular system. The behavior covered ranges from atomic to molecular, cellular, and multicellular length scales. A variety of experimental and theoretical approaches are used by the authors to study the dynamic interactions between cells: from advanced optical microscopy methods, to transfected cell systems, animal models, and computer simulations. Novel biomolecular surfaces are described that have been engineered either to control or to measure cell function. One recurring theme is that of inflammation and cellular immune response; remarkable progress has been made in recent years on this challenging problem, where the participating cells rapidly upregulate and downregulate their adhesion molecule expression in response to molecular cues, and demonstrate an exquisite balance between chemical adhesion and well-defined fluid shear forces. Those working in the fields of inflammation and vascular biology will be particularly interested in the chapters devoted to these topics, but these chapters also represent model problems demonstrating how engineering concepts and tools may be effectively applied to complex systems in biomedicine.

Principles of Cellular Engineering is organized into four parts. Part I is concerned with the firm adhesion of anchorage-dependent cells to substrates. Chapters are devoted to measurement of the forces that cells actively exert on their substrates (Chapter 1), mechanotransduction and the mechanical coupling between the luminal and basal cell surfaces (Chapter 2), and a spinning disk assay developed to quantify the strength of integrin–ligand adhesive bonds (Chapter 3). Chapter 4 presents evidence in support of the tensegrity, "prestress" model for the cell cytoskeleton. Part II focuses on three important aspects of the cell surface: the relationship between compression of the cell interface and receptor–ligand bond formation (Chapter 5), the lateral mobility of cell surface adhesion receptors (Chapter 6), and the cell glycocalyx layer and its influence on adhesive interactions with other cells (Chapter 7). Part III focuses on artificial systems designed to mimic the biomolecular interface of living cells. Specific examples include molecular dynamics simulations of protein adsorption to ceramic biomaterials (Chapter 8), supported lipid bilayers as models of the cell membrane (Chapter 9), and the use of functionalized glass micropipettes to model cell interactions in the capillary bed (Chapter 10). The final part of the book is dedicated to the roles that fluid shear forces play in the adhesion between reactive surfaces and circulating cells, specifically, the effects of complex hemodynamics (Chapter 11), pairwise interactions between multiple suspended cells (Chapter 12), and the unique physics of platelet-shaped cells (Chapter 13). I thank all of the contributing authors for lending their valuable time and expertise, the editors at Academic Press (in particular, Luna Han) who helped in the planning and production of this book, and my wife Cindy for her support and scientific insights. Finally, I express my thanks to the readers of this book, and the hope that they may find ideas or inspiration in this exciting field of cellular engineering.

Michael R. King
Rochester, New York

I

CELL–SUBSTRATE ADHESION

1

TRACTION FORCES EXERTED BY ADHERENT CELLS

CYNTHIA A. REINHART-KING and DANIEL A. HAMMER

Department of Bioengineering, University of Pennsylvania, Philadelphia, Pennsylvania, USA

The mechanical interaction between a cell and its substrate can affect and control cell behavior and function. As a cell adheres to and moves across a substrate, it exerts traction forces. Several laboratories have put forward methods to investigate and quantify these traction forces. Here, we review the various approaches developed to measure traction forces and some of the most significant findings resulting from the application of these techniques. Additionally, we discuss recent results from our laboratory quantifying traction forces exerted by endothelial cells, to better understand the process of two-dimensional tissue formation using angiogenesis as a model. As the interaction between a cell and its substrate is crucial for transmission of mechanical cues from the cell to the substrate and from the extracellular milieu to the cell, further investigation into cellular traction forces will continue to elucidate the role of mechanical forces in cell function and tissue formation.

I. INTRODUCTION

Cell migration is a chemically and mechanically integrated process that plays a crucial role in a number of biological processes. During embryogenesis, specifically gastrulation, cells migrate in the early embryo to form the germ layers that later become the specific tissues of the body. Later in embryogenesis, cells within these layers migrate to specific regions to give rise to specialized organs and tissues of the adult organism, such as the heart, liver, nervous system, and skin. The role of cell migration extends into adulthood as well. For example, migration is critical to wound healing: fibroblasts and endothelial cells must migrate from their existing space and into the wound bed to heal the injury site. Additionally, migration plays a key role in the inflammatory response: leukocytes migrate into sites of inflammation and mount an immune response. Defects in migration also play an important pathophysiological role. Uncontrolled leukocyte migration results in chronic inflammation and diseases such as rheumatoid arthritis and asthma. Tumor metastasis results when the cells of a tumor enter the circulatory system and then migrate to new sites to form new tumor masses. Clearly, cell migration persists from the time of embryogenesis throughout the life of an organism. Therefore, a more thorough understanding of migration could lead to insights into the physiology and pathological development of numerous disease states.

Cell migration results from complex chemical and mechanical interactions integrated into one concerted signal to move a cell forward. It has been described as a cyclical process occurring in five macroevents that distinguish the front of the cell from the rear of the cell and lead to translocation of the cell body (Lauffenburger and Horwitz, 1996).

1. The cell extends its membrane, first using filopodia, then by extending the leading edge of the lamellipodium. This first step in motility is often initiated by chemotactant factors located in the extracellular milieu. The cell senses these chemoattractants and extends its membrane in the direction of the signal. Membrane extension is thought to be physically driven by actin polymerization (Condeelis et al., 1988).

2. The new membrane extension forms firm attachments with the substrate. The attachments are initiated by binding of the membrane receptors, integrins, to substrate-bound ligand and are strengthened by the recruitment of structural and signaling molecules to the intracellular portion of the integrin. The strength and size of these attachments are related to the amount of substrate-bound ligand presented to the cell.

3. Contractile force is generated within the actin cytoskeleton and transmitted to these adhesion sites. This contractile force is believed, in most cases, to be driven by the interaction of the actin cytoskeleton with myosin.

4. The contractile force results in detachment of the rear of the cell. This allows for retraction of the tail of the cell.

5. Detachment of the rear of the cell results in a front-versus-rear asymmetry that propels the cell forward. This process repeats itself, resulting in cell translocation along its substrate.

Each of these five processes is mediated by the integration of numerous, interconnected molecular-level interactions governed by physicochemical (both mechanical and chemical) properties of the system in space and time. Therefore, to fully understand how a cell migrates, it is necessary to identify the key molecular and physical parameters that govern cell motility and develop an accurate model that describes their specific interactions leading to motion of a cell.

There have been significant advances in our understanding of cell adhesion and migration. Cell behavior has been shown to be affected by a number of extracellular cues, including the presence of chemoattractants and growth factors, substrate ligand density, substrate topography, and external mechanical signals such as shear stress, osmotic pressure, and stretch (Geiger and Bershadsky, 2002). It should be noted that the ability of a cell to respond to mechanical forces and chemical signals is not limited to sensory cells, but is exhibited by a variety of cell types that are subjected to physiological forces. As such, a number of methodologies have been developed to probe and measure cell response to these extracellular cues. One of the most significant advances has been the development of methods that characterize the mechanical interaction between a cell and its substrate during adhesion and migration.

To adhere and migrate, a cell must exert force tangential to its substratum, pushing the cell membrane outward and propelling the cell forward. These traction forces exerted against the substrate are a result of both the active forces exerted at the cell–substratum attachment sites and the nonspecific force generated from nonspecific frictional interactions between the cell and substratum (Oliver et al., 1995). Traction forces, in conjunction with the adhesion–contraction events described above, drive cell adhesion and locomotion. Physiologically, traction forces may also play a role in tissue remodeling. Cell traction forces transmitted to the substrate have been shown *in vitro* to contract collagen gels and realign extracellular matrix fibrils (Tranquillo et al., 1992). Cell-mediated extracellular matrix remodeling can change both the chemical composition and the mechanical composition of the extracellular matrix. It is well established that these changes affect cell behavior, specifically the degree of cell spreading, the direction of migration, and the rate of proliferation. However, the relative contribution of traction to cell migration in comparison to matrix remodeling remains unclear. It has been hypothesized that cells may generate traction to reorganize their matrix rather than to migrate (Elson et al., 1999). The ability to quantify cellular traction forces is critical to understanding the mechanisms by which cells generate tractions and the mechanism by which these tractions can alter the extracellular matrix environment and cell motility.

Unlike migration metrics like cell speed and persistence length, traction force generation is not easily detected or measured using standard microscopy tools, namely because of the nature of the force (i.e., the small forces over the area of a single cell) (Beningo and Wang, 2002). However, several cell traction measurement techniques have emerged. In this chapter, we review the major technological advances in the measurement of cell traction forces, highlight the major findings in the area of cell traction and how those findings relate to a better understanding of cell physiology, and lastly, make predictions on future studies of cell–substrate mechanics.

II. THE BIOLOGY OF CELL–SUBSTRATUM INTERACTION AND TRACTION GENERATION

A number of different forces exist within a cell, including cell adhesion, membrane extension, actin polymerization, and organelle translocation. The dominant force-producing mechanism in most adherent cells is the actomyosin–cytoskeleton complex. The primary components of this complex are stress fibers, containing mainly bundles of actin in addition to myosin, tropomyosin, and numerous other cytoskeletal proteins (Burridge et al., 1988). Force is generated as the myosin head interacts with actin in an ATP-dependent fashion to slide filaments past each other. In separate experiments, both the cell cortex and stress fibers have been found to be contractile. Initial experiments using detergent-extracted cells, which contain only stress fibers and the cell cortex, contracted on addition of Mg^{2+} and ATP (Kreis and Birchmeier, 1980). It was later found that the cell cortex alone is contractile (Mochitate et al., 1991; Halliday and Tomasek, 1995), and single isolated stress fibers shorten on addition of Mg-ATP in a myosin-dependent fashion (Katoh et al., 1998).

Stress fibers and the cytoskeleton complex are anchored to the substratum through focal adhesion sites. At these adhesions, integrins span the cell membrane, connecting the extracellular matrix to the actin cytoskeleton. Focal adhesions contain a large number of both structural and signaling proteins (Geiger and Bershadsky, 2001). They develop from small (<1 μm) dotlike adhesions called focal complexes, which form from integrin–extracellular matrix (ECM) binding occurring in lamellipodium extensions. Both focal adhesions and focal complexes are associated with the actin meshwork; however, the formation of stress fibers is one of the markers indicating the development of focal complexes into focal adhesions (Heath and Dunn, 1978). Stress fiber contractility is transmitted to the ECM through these adhesion sites. It is this contractility that is measured through various traction force measurement techniques.

III. CONTINUUM-BASED METHODS TO DETECT AND MEASURE CELLULAR TRACTION FORCES

According to Newton's law, the force that a non-accelerating cell exerts on a substrate is equal and opposite to the force exerted by the substrate on the cell. Because direct measurements of cell traction forces are not possible, currently used techniques calculate tractions based on the deformations in the substrate caused by cellular force generation. As an example, solving this inverse problem is analogous to analyzing footprints in the sand to characterize the motions of a pedestrian. In both cases, information about movement is extracted from changes detected in the substrate.

In calculating the traction force exerted by the cell through examination of displacements in the substrate, it should be noted that it is possible that the force a cell generates intracellularly may actually be greater than the force detected based simply on deformations in the substrate. Some of the intracellularly generated force may be dissipated within cellular deformations. Additionally, the amount of force exerted by the cell on the substrate, and vice versa, is dependent on the strength of the bonds between the cell and the substrate. For the force to be transmitted from one to the other, the linkages between the two must be capable of withstanding the force rather than breaking. Therefore, it is possible that by analyzing changes in the substrate to calculate traction forces, we lose some information about the total force a cell is capable of exerting.

For a cell to move, the sum of the traction forces must be equal to or greater than the hydrodynamic drag force on the cell, which is the only external force resisting translation of the cell. We can easily approximate the hydrodynamic drag as follows: Let the velocity of the cell (v) be on the order of 0.01 μm/s. Given a characteristic length scale (L) over which the velocity attenuates to zero equal to the radius of the cell ($\sim$10 μm), the shear rate equals $v/L = \sim10^{-3}$ s^{-1}. Assuming the viscosity of the medium to be approximately the viscosity of water (μ) and equal to 10^{-3} N-s/m^2, the shear stress equals the shear rate multiplied by μ: approximately equal to 10^{-6} N/m^2. The hydrodynamic drag on the cell is equal to shear stress multiplied by cell area. If the cell area is approximated to be 1000 μm^2, the hydrodynamic drag force is approximately equal to 10^{-10} dyn. This drag force is well below the detection limit of any traction measurement technique available, and therefore the sum of the forces across a cell can be assumed to equal zero. This assumption is important, as it is made in many of the techniques used to measure cellular tractions across the cell.

A. The Wrinkling Substratum

The idea of using deformable substrates to detect and measure traction forces was first conceived by Harris and coworkers (1980). In their landmark article, Harris and coworkers laid the foundation for all of the current methods used to quantify traction. Prior to this study, many groups had studied actomyosin contractility, but no group had been able to successfully map the forces to locations on the cell body or to quantify the magnitude of these tractions. Harris et al. used the idea of calculating tractions based on substrate deformations to create the first method to study forces exerted during cell locomotion (Fig. 1.1).

The forces exerted by the cell on its substrate are estimated based on the size and distribution of wrinkles of the substrate created by the adherent cell. Harris et al. not only developed and described a method to detect cellular traction forces, but also addressed several important scientific questions.

In developing this technique, several experimental parameters needed to be established. First, the substratum needs to be weak enough that forces exerted by the cell produce visible distortions in the substrate material. Additionally, the substratum needs to be nontoxic to cells, relatively inert to prevent undesirable biochemical changes, and visibly transparent to allow for optical microscopy. With these considerations in mind, Harris et al. created silicone rubber substrata to investigate cellular tractions. Prior to these experiments, silicon had been used in medical devices and implants, indicating that crosslinked silicon is nontoxic to cells. Additionally, the strength of silicon can be tuned based on the viscosity of the silicon fluid prior to polymerization and the duration of heat exposure leading to polymerization. Lastly,

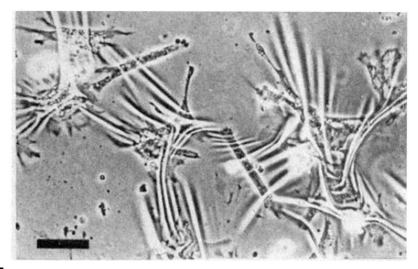

FIGURE 1.1 Harris et al. (1980) developed the first technique to measure cellular traction forces. Forces are measured based on wrinkles created in a silicon substrate by the force exerted by adherent fibroblasts. Bar = 100 µm. Adapted from Harris et al., 1980, by permission of the American Association for the Advancement of Science.

unlike collagen networks or other protein sheets, silicon cannot be bio-chemically remodeled by the cells. Therefore, chemically, the surface remains the same throughout culture, and any changes in the substrate are due to mechanical forces rather than chemical forces.

The chemical inertness of silicon allowed Harris et al. to test the long-held hypothesis that the radial orientation of clots around explants is due to dehydration of the protein network by cells, causing shrinkage and stretching of the network (Weiss, 1934). Because silicon cannot be dehydrated by the adherent cells, any physical changes in the substrate would be due to physical forces exerted by the substrate. Perhaps the most significant finding from this study (Harris et al., 1980), other than development of a technique to observe cellular traction forces, is that traction forces, and not dehydration, are responsible for the pattern seen around plasma clots. Cells create wrinkles in their substrate, which then cause the cells to reorient by contact guidance. This study was one of the first to show that cellular traction forces produced by cell adhesion and migration may reorganize the ECM, playing a role in development and tissue morphogenesis.

In addition to observing the development of traction forces in the cell substratum, Harris et al. also attempted to quantify the magnitude of the forces exerted by the cell. They used calibrated microneedles to push on the silicon substrate with known forces, attempting to reproduce the same degree of distortion in the substrate that is created by a cell. Using this technique, they estimated that the shear force exerted by the cell at its advancing edge is in excess of 0.001 dyn/μm. However, attempting to obtain reproducible, accurate traction measurements from wrinkles proved to be difficult due to the intrinsic nonlinearity of wrinkle formation. The wrinkles form slowly and are often larger than the cell, resulting in poor spatial and temporal resolution. Despite the difficulty in quantifying the forces exerted by cells on their substrate, the experiments of Harris et al. laid the foundation for all of the techniques currently used to observe and measure traction forces.

Notwithstanding its limitations, the wrinkling substratum method is still used by some laboratories. Further modifications on the method include attempts to modulate the compliance of the substrate to obtain better spatial resolution in the size and spacing of the wrinkles (Burton and Taylor, 1997). This method is very time-intensive, because the compliance of each substrate must be tested individually due to variations occurring in the crosslinking procedure. As in previous applications, this method remains limited in the accuracy of the force measurements as calculated from the wrinkle size and location.

B. Silicon Embedded with Beads

To overcome the difficulty associated with deriving quantitative information from the size and location of the wrinkles in the system described by Harris et al. (1980), Lee et al. (1994) modified the silicon substratum by creating a nonwrinkling silicon surface embedded with 1-μm beads. In this method, traction forces are calculated based on the in-plane displacements of the beads, which are much easier to detect and track than wrinkles. As the cell adheres and migrates on the substrate, the cell deforms the substrate, producing movements within the bead field. The bead field under the cell is compared with the bead field after the cell has moved away from that region of the substrate, when the beads are no longer subjected to the force of the cell, to calculate a strain field. On the basis of this strain field and the compliance of the silicon, the authors attempted to calculate the force exerted by the cell, where the traction force exerted on each bead equals the displacement of the bead multiplied by substrate compliance. As was done in the case of the wrinkling substratum technique, substrate compliance was calculated using calibrated microneedles. By using embedded beads, Lee et al. showed that significantly more accurate information about the location, magnitude, and direction of traction forces can be obtained than in the case of the wrinkling substratum technique.

To further simplify the traction calculations in comparison to the Harris et al. (1980) study, Lee et al. (1994) investigated the traction forces exerted by a relatively simple cell type — keratocytes. Fish keratocytes are simple, crescent-shaped cells in culture that maintain their shape throughout migration, unlike, for instance, fibroblasts, which are continually changing their shape to propel themselves forward. Therefore, by capture of a single moment in the movement of a keratocyte, the steady-state tractions of the cell can be calculated.

The study by Lee et al. (1994) was significant for a number of reasons. Using the modified silicon technique, Lee et al. were able to begin to elucidate components of the relationship between traction generation and cell locomotion. Perhaps the most striking result is that traction forces are not detected in the lamellipodia of keratocytes, but rather only along the sides of the crescent, pointing inward. This contrasts with the results of Harris et al. (1980) that seem to indicate that fibroblasts move by exerting rearward pointing traction at the front of the cell that propels the cell forward. Additionally, Lee et al. (1994) found that the traction forces exerted by keratocytes were an order of magnitude smaller than those found for fibroblasts by Harris et al. (1980). Lee et al. proposed that this pattern of tractions in keratocytes contributes to their fast migratory ability in comparison to fibroblasts. In keratocytes, there exist close cell–substratum contacts and negligible contractility at the front of the cell, and no close cell–substratum adhesions and high contractility at the rear of the cell. This asymmetry between the front and rear of the cell allows for rapid lamellipodium extension and rapid rear retraction, resulting in fast locomotion. In contrast, according to Harris et al. (1980), fibroblasts exhibit high contractility at both the front and the rear of the cell, which could compete and result in slow lamellipodium extension and rear retraction. Clearly, the mode of migration can vary based on cell type.

Historically, Lee et al. (1994) were the first to implement a nonwrinkling substrate and the use of fiducial markers to measure displacements within the plane of the substrate. This technique aids in the quantification of the forces exerted by the cell, but it also allows for determination of the location and direction of the forces. It should, however, be noted that there are several limitations to this technique. The compliance of the substrate can be altered so that even cells capable of exerting very little force can produce detectable bead displacements. However, the authors found that silicon substrates fabricated to be hypercompliant do not exhibit linear behavior. That is, the beads return only 30–60% of the way to their original position and take several seconds to do so, indicating the silicon exhibits both plastic and viscoelastic behavior. Therefore, when these hypercompliant substrates are used, the bead displacements cannot be directly translated into the forces exerted by the cell.

As such, this technique is best suited for stiffer substrates that do not exhibit plastic behavior.

Given a nonwrinkling, elastic substratum that exhibits linear elastic (rather than viscoelastic) behavior, Dembo et al. (1996) developed a computational method for determining both the magnitude and direction of the forces exerted by an adherent cell based on the theory of elasticity. This method has been modified to apply to the study of a variety of cell types, but the basic methodology remains the same. First, an image of the cell (a bovine aortic endothelial cell is pictured in Fig.1.2A) is taken followed by an image of the beads directly beneath the cell (Fig. 1.2B). The cell is removed using trypsin–EDTA and a second image of the unstressed bead field is taken. The bead displacements created by the cell within the top plane of the substratum are determined by comparing the two bead fields. Early in the development of the elastic substratum method, this was done manually, but in later years an algorithm was developed to detect bead displacement on the basis of comparison of two images of the bead field using a correlation-based optical flow method (Marganski et al., 2003). The projected area of the cell is divided into a mesh (Fig. 1.2C), and the traction vectors are calculated for each node of the mesh (Fig. 1.2D) based on the bead displacements, the material properties of the substrate, and the cell outline. Dembo et al. (1996) calculated the traction field by solving the inverse problem, where the displacement field is expressed as an integral over the traction field. This transform cannot be solved analytically and relies on Bayesian statistics to calculate the most likely traction vectors that can most accurately describe the given displacement field. Despite being computationally intensive, the algorithm developed by Dembo and coworkers can determine the magnitude of the forces exerted by the cell, as well as map tractions to specific locations on the cell (Fig. 1.2E).

C. Polyacrylamide Embedded with Fluorescent Beads

As described in the preceding section, silicon substrates proved valuable in measuring the forces exerted by fast-moving keratocytes. However, these same substrates are much less useful for studying most mammalian cells. For tractions to be accurately calculated, the substrate must be tuned to match the motility and force generation of a given cell type. It is difficult to accurately produce a silicon substrate of desired compliance for slower-moving cells capable of exerting higher forces. To overcome this limitation, Dembo and Wang (1999) used a substrate of polyacrylamide embedded with submicrometer-sized fluorescent beads. The compliance of polyacrylamide substrata can be tuned chemically by varying the monomer and crosslinker concentrations (Pelham and Wang, 1997). Polyacrylamide offers several additional advantages over silicon substrata. Over a wide range of deformations, it exhibits linearly elastic behavior. Also, polyacrylamide is not typically amenable for cell binding on its own, without the conjugation of specific cell adhesion ligands (Nelson et al., 2003). Therefore, it is a perfect scaffold for studying cell adhesion and behavior in a controllable, defined way.

The computational method by which deformations in the substrate are used to determine the tractions exerted by the cell are very similar to the methods used on the aforementioned stressed silicon substrata. However,

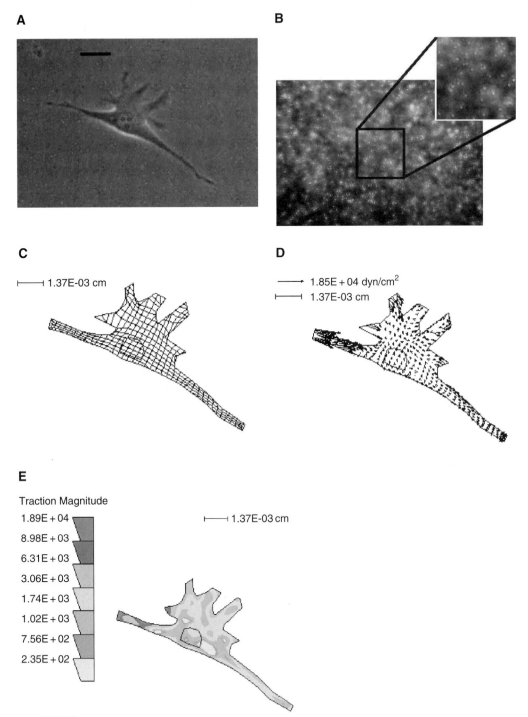

A

B

C

├──────┤ 1.37E-03 cm

D

──────▶ 1.85E + 04 dyn/cm^2

├──────┤ 1.37E-03 cm

E

Traction Magnitude

1.89E + 04

8.98E + 03

6.31E + 03

3.06E + 03

1.74E + 03

1.02E + 03

7.56E + 02

2.35E + 02

├──────┤ 1.37E-03 cm

FIGURE 1.2 Traction force microscopy calculates forces exerted by a cell on its substrate based on the displacements of marker beads within the surface of the substrate. (A) An image of a single cell is taken (bar = 20 μm) followed by (B) images of the bead field in its stressed state, due to the adherent cell (red), and its unstressed state, after the cell is removed (green). Yellow beads have not displaced from one image to the next. Inset: Magnified view of the bead field with the background subtracted for visual clarity (image processing courtesy of M. Mancini). (C) The cell is divided into a mesh, and (D) traction forces are calculated at each node of the mesh. (E) Color contour plot of the traction field. Note: Substrate depicted is polyacrylamide. (See Color Plate 1)

using fluorescent markers has greatly improved the tracking method and the ability to calculate an accurate strain field.

Dembo and Wang published several studies using the resulting technique, traction force microscopy, which elucidate the mechanisms of fibroblast migration. Specifically, they have shown that the lamellipodia of the cell provide almost all of the force required for forward locomotion (Munevar et al., 2001a). Their results indicate that the lamellipodium is a mechanical entity distinct from the rest of the cell body. Interestingly, this same mechanical division within the cell does not seem to exist in H-ras transformed cells, perhaps explaining the difference in their motile behavior. Additionally, Beningo et al. (2001) investigated the role of focal adhesions in regulating traction generation and found that the size of focal adhesions is inversely related to the amount of force generated. Moreover, the distribution of adhesions does not correspond well with the distribution of traction forces. The authors conclude that these results may indicate that early focal complexes are responsible for strong propulsive forces, and maturation of these adhesion sites results in a change into passive anchorage sites — a conclusion that has been widely discussed in the literature. Additionally, Dembo and coworkers investigated the dynamic roles that front-versus-rear adhesions (Munevar et al., 2001b), myosin IIb (Lo et al., 2004), focal adhesion kinase (Wang et al., 2001), and stretch-activated Ca^{2+} channels play in fibroblast migration (Munevar et al., 2004). Using traction force microscopy, Dembo and coworkers have made significant progress in understanding the role of force generation in fibroblast migration.

One of the most significant technical advances using polyacrylamide gel is the ability to reliably control the compliance of the cellular substrate without changing the ECM density. Tuning the compliance of the substrate was a critical turning point in the development of traction force microscopy, as it allowed for the investigation of almost any cell type and an understanding of cell behavior as a function of the mechanical environment. Prior to the study by Pelham and Wang (1997), most studies investigating cell migration and adhesion focused on cell migration in response to its soluble chemical environment (chemotaxis) or in response to the ligand conjugated to the substrate (haptotaxis). Additionally, studies involving the cell's mechanical environment focused on the response due to imposed forces such as fluid shear stress and mechanical stretch. However, by changing the stiffness of the substrate, Pelham and Wang (1997) created a significant shift in the way researchers approach cellular response and mechanotransduction. Using polyacrylamide substrata, Pelham and Wang kept the ECM density on the substrate constant while altering the mechanical compliance. They demonstrated that fibroblasts are capable of actively responding to the mechanical compliance of their substrate. Cells on stiffer gels are more spread and migrate more slowly than cells on more compliant gels. Moreover, the ability of cells to sense the mechanical compliance of their substrate is reflected in their ability to change the phosphorylation state of numerous proteins contained within the focal adhesion structure. Focal adhesions on stiff substrates are larger, more elongated, and more stable, whereas focal adhesions on more compliant substrates contain less phosphorylated $pp125^{FAK}$ and paxillin and appear much more irregularly. These results were the first to suggest that mechanical ECM cues may be just as important as chemical cues in regulating cell adhesion.

Since the seminal article of Pelham and Wang (1997), a number of studies have investigated the effects of compliance on cell behavior. Lo et al. (2000) used polyacrylamide chemistry to create a substrate containing a step in stiffness — a central region of the substrate where two substrates of different compliances meet. They demonstrated a behavior called *durotaxis,* by which the cells were able to actively detect and respond to changes in the compliance of the substrate. Cells that were migrating on the soft substrate, on hitting the boundary of the stiff–soft transition, would cross onto the stiff substrate, whereas cells on stiff substrates exhibited higher tractions and more spread area, and either retracted or changed directions in response to the stiff–soft boundary. Later, Wong et al. (2003) investigated the ability of fibroblasts to migrate on polyacrylamide hydrogels containing gradients of compliance, rather than a step, as was done by Lo and co–workers. They found that vascular smooth muscle cells tend to migrate faster on softer substrates than on stiffer substrates (15 kPa vs 25 kPa), and that cells tend to accumulate on stiffer substrates. Moreover, the migration pattern on gradient-compliant gels appeared to be directed toward the stiffer gel regions rather than exhibiting the typical random walk pattern characteristic of cell migration. Engler et al. (2004) further investigated cell response to compliant gels and showed that the response is in large part mediated by the assembly of the actin cytoskeleton. By testing for changes in the cytoskeleton, Engler and coworkers were able to show that slight overexpression of actin in the cell can compensate for the loss of spreading seen in soft gel responses. Additionally, Yeung et al. (2005) showed that the sensitivity threshold for compliance sensing is cell type specific and that cell–cell contacts can also help rescue the morphology changes seen in the soft substrates to more closely resemble the morphology of cells on stiffer substrates (Yeung et al., 2005). Overall, the study of durotaxis is still relatively young, and much remains to be learned about how a cell mechanically senses and responds to the material properties of its substrate and environment.

D. Micropatterned Elastomeric Substrates

The elastic substratum method developed by Dembo and Wang (1999) is not without limitations. Because beads are located throughout the gel, and not just at the upper surface, out-of-plane fluorescence and bead clustering can make it difficult to accurately track bead displacements using an automated system without significant noise. Balaban et al. (2001) attempted to overcome this difficulty by using soft lithography techniques to stamp well-defined patterns on the surface of the substrate rather than using marker beads. As a cell exerts tractions on the substrate, the patterns become deformed, and tractions are calculated based on these deformations. A well-defined pattern is advantageous over randomly placed beads because the deformations are much easier to detect and predict. The subsequent calculation is similar to that of Dembo and Wang (1999), except that rather than mapping forces across the cell body, Balaban et al. (2001) assumed that all forces must be transmitted at sites of focal adhesion, and therefore all forces are mapped to fluorescently tagged focal adhesions. Solving this inverse problem is much the same as described for other elastic substratum methods and requires regularization to create a good fit.

Using this method, Balaban et al. (2001) investigated the correlation between focal adhesion size, orientation, and force generation in human fore-skin fibroblasts and cardiac myocytes. They showed that force is transmitted primarily in the direction parallel to the long axis of the focal adhesion. Additionally, they demonstrated that the size of the focal adhesion corresponds to the amount of force generated at the site. As the focal adhesion grows in time, the amount of force it exerts increases proportionately. These data are inconsistent with the observations of Beningo and co–workers (2001) mentioned earlier in the chapter that showed smaller focal complexes are responsible for large forces. These differences may be attributable to differences in the cell types studied. It should also be noted that Balaban et al. (2001) investigated stationary cells, whereas Beningo and co–workers studied locomoting cells. Because locomoting cells must generate large propulsive forces at the leading edge using newly formed adhesions and must release the adhesions at the rear of the cell to move forward, it is reasonable that the force distribution and generation at focal adhesions may differ.

The technique presented by Balaban et al. (2001) has potential technical drawbacks. First, to create the pattern, the stamp indents the substrate several micrometers. While cells do have the ability to respond to topological cues, the size of features necessary to elicit a response is not well established. Therefore, it is possible that the microstamped pattern elicits a change in cellular behavior that is unaccounted for in the analysis of cell response. A second potential drawback is in the assumption that all forces must stem from focal adhesions as marked by GFP-zyxin. The authors stated that this assumption is valid because they did not detect substrate deformations in areas without focal adhesions. As is discussed in more detail later in the chapter, this assumption may not always be valid, as there have been cases where forces were detected and yet no focal adhesions were evident. Overall, however, the technique described by Balaban et al. (2001) improves the method of displacement detection, thereby simplifying the displacement calculation and making it a satisfactory system for looking at force generation.

IV. DISCRETE METHODS FOR MEASURING CELLULAR TRACTION FORCES

The methods addressed thus far for mapping cellular tractions have focused on looking at a continuum of force distributions over the substrate, in which the displacements across the substrate are all coupled. In addition to these continuum methods, several techniques have been developed to meas-ure forces exerted at discrete locations on the cell rather than over a contin-uum. In these cases, (1) the cell–substrate interface is restricted to discrete locations so that force can be assumed to be generated only at locations where the cell is touching the substrate, or (2) measurements are taken only at specific locations on the cell. To date, these techniques appear less fre-quently in the literature; the advantages and disadvantages of the techniques are discussed here.

A. Micromachined Cantilevers

With the advent of microfabrication technologies, several groups have created novel devices to measure cellular traction force. Galbraith and Sheetz (1997) developed a technique by which forces are detected based on deflections in substrate-embedded cantilevers. Briefly, as shown in Fig. 1.3A, a silicon substrate containing embedded cantilevers ranging in size from 4 to 25 μm² was designed. As cells migrate over the substrate, they deflect the cantilevers from side to side. These deflections can be detected, measured, and calibrated to determine the amount of force the cell has exerted to deflect the cantilever. Because the cantilever can be deflected only along one axis, the force that a cell is capable of exerting is attenuated if the cell does not cross the cantilever at a 90° angle. To calculate the total force the cell is exerting, the displacement of the cantilever is multiplied by the stiffness of the cantilever, and the resulting force is multiplied by the sine of the angle between the cantilever and the long axis of the cell. This calculation is based on the assumption that fibroblasts generate most of their force parallel to the direction of motion, a conclusion reached by Harris et al. (1980).

Measuring traction forces based on cantilever deflection has several advantages over the continuum methods presented earlier in this chapter. Most notably, by isolating force measurements to individual portions of the cell based on the size and distribution of cantilevers, Galbraith and Sheetz were able to isolate the contributions of the leading versus trailing edge of the cell. Galbraith and Sheetz found that in fibroblasts, tractions did not form at the leading edge. Rather, tractions began to form 5–25 μm behind the leading edge, which is the same location where they observed an increase in cytoskeletal and myosin organization. The forces tend to point inward at the front of the cells, whereas beneath the nuclear region the tractions change direction, and at the tail of the cell, the forces are pointing forward in the direction of migration. Additionally, at the tail end of the cell, where the cell area is smaller, forces appear to be larger. This may indicate that this smaller

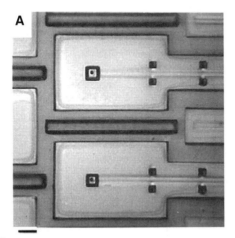 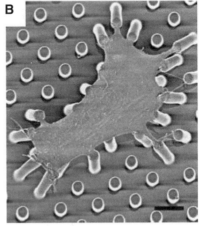

FIGURE 1.3 Several methods have been developed to measure forces at discrete locations on the cell. (A) Galbraith and Sheetz (1997) fabricated a substrate embedded with cantilevers that deflect as cells migrate across. Pictured here are two of the cantilevers. Image adapted from Galbraith and Sheetz (1997). Copyright ©1997, National Academy of Sciences, USA. Reproduced by permission from the National Academy of Science. (B) Tan et al. (2003) developed a lithographically created micoarray of posts that can be deflected by cellular force. Image adapted from Tan et al. (2003). Copyright © 2003, National Academy of Sciences, USA. Bar = 10 μm in both figures.

area of the cell most likely contains fewer contacts, each of which is under greater tension than the contacts at the front of the cell. On the basis of these data, Galbraith and Sheetz helped to confirm the model of migration, which states that a cell migrates by breaking these rear adhesions, allowing the cell body to move forward in response to the tractions exerted at the front of the cell. By isolating the discrete contributions of various components of the cell, Galbraith and Sheetz were able to partially dissect their contributions to traction generation and cell migration.

Calculating the force generated by a cell using a cantilever-embedded substrate has one significant advantage over the elastic substratum methods. Calculating the traction force exerted on a cantilever is significantly easier than the calculations based on a continuum. Because the force sensors are isolated and independent of each other, the force calculation is fairly straight-forward. However, while the force calculation is relatively easy, the fabrication of the substrate containing micrometer-sized cantilevers is significantly more complicated than the fabrication of elastic substrates. The process involves a series of microetching and coating steps that result in submicrometer-sized cantilevers embedded in-plane with the substrate surface, anchored at one end in the substrate and capable of deflecting at the opposite end. This process is labor-intensive and expensive. Additionally, because of limitations in the micromachining procedure, the spacing and size of the cantilevers are limited, thereby limiting the resolution of the measurements. Overall, the bed of cantilevers is an effective way of isolating force, but since its publication in 1997, simpler, less costly, and less labor-intensive methods have been developed.

B. Microfabricated Elastomeric Posts

As described in the previous section, one of the advantages of the discrete methods for quantifying cellular traction forces is the ability to isolate and map forces to specific locations on the cell. Tan et al. (2003) built on the idea first established by Galbraith and Sheetz (1997) to measure force at discrete locations, and built a simpler system to isolate force. They designed a substrate containing an array of evenly spaced microneedles made of silicon elastomer (Fig. 1.3B). The tips of the post can be coated with ECM proteins to allow for cell adhesion, and the mechanical compliance altered by varying the height of the posts. Cells adhere to the top of the posts and because the posts are elastomeric, traction forces exerted by the cell cause the posts to bend. On the basis of simple bending mechanics, the force exerted by the cell on an individual post can be calculated by tracking the displacement of the tip of the post from its original position.

Using elastomeric posts to measure force generation, Tan et al. (2003) investigated the relationship between adhesion size and the amount of force exerted. They were able to show that two distinct types of adhesions exist. Those adhesions that are larger than 1 μm^2 exert force that is proportional to their area. However, adhesions that are smaller appear to exert disproportionately larger forces. The authors concluded that these larger forces may be focal complexes, rather than focal adhesions. These results would coincide somewhat with the findings of Beningo et al. (2001) that showed small nascent adhesions were capable of exerting relatively large forces. In addition to examining how adhesion size correlates with force, Tan et al.

(2003) also examined how cell size correlates with force. The authors found that when cell spreading was restricted by reducing the number of ECM-coated posts in a given area, the cells exerted less force. However, when the cells were transfected with a constitutively active RhoA, they regained the ability to exert force. These findings directly link cell morphology to the biochemical changes that result in changes in contractility.

In the study by Tan et al. (2003), the spatial resolution is on the order of 9 μm, based on the post spacing. However, it is not well established whether the spacing affects cell adhesion and migration in comparison to behavior on a planar continuum. To increase the resolution of the technique and more closely mimic a planar substrate, du Roure et al. (2005) improved the micro-fabrication technique to create arrays of elastomeric posts with 2- to 4-μm spacing. By use of this more closely packed array that better mimics a continuous surface, a spread cell covers significantly more posts, improving the number of force measurements that can be taken. However, this technique does have one significant limitation. Because the posts are positioned closely together, force exerted by the cell occasionally results in two posts coming into contact. In these cases, the force cannot be measured accurately.

Du Roure et al. (2005) used the micrometer-sized array of posts to investigate forces exerted by epithelial cell monolayer assembly in comparison to forces exerted by single isolated cells. They showed that the forces exerted at the edges of the assembly are significantly greater than the forces exerted by individual cells. This may indicate that the ability of a cell to exert force can be affected by cell–cell junctions and multicellular assemblies. These results are significant in that they are the first in which traction measurements were used to look at tissue assemblies. In the future, such experiments will be critical in understanding tissue assembly.

V. INVESTIGATION INTO THE MECHANISMS OF ENDOTHELIAL CELL TRACTION GENERATION

Our laboratory has investigated the mechanics of cell adhesion using angiogenesis as a model. More specifically, we have studied how substrate ligand density and substrate compliance affect the ability of endothelial cells to adhere, spread, and migrate, to ultimately understand the balance of factors that result in tissue formation. In collaboration with Micah Dembo, we have characterized endothelial cell response to chemical and mechanical changes in the substrate using traction force microscopy. Traction force microscopy has been useful in elucidating basic principles of endothelial cell adhesion and cell–cell cohesion.

Our initial studies of endothelial cell adhesion focused on investigating the effects of ligand density on the ability of cells to adhere and exert traction forces on the substrate (Reinhart-King et al., 2003). Because we sought to understand endothelial cell adhesion at a very fundamental, biophysical level, we designed and used a nine-amino-acid peptide containing an arginine–glycine–aspartic acid sequence. This sequence is a well-characterized component of a number of ECM proteins and has been shown to bind in a very specific way to integrin cell surface receptors. Using polyacrylamide substrates conjugated with varying densities of RGD-peptide, changes in cell adhesion and traction force generation were measured. Controlled, defined variation in the density of peptide conjugated to the gel results in changes in cell morphology. More specifically, increasing peptide density causes an increase in the average cell area. The polyacrylamide chemistry in conjunction with the RGD-peptide provides a well-defined system to control cell area.

Using polyacrylamide gels with well-defined constant compliance (2500 Pa), we investigated the effects of substrate ligand density on the ability of cells to exert tractions. Regardless of ligand density or cell area, the same general pattern of traction forces existed for all of the cells studied. Traction forces tend to point inward, with the most significant traction localized to the tips of the pseudopodia. In general, the traction forces under the nucleus are negligible in comparison to those at the cell periphery. A comparison of cell force with cell area reveals a linear, increasing relationship between the two; a large cell exerts proportionately more force than a smaller cell. It is interesting to note that the cell is able to maintain this force/area ratio over a wide range of ligand densities. This led us to hypothesize that perhaps the cell membrane is analogous to a rubber band that passively resists spreading. Tractions are the forces that drive the membrane outward to achieve spreading, and the further the membrane is driven outward, the more force that must be exerted to maintain the spread area.

Our finding that force and area are related led us to question how this relationship develops and whether it is related to focal adhesion development and size, as has been suggested by several other studies (Balaban et al., 2001; Beningo et al., 2001; Tan et al., 2003). To gain insight into the basis for the relationship between force and area, we investigated how forces develop while a cell is spreading (Reinhart-King et al., 2005). When a cell first contacts a substrate, it begins rounded, and over time it spreads to some steady-state area. Investigating how the cytoskeleton assembles, how focal adhesions form, and how tractions are generated during cell spreading is essential in understanding the mechanisms that drive traction force generation and endothelial cell adhesion.

Endothelial cells were plated on polyacrylamide gels conjugated with varying densities of RGD-peptide and monitored from the time of plating until they had reached their maximum spread area. Substrate ligand density is a strong determinant of the dynamics of endothelial cell spreading. In general, the maximum spread area is dependent on ligand density, but additionally, the *rate of spreading* is also dependent on ligand density; cells plated on higher ligand densities spread faster than those on substrates containing less RGD-peptide. Additionally, the mode of spreading appears to differ across ligand densities. At higher ligand densities, cells tend to spread isotropically, extending the membrane uniformly outward from the center. At lower ligand densities, cells tend to spread by extending thin pseudopodia. This very fundamental difference in the shape changes a cell undergoes during spreading indicates that even very early in the adhesion process, a cell has the ability to actively respond to differences in substrate chemistry. We hypothesized that a cell at lower ligand density extends more thin pseudopodia in search of ligand, whereas at higher ligand density, the cell's receptors are close to saturated during spreading and so there is no need for the cell to send out additional sensory extensions in search of additional ligand.

In our investigations of endothelial cell adhesion and spreading, traction force microscopy has enabled us to measure the forces a cell exerts as it undergoes the dynamic shape changes associated with spreading. Interestingly, endothelial cells are capable of exerting significant, measurable forces within only a few minutes after contacting a surface. As seen in Fig. 1.4, within the first 15 min of plating, the cell has begun to spread and is exerting detectable, well-organized forces on the substrate. Interestingly, these forces are oriented very similarly to cells that have completely spread and adhered. The most significant forces are present at the cell periphery and decrease as one approaches the nucleus. Additionally, the forces all tend to point inward

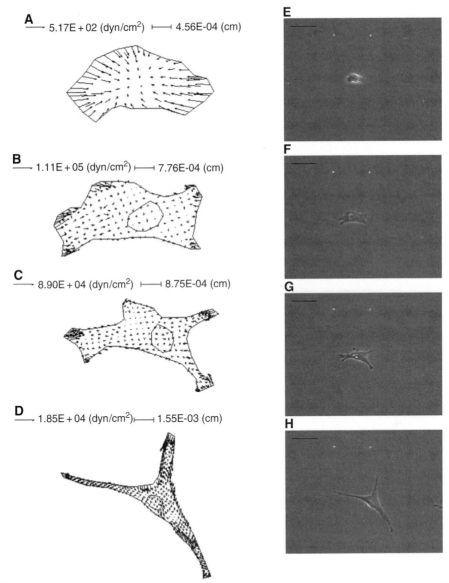

FIGURE 1.4 Using traction force microscopy, endothelial cell traction forces were measured during cell spreading. Traction forces are evident early in spreading, and typically point inward throughout the entire process of spreading. Phase-contrast images are presented adjacent to their corresponding traction maps. (A) and (E) were taken 15 min after plating; (B) and (F) were taken 30 min after plating; (C) and (G) were taken 1 h after plating; (D) and (H) were taken 4 h after plating.

throughout the process of spreading, despite the movement of the membrane outward. This may indicate that a cell uses the substrate to stabilize membrane extensions while continuing to spread. Actin staining with phalloidin at various time points during spreading reveals that despite the presence of force and membrane extension, no stress fibers are seen at early time points (before 1 h). So while actin stress fibers have been shown to be contractile (Katoh et al., 1998), stress fibers are not necessary to produce contractile stresses in spreading cells.

Perhaps one of the most interesting findings from our investigations into the dynamics of endothelial cell spreading is that the presence of vinculin clustering may not be necessary for force generation. Vinculin has been shown to be one of the earliest recruited proteins of a focal adhesion complex (Galbraith et al., 2002). Therefore, by looking for vinculin clustering, the initiation of focal adhesion formation can be determined. Endothelial cells were fixed at various times during spreading and stained with an anti-vinculin antibody. Despite the presence of force generation, no vinculin clustering was detected before 2 h in cells plated on 0.1 and 0.01 mg/ml RGD-peptide and not detected at all in cells on gels coated with 0.001 mg/ml peptide, even 24 h after plating. It should be noted that it is possible that clusters smaller than the detection limit of our optical technique might exist. However, given the possibility that focal adhesion formation is not a requisite for force generation, techniques that map forces to discrete focal adhesions may not be considering all the sources of force generation (Balaban et al., 2001; Tan et al., 2003).

As mentioned previously in the chapter, cells are capable of actively responding to tension created in the membrane by a microneedle (Lo et al., 2000). Additionally, we have shown that cells are capable of exerting tension on a substrate. Therefore, it is possible that cells may be able to respond to adjacent cells through tension created in the substrate. We have initiated investigation into the effects that cell–cell contact and substrate mechanics have on adhesion, migration, and traction generation. Looking at two cells in contact and cells as they are coming into contact should provide key insights into the mechanisms of tissue formation and biomaterial design.

In general, as cells come into contact on polystyrene, a very rigid surface in comparison to polyacrylamide, they tend to bump into each other and push past each other relatively quickly. On the other hand, on polyacrylamide, cells tend to stay in contact with each other much longer. As they eventually move apart from each other, the cells tend to form long tethers, rather than pushing past each other, as is the case on polystyrene. Using traction force microscopy to map the forces between two cells revealed that cells maintain the same general pattern of traction forces seen previously in single cells, with the most significant forces pointing inward and localized to the periphery of the cells. It appears that the forces under the points of contact are much weaker than those at the cell edges opposite the contact edge. Additional studies should help reveal the mechanism for the migration and adhesion differences seen on stiff (polystyrene) versus soft (polyacrylamide) substrates. As the mechanism by which substrate compliance affects the ability of cells to form intercellular contacts is elucidated, better biomaterial and tissue engineering therapeutics may result.

VI. SUMMARY AND FUTURE WORK

Looking at the current trend in cell migration research, scientists tend to focus on studying migration and adhesion in three-dimensional matrices, in comparison to the two-dimensional substrates that have been investigated previously. While most tissues have three-dimensional architecture, most of the research and subsequent significant findings that pertain to cell adhesion and migration have been discovered as a result of two-dimensional experiments. Perhaps the most obvious explanation for the preponderance of two-dimensional (2D) explorations preceding three-dimensional (3D) studies is the ease of experimentation. Cells are easier to culture, observe, and assay in two dimensions. However, in two dimensions, cells are forced to establish an artificial polarity based on the ability to bind immobilized ligand only on its ventral surface. On the other hand, cells in a 3D environment are completely surrounded by ECM, enabling receptors on all sides of the cell to bind proteins in the extracellular environment. Because of this very basic but important difference in receptor binding, cell behavior and morphology can be different in three dimensions than in two dimensions. A natural extension of understanding cell migration in three dimensions is the study of traction forces exerted in three dimensions.

Several studies have begun to address traction forces in pseudo-3D environments. However, to calculate forces in a completely 3D environment, several technical challenges must be overcome. One of the most significant hurdles is the difficulty associated with tracking displacements surrounding the cell in all three dimensions. To do so accurately would require image z-slices taken throughout the focal planes of the cell. Vanni and coworkers (2003) were able to circumvent this experimental and technical difficulty by noting that within a 3D matrix, it is possible to find individual cells randomly oriented within a 2D plane. Therefore, using plane stress analysis, one can neglect the out-of-plane forces, while neglecting circumferential strain in the calculations. Unlike many of the 2D techniques discussed earlier in the chapter, Vanni et al. do not rely on marker beads to track substrate displacements, but rather use differential interference contrast (DIC) microscopy to track movements in the fibers constituting the collagen matrix. Therefore, visualizing the displacements is much easier as they do not require the same refocusing issues encountered in many of the other displacement tracking methods. Although this technique begins to address traction generation in 3D matrices, it neglects z-displacements and does not calculate the forces exerted on all sides of the cell.

Rather than embedding the cells in a 3D matrix, Beningo et al. (2004) sandwiched the cells between two 2D gels. Both gels were derivatized with ECM, thereby permitting the cells to bind both their dorsal and ventral receptors. In this sandwiched culture, fibroblast morphology more closely resembles cells plated in 3D culture rather than those plated on 2D substrates. In addition to morphological changes, the cells exhibited fewer large actin bundles and few focal adhesions. By embedding both sides of the sandwich with submicrometer-sized beads, they used traction force microscopy to measure the forces exerted by the cell on both its dorsal and ventral surfaces. Computationally, the technique is similar to that developed by Dembo and Wang (1999) several years prior; however, displacements and tractions for each surface are solved separately. While Beningo et al. (2004) created a system where the cell can bind on both its dorsal and ventral surfaces and the forces exerted by the cell on the substrate can be calculated, this technique does not fully recapitulate an *in vivo*-like 3D environment.

To fully understand how a cell interacts in its native 3D environment, traction algorithms must be adapted for 3D calculations. However, a number of important problems remain to be addressed using 2D systems. For instance, it is still not well understood how focal adhesions contribute to force generation and which focal adhesion proteins are necessary. While the traction techniques described in this chapter have made significant advances in this field, the signaling pathways that lead to force generation are still not fully explained. Ultimately, by elucidating the mechanisms of traction generation, a better understanding of cell adhesion and migration will result in critical insights into development, wound healing, and other physiological processes, in addition to better design of biomaterials for tissue engineering applications.

SUGGESTED READING

Beningo, K. A., and Wang, Y. L. (2002). Flexible substrata for the detection of cellular traction forces. *Trends Cell Biol.* **12**(2): 79–84.

This review takes a brief look at various tools to measure cellular tractions, highlights some of the major findings involving fibroblast migration, and looks forward to the future of studies involving cellular tractions.

Bershadsky, A. D., Balaban, N. Q., and Geiger, B. (2003). Adhesion-dependent cell mechanosensitivity. *Annu. Rev. Cell Dev. Biol.* **19**: 677–695.

Bershadsky et al. discuss the phenomenon of mechanosensitivity and review what is known about the signaling components at adhesion sites that aid in the ability of cells to sense and respond to mechanical changes in their surroundings.

http://www.cellmigration.org (accessed 2005).

This web page was assembled by a consortium of more than 30 investigators to tackle the problem of cell migration using a collaborative, multidisciplinary approach. The web site provides details on novel reagents, techniques, and findings in cell migration research.

Lauffenburger, D. A., and Horwitz, A. F. (1996). Cell migration: A physically integrated molecular process[review]. *Cell* **84**: 359–369.

This review is widely cited as it provides an analysis of cell migration and adhesion from molecular, cellular, and whole tissue perspectives, focusing mostly on the molecular level signals that result in larger-scale phenomena.

Li, S., Guan, J. L., and Chien, S. (2005). Biochemistry and biomechanics of cell motility. *Annu. Rev. Biomed. Eng.*, **7**: 105–150.

A very thorough review of what is known about the molecular mechanisms that lead to cell adhesion and migration. This review breaks down some of the signaling cascades and attempts to integrate them into a larger picture that describes the cell's response to its chemical and mechanical environment.

REFERENCES

Balaban, N. Q., Schwarz, U. S., Riveline, D., Goichberg, P., Tzur, G., Sabanay, I., Mahalu, D. Safran, S., Bershadsky, A., Addadi, L., and Geiger B. (2001). Force and focal adhesion assembly: A close relationship studied using elastic micropatterned substrates. *Nat. Cell Biol.* **3**: 466–472.

Beningo, K. A., M. Dembo, I. Kaverina, J. V. Small and Y. L. Wang (2001). Nascent focal adhesions are responsible for the generation of strong propulsive forces in migrating fibroblasts. *J. Cell Biol.* **153**: 881–888.

Beningo, K. A., M. Dembo and Y. L. Wang (2004). Responses of fibroblasts to anchorage of dorsal extracellular matrix receptors. *Proc. Natl. Acad. Sci. USA* **101**: 18024–18029.

Beningo, K. A. and Y. L. Wang (2002). Flexible substrata for the detection of cellular traction forces. *Trends Cell Biol.* **12**: 79–84.

Burridge, K., K. Fath, T. Kelly, G. Nuckolls and C. Turner (1988). Focal adhesions: Transmembrane junctions between the extracellular matrix and the cytoskeleton. *Annu. Rev. Cell Biol.* **4**: 487–525.

Burton, K. and D. L. Taylor (1997). Traction forces of cytokinesis measured with optically modified elastic substrata. *Nature* **385**: 450–454.

Condeelis, J., A. Hall, A. Bresnick, V. Warren, R. Hock, H. Bennett and S. Ogihara (1988). Actin polymerization and pseudopod extension during amoeboid chemotaxis. *Cell Motil. Cytoskel.* **10**: 77–90.

Dembo, M., T. Oliver, A. Ishihara and K. Jacobson (1996). Imaging the traction stresses exerted by locomoting cells with the elastic substratum method. *Biophys. J.* **70**: 2008–2022.

Dembo, M. and Y. L. Wang (1999). Stresses at the cell-to-substrate interface during locomotion of fibroblasts. *Biophys J.* **76**(4): 2307–2316.

Du Roure, O., A. Saez, A. Buguin, R. H. Austin, P. Chavrier, P. Siberzan and B. Ladoux (2005). Force mapping in epithelial cell migration. *Proc. Natl. Acad. Sci. USA* **102**: 2390–2395.

Elson, E. L., S. F. Felder, P. Y. Jay, M. S. Kolodney and C. Pasternak (1999). Forces in cell locomotion. *Biochem. Soc. Symp.* **65**: 299–314.

Engler, A., L. Bacakova, C. Newman, A. Hategan, M. Griffin and D. Discher (2004). Substrate compliance versus ligand density in cell on gel responses. *Biophys. J.* **86**(1, Pt. 1): 617–628.

Galbraith, C. G. and M. P. Sheetz (1997). A micromachined device provides a new bend on fibroblast traction forces. *Proc. Natl. Acad. Sci. USA* **94**: 9114–9118.

Galbraith, C. G., K. M. Yamada and M. P. Sheetz (2002). The relationship between force and focal complex development. *J. Cell Biol.* **159**: 695–705.

Geiger, B. and A. Bershadsky (2001). Assembly and mechanosensory function of focal contacts. *Curr. Opin. Cell Biol.* **13**: 584–592.

Geiger, B. and A. Bershadsky (2002). Exploring the neighborhood: Adhesion-coupled cell mechanosensors. *Cell* **110**: 139–142.

Halliday, N. L. and J. J. Tomasek (1995). Mechanical properties of the extracellular matrix influence fibronectin fibril assembly *in vitro*. *Exp. Cell Res.* **217**: 109–117.

Harris, A. K., P. Wild and D. Stopak (1980). Silicone rubber substrata: a new wrinkle in the study of cell locomotion. *Science* **208**: 177–179.

Heath, J. P. and G. A. Dunn (1978). Cell to substratum contacts of chick fibroblasts and their relation to the microfilament system: A correlated interference-reflexion and high-voltage electron microscope study. *J. Cell Sci.* **29**: 197–212.

Katoh, K., Y. Kano, M. Masuda, H. Onishi and K. Fujiwara (1998). Isolation and contraction of the stress fiber. *Mol. Biol. Cell.* **9**: 1919–1938.

Kreis, T. E. and W. Birchmeier (1980). Stress fiber sarcomeres of fibroblasts are contractile. *Cell* **22**(2 Pt 2): 555–561.

Lauffenburger, D. A. and A. F. Horwitz (1996). Cell migration: A physically integrated molecular process. *Cell* **84**: 359–369.

Lee, J., M. Leonard, T. Oliver, A. Ishihara and K. Jacobson (1994). Traction forces generated by locomoting keratocytes. *J. Cell Biol.* **127**(6, Pt. 2): 1957–1964.

Lo, C. M., D. B. Buxton, G. C. Chua, M. Dembo, R. S. Adelstein and Y. L. Wang (2004). Nonmuscle myosin IIb is involved in the guidance of fibroblast migration. *Mol. Biol. Cell.* **15**: 982–989.

Lo, C. M., H. B. Wang, M. Dembo and Y. L. Wang (2000). Cell movement is guided by the rigidity of the substrate. *Biophys. J.* **79**: 144–152.

Marganski, W. A., M. Dembo and Y. L. Wang (2003). Measurements of cell-generated defor-
    mations on flexible substrata using correlation-based optical flow. *Methods Enzymol.*
    361: 197–211.

Mochitate, K., P. Pawelek and F. Grinnell (1991). Stress relaxation of contracted collagen gels:
    Disruption of actin filament bundles, release of cell surface fibronectin, and downregulation
    of DNA and protein synthesis. *Exp. Cell Res.* **193**: 198–207.

Munevar, S., Y. Wang and M. Dembo (2001a). Traction force microscopy of migrating normal
    and H-ras transformed 3T3 fibroblasts. *Biophys. J.* **80**: 1744–1757.

Munevar, S., Y. L. Wang and M. Dembo (2001b). Distinct roles of frontal and rear cell–substrate
    adhesions in fibroblast migration. *Mol. Biol. Cell.* **12**: 3947–3954.

Munevar, S., Y. L. Wang and M. Dembo (2004). Regulation of mechanical interactions between
    fibroblasts and the substratum by stretch-activated Ca^{2+} entry. *J. Cell Sci.* **117**(Pt. 1):
    85–92.

Nelson, C., S. Raghavan, J. Tan and C. Chen (2003). Degradation of micropatterned surfaces by
    cell-dependent and independent processes. *Langmuir* **19**: 1493–1499.

Oliver, T., M. Dembo and K. Jacobson (1995). Traction forces in locomoting cells. *Cell Motil.
    Cytoskel.* **31**: 225–240.

Pelham, R. J., Jr. and Y. Wang (1997). Cell locomotion and focal adhesions are regulated by
    substrate flexibility. *Proc. Natl. Acad. Sci. USA* **94**: 13661–13665.

Reinhart-King, C. A., M. Dembo and D. A. Hammer (2003). Endothelial cell traction forces on
    RGD-derivatized polyacryalmide substrata. *Langmuir* **19**: 1573–1579.

Reinhart-King, C.A., M. Dembo and D.A. Hammer (2005). The dynamics and mechanics of
    endothelial cell spreading. *Biophys. J.* in press.

Tan, J. L., J. Tien, D. M. Pirone, D. S. Gray, K. Bhadriraju and C. S. Chen (2003). Cells lying on
    a bed of microneedles: an approach to isolate mechanical force. *Proc. Natl. Acad. Sci. USA*
    100: 1484–1489.

Tranquillo, R. T., M. A. Durrani and A. G. Moon (1992). Tissue engineering science: Conse-
    quences of cell traction force. *Cytotechnology* **10**: 225–250.

Vanni, S., B. C. Lagerholm, C. Otey, D. L. Taylor and F. Lanni (2003). Internet-based
    image analysis quantifies contractile behavior of individual fibroblasts inside model tissue.
    Biophys. J. **84**: 2715–2727.

Wang, H. B., M. Dembo, S. K. Hanks and Y. Wang (2001). Focal adhesion kinase is involved in
    mechanosensing during fibroblast migration. *Proc. Natl. Acad. Sci. USA* **98**: 11295–11300.

Weiss, P. A. (1934). In vitro experiments on the factors determining the course of the outgrowing
    nerve fiber. *J. Exp. Zool.* **68**: 393–448.

Wong, J., A. Velasco, P. Rajagopalan and Q. Pham (2003). Directed movement of vascular
    smooth muscle cells on gradient-compliant hydrogels. *Langmuir* **19**: 1908–1913.

Yeung, T., P. C. Georges, L. A. Flanagan, B. Marg, M. Ortiz, M. Funaki, N. Zahir, W. Ming,
    V. Weaver and P. A. Janmey (2005). Effects of substrate stiffness on cell morphology,
    cytoskeletal structure, and adhesion. *Cell Motil. Cytoskel.* **60**: 24–34.

2

CONTROL OF ENDOTHELIAL CELL ADHESION BY MECHANOTRANSMISSION FROM CYTOSKELETON TO SUBSTRATE

ROSALIND E. MOTT and BRIAN P. HELMKE

Department of Biomedical Engineering and Cardiovascular Research Center, University of Virginia, Charlottesville, Virginia, USA

The physical interactions among cells and extracellular matrix guide the physiological behavior of tissues. The success of reparative medicine and tissue engineering depends on the ability to promote migratory cell behavior for wound healing and vascular development and a quiescent phenotype for tissue and blood vessel stabilization. Thus, an improved understanding of how cells sense mechanochemical cues in their local environment is a critical challenge facing biomedical engineers. Because the cytoskeleton represents an interconnected mechanical structure that dynamically interacts with signaling proteins throughout the cell, it is a primary candidate to act as a central controller to integrate and differentiate complex signaling networks that determine cellular phenotype. This chapter explores mechanisms of physical interaction and signaling from the cytoskeleton to the extracellular matrix, using endothelial responses to hemodynamic forces as an example to suggest that a systems modeling approach will open new directions for vascular bioengineering.

I. INTRODUCTION

The German scientist Julius Wolff first established the idea that biological form adapts in response to mechanical stimuli. Although Wolff's theory focused on the functional remodeling of bone, work during the last 25 years demonstrates that most biological tissues adapt their genetic, proteomic, and morphological profiles in response to a diverse set of mechanical stimuli, including gravity, blood flow, and contractile forces generated within the tissue itself. In addition to recognizing cell physiology to be a complex system of genomic and proteomic signaling networks, understanding the molecular basis of force transmission and mechanotransduction is critical to the successful development of new therapeutic approaches.

Some of the most profound observations of tissue adaptation to mechanical force have been observed in the cardiovascular system. In particular, vascular inflammatory diseases such as atherosclerosis involve pathological remodeling of composition and structure in the blood vessel wall that occurs regionally near locations of spatially and temporally complex blood flow profiles (Dewey et al., 1981; Nerem et al., 1981). At the blood–tissue interface, endothelial permeability to macromolecules is increased in regions of atherosclerotic lesion formation, and the distribution of both cytoskeletal structures and intercellular junctions is significantly different from that in endothelial cells located in nonlesion artery regions. Furthermore, the gene expression profile in regions of atherogenesis reflects a pro-inflammatory genotype, and the expression of proteins associated with leukocyte adhesion and transmigration is upregulated. The endothelium overlying lesions contributes to remodeling of the extracellular matrix (ECM) composition from a physiological basement membrane consisting primarily of collagen IV and laminin to one that is enriched in fibronectin, fibrinogen, thrombospondin, and other collagen subtypes (van Zanten et al., 1994). Most notably, enrichment of collagen I and III has been implicated in increased thrombogenicity in these regions. Because the endothelium serves as the primary interface between blood flow and vessel wall tissue cells, endothelial cells have been implicated in regulating vessel wall remodeling by transducing hemodynamic stress into biochemical signaling events associated with lesion formation. Conversely, restoration of physiological function of the endothelium after vascular interventions such as balloon angioplasty and stent placement is likely to be critical for preventing downstream clinical events such as stroke and thrombosis. In addition, promoting endothelial function in tissue-engineered blood vessels will improve patency after graft placement. In both cases, an improved understanding of the mechanisms of endothelial mechanotransduction that guide cell adhesion and migration and monolayer assembly would enable the rational design of pharmacological and physical cues to engineer a healthy endothelium.

Hemodynamic shear stress, mechanical stretch, and hydrodynamic pressure constitute the primary mechanical forces acting on the artery wall. As early as 1976, a significant increase in collagen I and III production by vascular smooth muscle cells was measured after *in vitro* exposure to 10% cyclic stretch of the substrate for a period of 2 days (Leung et al., 1976). This observation was consistent with the correlation between enhanced elastin and collagen synthesis and increased medial tension *in vivo*. Such force-induced ECM remodeling has been hypothesized to alter the mechanical properties of the arterial wall and to contribute to vascular hypertrophy. Five years later, Nerem et al. (1981) reported that the shapes of endothelial cells in the rabbit descending aorta varied with distance from flow dividers at bifurcations. Cells were elongated, with their major axis aligned parallel to the direction of blood flow, but did not exhibit this preferred orientation closer to the ostia. Endothelial cells exposed to unidirectional laminar shear stress *in vitro* exhibited a similar ellipsoidal shape oriented parallel to the flow direction, whereas cells in static fluid culture appear rounded or polygonal and without preferred orientation (Dewey et al., 1981). Since these initial studies were performed, endothelial mechanotransduction of the hemodynamic shear stress profile to modulate arterial wall physiology and pathology has been implicated in the adaptation of gene and protein expression, structure and morphology, and communication with adjacent cell types. A similar array of adaptive responses by the endothelium to substrate stretch suggest that cells are capable of sensing multiple mechanical cues simultaneously. More recent reductionist models have focused on decoupling endothelial cell responses to stretch and shear stress to decipher how these stimuli are integrated in a complex *in vivo* environment (Qiu and Tarbell, 2000).

Cytoskeletal adaptation to hemodynamic forces has clear implications for vascular function. For example, mice lacking the vimentin gene exhibit a reduction in flow-dependent vasodilation and arterial remodeling (Schiffers et al., 2000) that is in part attributable to impaired cellular contractility and ability to migrate. Cytoskeletal disruption in cultured endothelial cells inhibits the flow-induced decrease in expression of endothelin-1, a potent vasoconstrictor (Malek et al., 1999). Intravascular disruption of microtubules *in vivo* abrogates flow-dependent vasodilation, but not vascular smooth muscle cell relaxation by sodium nitroprusside treatment, indicating that an intact cytoskeleton is required for endothelium-dependent regulation of blood pressure and flow distribution (Sun et al., 2001). Thus, the ability of the endothelium to sense and adapt to changes in the local hemodynamic environment is critical to maintaining vascular homeostasis *in vivo*, but the mechanisms of mechanotransduction that promote adaptation remain difficult to decipher. Identification of the upstream mechanosensory trigger has remained elusive, as mechanical stimuli induce multiple biochemical events at short time scales at different locations in the cell. Activation of ion channels in the plasma membrane, heterotrimeric G-proteins in the plasma membrane and near intercellular junctions, and focal adhesion scaffold–associated proteins occurs within seconds of the onset of steady unidirectional laminar shear stress, and the time course of activity may depend on the temporal profile of applied force (White et al., 2001). The cytoskeleton is a likely candidate for the redistribution of intracellular forces to initial locations of signaling network activation. Structures that potentially transmit force from the extracellular milieu to the cytoskeleton include components of the cell surface glycocalyx (Weinbaum et al., 2003; see Chapter 7), and focal adhesion complexes linked via integrin receptors to the ECM (Wang et al., 1993). However, a critical question remains to be answered: What are the mechanisms that modulate a spatially and temporally complex mechanical environment to enable cells to choose which stimuli should trigger adaptive responses and which should promote quiescent behavior? This chapter examines the composition and dynamic regulation of cytoskeletal and adhesion-associated scaffolds that determine endothelial interactions with the ECM.

II. MECHANOADAPTATION OF ADHESION-ASSOCIATED STRUCTURES

Early studies measuring the relationship between endothelial cell shape and location in the artery wall demonstrated parallel trends in intracellular structural morphology. Continuous exposure of endothelial cells to hemodynamic force profiles for 6 to 24 h *in vitro* induces morphological adaptations that mimic those observed *in vivo*. In general, the remodeling process affects most cellular structures, including the cytoskeleton, adherens junctions, focal adhesion sites, and ECM. In many cases, redistribution of signaling molecules associated with these structures also occurs, suggesting that cell remodeling serves, in part, as a negative feedback mechanism that helps to desensitize biochemical networks to continuous changes in a complex array of extracellular mechanical cues.

A. Cytoskeleton

Intracellular structural remodeling serves to reorganize the spatial distribution of all three major cytoplasmic cytoskeleton networks. Early observations of elongated endothelial cell morphology aligned parallel to fluid shear stress corresponded with disassembly of F-actin dense peripheral bands and formation of bundled stress fibers that were oriented approximately parallel to the major axis of cell elongation (Dewey et al., 1981; Levesque and Nerem, 1985). In confluent monolayers of endothelial cells exposed to a wall shear stress of 15 dyn/cm^2, intermediate filaments and microtubules also exhibit a gradual redistribution from rounded filament network shapes to a preferred orientation corresponding to the direction of cell elongation (Galbraith et al., 1998). Measurement of F-actin distribution over a range of time points following onset of steady unidirectional shear stress demonstrates that the actin cytoskeleton is a primary determinant of cell shape during adaptation (Noria et al., 2004). Transient inactivation of the small GTPase RhoA is likely to facilitate actin dynamics (Tzima et al., 2001), but a mechanism of temporal amplification that allows reorganization to continue for longer periods remains unknown. An intact microtubule network is required for both F-actin reorganization and cell shape elongation to occur (Malek and Izumo, 1996), suggesting that a balance of cytoskeletal components exists to modulate cellular behavior during this process. The role of intermediate filaments in this process is less clear, as disruption with acrylamide did not prevent adaptation (Malek and Izumo, 1996). However, the vimentin network is required for cell migration and maintenance of mechanical integrity during adaptation of the artery wall to flow *in vivo* (Eckes et al., 1998).

The endothelial cell cytoskeleton also remodels in a manner consistent with cell elongation in response to chronic substrate stretch. In response to cyclic uniaxial stretch (1 Hz), endothelial cells exhibited an elongated shape that was nearly perpendicular to the stretch axis. The orientation angle of actin stress fibers varied as a function of stretch magnitude (Takemasa et al., 1997) and location in the artery tree from which the primary cells were isolated (Sugimoto et al., 1995). This morphological and cytoskeletal adaptation occurred not just *in vitro,* but also in intact arteries subjected to axial stretch *ex vivo* (Sipkema et al., 2003). Adaptation of the endothelium to axial stretch served to decrease reactivity to acetylcholine, which induces activation of endothelial nitric oxide synthase (eNOS). This observation is consistent with the decreased eNOS activity and vasodilator function observed in endothelium overlying atherosclerotic lesions, where actin cytoskeleton structure is more randomly oriented (Nerem et al., 1981). Because eNOS activity depends on the polymerization state of the actin cytoskeleton (Su et al., 2003), the force-mediated regulation of eNOS may be directly linked to cytoskeletal remodeling. Cyclic strain also modulates mitochondrion-dependent production of reactive oxygen species (ROS) in response to cyclic strain in a manner that may depend on interactions between mitochondria and the actin cytoskeleton (Ali et al., 2004). Taken together, these observations indicate that mechanically induced structural remodeling, especially in the actin cytoskeleton, is important not only for mechanical stability of the endothelium but also for maintenance of physiological functions.

Substantial changes in cytoskeletal structure induced by hemodynamic forces are translated into alterations in mechanical properties. Adaptation of cytoskeletal structure to steady unidirectional arterial shear stress for 24 h resulted in a twofold increase in both the elastic modulus and the viscosity of the cells (Sato et al., 1996). The increase in viscoelastic moduli directly corresponded to the degree of actin polymerization in the cells. Theoretical explanations of these phenomena are often based on the idea that adherent cells exist in a state of prestress that is characterized by a balance between tension and compression forces in the cytoskeleton (Wang et al., 1993; see Chapter 4). Although the cell stiffening behavior that results from mechanical challenge depends on microfilaments and microtubules, the contribution of intermediate filaments in this model is less clear. As *in vitro* measurements of shear modulus of protein polymer gels demonstrate that vimentin is significantly less stiff than actin (Janmey et al., 1991), intermediate filaments may not act as principal force-bearing elements in the cytoplasm. This hypothesis is supported by both theoretical models of intact cells and experimental measurements of cell stiffening in cells lacking vimentin (Wang and Stamenovic, 2000). However, because intermediate filament networks remodel without disassembly during adaptation to shear stress (Helmke et al., 2000), they may play a critical role in maintaining the mechanical integrity of the cells during mechanical challenges.

B. Focal Adhesion Sites

For the mechanical adaptation in the cytoskeleton to impact the surrounding extracellular environment, forces associated with adaptation must be transmitted between the cytoplasm and the cell exterior. Focal adhesion sites, the intricate structures that define the interface between cytoskeleton and substrate (Fig. 2.1), also remodel during the application of external force, suggesting a direct mechanochemical link between the cytoskeleton and the ECM. Tandem scanning confocal microscopy first revealed both the coalescence and the alignment of focal adhesion structures in cells exposed to steady unidirectional laminar shear stress (Davies et al., 1994). Throughout the extensive remodeling process, cells maintained a constant total adhesion area, even though the number of individual focal adhesion sites decreased.

After adaptation to steady shear stress for 24 h, focal adhesion sites, recognizable by the immunofluorescence labeling of vinculin, were redistributed to the cell periphery and became elongated with their shape aligned parallel to the flow axis (Galbraith et al., 1998). Vinculin colocalized with the ends of newly formed actin stress fibers, suggesting a direct link between adhesion site remodeling and redistribution of tension in the actin cytoskeleton. In other studies, both vinculin and the integrin receptor $\alpha v \beta 3$ localized predominantly to the upstream end of cells in streaklike adhesions (Girard and Nerem, 1995). Similarly, focal adhesion kinase (FAK) and paxillin translocated from the cytosol to adhesion sites near the cell periphery after adaptation of human pulmonary endothelial cells to flow (Shikata et al., 2005). Shear stress induces a dramatic increase in binding of the adapter protein Shc to $\alpha v \beta 3$ and $\alpha 5 \beta 1$ (Jalali et al., 2001). In contrast, the fibronectin receptor $\alpha 5 \beta 1$ and the large scaffold protein talin also redistributed into streaklike, aligned adhesions, but were present throughout the cell

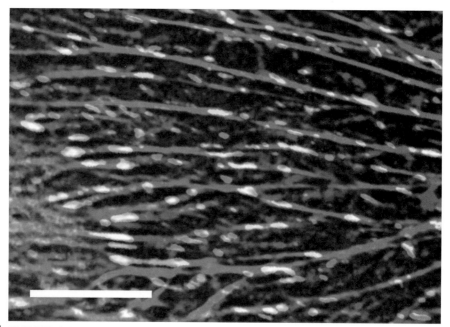

FIGURE 2.1 Adhesion sites connect the intracellular architecture to the ECM to serve as mechanochemical links between the inside and outside of the cell. In a subregion of an endothelial cell, discrete adhesion sites containing vinculin (green) orchestrate a continuous connection from the actin cytoskeleton (red) to the fibronectin matrix (blue). Bar = 10 μm. (See Color Plate 2)

(Girard and Nerem, 1995). The differences between spatial redistribution of specific molecular complexes may indicate that shear stress promotes the segregation of various adhesion molecules to create distinctly different adhesive regions of the cell. One type of adhesion may be more dynamic and biochemically active than another, thus allowing for spatially defined signaling heterogeneity. The importance of focal adhesion-associated signaling events linked to FAK, Shc, and paxillin is revisited later in this chapter.

Although mechanisms modulating the rate of focal adhesion remodeling during adaptation to hemodynamic forces are not yet clear, both ECM composition and mechanotransmission between the cytoskeleton and the matrix may serve as primary factors in this process. Cells plated on a 6-nm–thick gelatin layer exhibited a flow-induced remodeling rate that was reduced by nearly 47% in comparison to that of uncoated glass (Davies et al., 1994), suggesting that either integrin-matrix adhesion strength or cytoskeletal reinforcement directly impacts focal adhesion remodeling. This idea is further supported by comparing endothelial cell monolayers at different stages after reaching confluence. Morphological changes occur much more rapidly in cells that have just reached confluence than in cells that have been confluent for at least 2 days (Noria et al., 2004). After adaptation, adhesion-associated proteins were redistributed to the ends of actin stress fibers near the cell periphery and changed from an ovoid to an extended linear shape (Fig. 2.2), suggesting that focal adhesion remodeling supports mechanical integrity of the endothelial monolayer by maintaining structural links among cells and the ECM. During the 24-h process of adaptation to shear stress, proteins such as tensin disappeared transiently from adhesion sites over a time scale consistent with the dynamics of actin disassembly and reassembly into stress

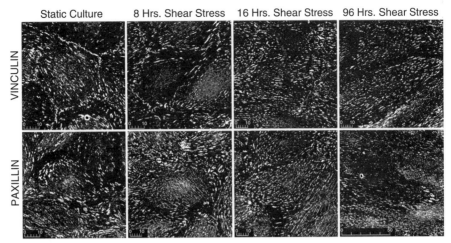

FIGURE 2.2   Adhesion structure adaptation to shear stress. Both paxillin and vinculin redistribute to ends of actin stress fibers near the cell periphery and change from an ovoid into a linear shape. Alignment of the adhesion sites becomes apparent after 16 h of shear stress exposure and is maintained until at least 96 h. This dramatic remodeling process supports mechanical integrity of the endothelium by physically linking cell structure to the ECM. Shear stress, left to right. Reprinted, with permission, from Noria et al. (2004).

fibers. During this period of active remodeling, the intermediate filament cytoskeleton serves as the mechanical support for the cytoplasm, because it links to $\alpha v \beta 3$ integrins in focal adhesions (Tsuruta and Jones, 2003) and does not disassemble. Recent theoretical predictions are consistent with the idea that maintenance of tension through focal adhesions by the cytoskeleton plays a major role in the mechanosensitivity of adhesion site remodeling (Nicolas et al., 2004). Overall, these studies clearly demonstrate that externally applied shear stress promotes both morphological and compositional adaptation of focal adhesion sites in endothelial cells.

Mechanical strain of the ECM substrate also induced reorganization of the focal adhesion sites in endothelial cells. The most curious difference between the effects of shear stress and strain on endothelial cells is that the former promotes structural alignment that is parallel to the direction of flow, whereas the latter promotes an alignment nearly perpendicular to the principal direction of stretch. Although it is likely that focal adhesion-mediated signaling plays a role in triggering these different responses, the mechanisms remain elusive. Cyclic strain results in the translocation of FAK and paxillin from the cytosol to adhesion sites in the cell periphery (Shikata et al., 2005), as in the response to shear stress. Furthermore, focal adhesion remodeling to strain also depends on the integrin type (Yano et al., 1997). After 4 h cyclic strain, a subset of adhesion sites aligned in a pattern parallel to the direction of stretch. Interestingly, this adhesion remodeling response was specific to the focal adhesions that contained $\beta 1$-integrins; $\beta 3$-integrins failed to reorganize into parallel adhesions. One hypothesis explaining this behavior is that the composition of the ECM conveys specificity in which integrin signals participate in structural remodeling. In this case, $\alpha 5 \beta 1$ binds to fibronectin, and $\alpha 2 \beta 1$ serves as a collagen receptor. In contrast, $\alpha v \beta 3$ serves as a receptor for an assortment of ligands, including vitronectin, fibrinogen, and fibronectin. Although the lack of $\alpha v \beta 3$ reorganization reported here appears to conflict with the polarized distribution after adaptation to shear stress (Girard and Nerem, 1995), neither of these studies examined the composition of the ECM after force-mediated adaptation. The reorganization of $\beta 3$-integrins in response to applied force is likely to be sensitive to the type of ligand present in the local ECM environment. These data strongly implicate the spatial distribution of ECM composition in regulating force-dependent signaling at the cell–matrix interface, but few quantitative measurements exist of compositional changes induced by mechanical stretch on focal adhesions. In addition, comparison of the responses to cyclic stretch and steady stretch would improve understanding of the relative effects of high pulsatility and hypertension in modulating adaptation of the artery wall to hemodynamic forces (Lehoux et al., 2005).

C. Extracellular Matrix

Alterations in hemodynamic forces induce remodeling outside of the cell as well as in the cytoplasm. Endothelial cells exposed to laminar shear stress *in vitro* adapt both the composition and morphology of deposited ECM (Thoumine et al., 1995). Remarkably, the observed changes varied for each matrix component examined. Shear stress induced a continuous increase in laminin expression, and both laminin and collagen IV were reorganized into fibril bundles but did not align in the direction of flow. Neither the expression level nor the spatial distribution of vitronectin changed during shear stress, in contrast to the redistribution of $\alpha v \beta 3$ vitronectin receptors to the upstream edges of the cells (Girard and Nerem, 1995). Fibronectin expression was transiently suppressed after onset of shear stress, but was significantly increased after 48 h, at which time bundled fibrils that were aligned parallel to the flow direction existed (Fig. 2.3) (Thoumine et al., 1995). The amount of fibronectin protein significantly decreased in the culture medium, ECM, and intracellular space within 12 h (Gupte and Frangos, 1990), and the global decrease was due to a shear-induced inhibition of *de novo* synthesis of fibronectin. As decreased levels of fibronectin on the abluminal surface of the cell and in the serum would inhibit the aggregation of platelets (see Chapter 13), this response could protect the endothelium from inflammatory processes during adaptation to shear stress. The shear-induced remodeling of the matrix composition and morphology on the basal surface is consistent with the assembly of a matrix similar to that which exists in atheroprotected regions of the artery wall.

Cyclic biaxial stretch of the substrate under endothelial cells in culture promotes an increase in fibronectin protein expression after 24 h (Gorfien et al., 1990). However, most measurements of ECM adaptation to hemodynamic forces have not specifically addressed the role of endothelial cells. In an organotypic culture model representing the fetal lung, mechanical strain induced an increase in fibronectin release into the culture medium (Mourgeon et al., 1999). The production of fibronectin was post-transcriptionally regulated and presumed to be derived from protein released from the cell layer or matrix rather than from intracellular stores. In a perfused arterial culture model, fibronectin protein content was increased in medial regions nearest the intima after exposure to high luminal pressure (Bardy et al., 1995).

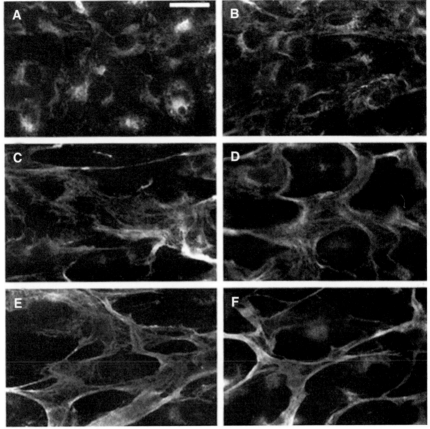

FIGURE 2.3 Shear stress promotes fibronectin fibril bundling and alignment parallel to the flow direction. Endothelial cells were either (A) maintained in static culture or exposed to 30 dyn/cm² for (B) 12 h, (C, D) 24 h, or (E, F) 48 h and then stained for fibronectin. Bar = 25 μm. Reprinted, with permission, from Thoumine et al. (1995).

In fibroblasts on stretched collagen matrices, secretion of both tenascin-C and collagen XII was increased compared with that from cells on relaxed collagen gels (Chiquet et al., 1996). This measurement is consistent with increased tenascin-C expression measured in an *in vivo* model of mechanical loading on skeletal muscle connective tissue (Fluck et al., 2000). Both the *in vitro* and *in vivo* models involve fibroblast-type cells in nonvascular connective tissues under physiological tensile loads. Nevertheless, these studies emphasize that external force directly impacts the composition of the adhesive matrix. It is interesting to note that the fibrillar structure of the matrix is also force-dependent. Fibroblasts initially plated on collagen attached to a fixed surface secreted fibronectin and assembled a fibrillar matrix, whereas cells on free floating collagen failed to form fibrils (Halliday and Tomasek, 1995). This result supports the idea that the ability of cells to generate traction force against their substrate modulates both the expression of matrix proteins and their assembly into fibrillar structures. Although most of the work examining the effects of stretch on matrix composition and structure has been performed using fibroblasts, endothelial cells are also subject to mechanical stretch *in vivo* and may induce ECM remodeling in a similar manner.

A large number of biomechanical studies indicate that both hemodynamic shear stress and substrate stretch directly impact the composition and form of the ECM. However, the effects of these force-induced matrix remodeling events on cellular signaling and function remain to be elucidated. Presumably, matrix adaptation occurs in parallel with cytoskeletal and focal adhesion reorganization. The structural reorganization may impart strain between these structures, and the compositional changes are likely to alter biochemical signaling near these locations. To gain insight into mechanisms that regulate cellular physiology, it is necessary to consider dynamic responses that imply force transmission as the cytoskeleton, adhesions and matrix simultaneously transform.

III. MECHANOTRANSMISSION BETWEEN CYTOSKELETON AND EXTRACELLULAR MATRIX

Most models of cellular mechanobiology examine morphology or signaling activity at discrete time points during the process of adaptation. In the cases described earlier, the cytoskeleton, focal adhesion sites, and ECM exhibited spatial reorganization that was qualitatively characterized by average morphological parameters. However, this approach lacks the time resolution necessary to fully understand the molecular mechanisms involved in force transmission and initiation of signaling. In response to body forces such as hemodynamic shear stress, pressure, and substrate strain, measurement of the dynamic process of reorganization reveals a spatial distribution of local mechanical events that cannot be detected in spatially averaged measurements. Recent approaches for applying local forces to regions of cells test hypotheses in which local activation results in distributed signaling or structural adaptation, and these techniques require time-resolved measurement of molecular displacement to draw conclusions regarding mechanotransmission from the exterior of the cell to the cytoplasm. Neither of these approaches is possible without advances in the dynamic measurement of behaviors on a subcellular length scale.

A. Dynamic Interactions among Cytoskeletal Components

Advances in live cell imaging using endogenously expressed fluorescent proteins such as green fluorescent protein (GFP) have enabled direct observation of dynamic intracellular processes. Cytoskeletal organization represented by morphological measurements on fixed endothelial cells suggests that the impact of unidirectional steady shear stress takes hours. In contrast, live cell measurements reveal regional displacement of the cytoskeleton only seconds to minutes following the onset of shear stress (Helmke et al., 2000). The rapid displacement induced by shear stress acting on the cell surface implicates the cytoskeleton in the redistribution of forces throughout the cytoplasm.

The first data to support this hypothesis arose from observations of GFP-vimentin that distributed to the endogenous intermediate filament cytoskeleton in aortic endothelial cells (Helmke et al., 2000). Intermediate filaments exhibited random wiggling and fluctuations in position in confluent cell layers under static fluid conditions. Within minutes of the onset of unidirectional steady laminar shear stress, significant directional displacement was measured in regions of the intermediate filament network. Quantitative spatiotemporal image analysis demonstrated that displacement magnitude was increased significantly in the downstream and apical regions of the cytoplasm, and maximum displacement was achieved during the first 3-min interval after onset of shear stress (Helmke et al., 2001). Tracking displacement of connections among intermediate filaments enabled computation of the strain tensor (Helmke et al., 2003). Thus, shear stress-induced structural dynamics in the cytoskeleton represented mechanical deformation in the cytoplasm. This analysis revealed cytoplasmic strain focusing, local maxima of strain that corresponded to locations consistent with cytoskeletal connections to focal adhesion sites and intercellular junctions. Such connections would provide a mechanism for redistribution of body forces acting on the cell surface to distinct structures known to participate in initial signaling events triggered by shear stress. However, direct measurement of force transmission among these physically connected structures remains to be performed.

Similar measurements using GFP-actin indicate that onset of shear stress also influences the rapid dynamics of this cytoskeletal component. In confluent monolayers of endothelial cells, shear stress induced displacement of actin stress fibers near the basal cell surface in a manner similar to that observed in the intermediate filament cytoskeleton. Cytoplasmic stress fiber connections were displaced up to 1 μm within 2 min of flow onset, suggesting that dynamic force-dependent interactions occur at these sites. Depolymerization of these stress fiber bundles often occurred on a time scale approaching 20 min, consistent with previous reports of partial disassembly of dense peripheral bands (Galbraith et al., 1998). In fact, actin polymerization mediated edge ruffle extension in confluent monolayers within 30 s of flow onset. This fast ruffling response was transient and preceded ruffle extension associated with cell migration in the downstream direction (Li et al., 2002). Edge extension during this directional cell migration requires activation of FAK and its recruitment into focal contacts near the leading edge.

Extracellular forces such as shear stress induce similar displacement distribution in both the actin and vimentin cytoskeletal networks, but the relative role of each component in redistributing intracellular forces remains unclear. Dynamic measurements of F-actin in the presence of the filament-severing protein gelsolin demonstrate that the shear modulus of the gel decreases with F-actin filament segment length (Janmey et al., 1988). Viscoelastic properties in the cytoplasm would therefore depend on the local state of actin polymerization, and regions near the edges of cells would be expected to become less effective at focusing force transmission as the dense peripheral bands disassemble under the influence of shear stress. An increase in polymerization dynamics locally in these regions also correlates with the increase in GFP-actin–mediated edge activity observed rapidly after onset of shear stress.

In contrast to polymer gels of F-actin or vimentin alone, a copolymer gel exhibits viscoelastic properties that closely mimic those of intact cells subjected to external force application. Steric entanglement of protein filaments partly explains the overall properties of the copolymer gel, but it is more likely that cross-linking protein interactions are the primary determinants of cytoskeletal constitutive properties. For example, plectin serves to link intermediate filaments to microtubules and microfilaments (Svitkina et al., 1996) in the vicinity of focal adhesion sites containing β4–integrins (Homan et al.,

2002). Thus, intermediate filaments are capable of supporting tensile forces during initial local depolymerization of microfilaments. In fact, onset of shear stress displaces intermediate filaments near the edges of adjacent endothelial cells in a monolayer in parallel directions by the same distance, and intermediate filaments do not disassemble on the time scale of actin depolymerization. This behavior occurs even though actin-mediated ruffling is increased in these regions, implicating the vimentin cytoskeleton in transmitting forces from one cell to the next and in providing mechanical stability during this initial period of remodeling.

Heterogeneous constitutive properties resulting in strain focusing near focal adhesion sites and intercellular junctions could elicit conformation changes in adapter proteins, integrins, or junction-associated proteins that potentiate biochemical signaling. As conformation also controls binding affinity to modulate assembly of these macromolecular complexes, signal amplification can occur through modulation of rates of assembly and clustering associated with mechanical stabilization at these sites. For example, integrin activation, binding to the adapter molecule Shc, and subsequent activation of the JNK signaling cascade occur only at newly-formed integrin–fibronectin or integrin–vitronectin binding sites (Jalali et al., 2001). Based on the redistribution of fibronectin and vitronectin receptors after adaptation to shear stress (Yano et al., 1997), the onset of tension through these new focal contacts after integrin ligation depends also on local force transmission to cytoskeletal components associated with these sites. Thus, the local mechanical properties of the cytoskeleton serve to promote activation of integrin-mediated signaling in response to hemodynamic forces acting at the cell surface.

Regional remodeling of mechanical properties also serves to modulate force transmission to the nucleus in a manner that depends on physical interactions among the cytoskeletal components. The shape of the endothelial cell nucleus deforms in response to substrate strain (Caille et al., 1998), demonstrating that the nuclear envelope is mechanically connected to focal adhesion sites. The nucleus shape also deforms after onset of fluid shear stress by a mechanism requiring an intact microtubule cytoskeleton (Stamatas and McIntire, 2001). In addition, microtubules serve to increase nuclear rotation and translation after onset of shear stress (Lee et al., 2005). These observations suggest that heterogeneous force redistribution by local cytoskeletal mechanics provides a mechanism for direct force-dependent regulation of gene expression, in addition to indirect regulation of transcription through cytoplasmic signaling networks.

B. Local Mechanical Cytoskeleton–ECM Interactions That Regulate Adhesion

Studies that apply external force at the whole cell and tissue levels have been critical in defining the general characteristics of adaptation. However, local application of forces to specific sites in or on the cell enables more careful dissection of the molecular mechanisms involved in triggering adaptation. Thus, several techniques have been developed to apply a local force to the cell, including modalities that involve microspheres or micropipets to probe molecular-level responses to force.

In recent years, microspheres have been used to construct detailed maps of the local effects of mechanical stress. One advantage of this approach is the ability to apply on the cell surface forces of a magnitude on the order of nanonewtons and with a spatial resolution of micrometers. Generally, a latex bead is displaced with an optical laser trap, or ferromagnetic beads on the cell surface are twisted with a controlled magnetic field. If the beads are pre-coated with adhesive ligands, molecular connections to the internal cytoskeleton can be established. These connections mimic the integrin-based adhesions to ECM, and adhesion site structure can be monitored as force is applied.

Using magnetic twisting cytometry with microspheres coated with integrin antibodies or the integrin ligand peptide arginine–glycine–aspartate (RGD), Wang et al. (1993) completed the first mechanical measurements of cytoskeletal reinforcement near the sites of applied force. When torque was applied to the microspheres after attachment to the cell surface, cytoskeletal stiffness increased in proportion to the applied tension. Subsequent measurements employed a microsphere coated with a fibronectin fragment containing type III repeat 7–10 domains (FNIII7-10), which specifically bind to $\alpha5\beta1$ integrins (Choquet et al., 1997). These latex microspheres adhered to the apical cell surface and underwent centripetal transport, which indicated attachment to the rearward-moving cytoskeleton. If the beads were held in place for more than 10 s with an optical trap force ranging from 5 to 60 pN, they escaped the restraining force of the optical trap, and subsequent attempts to retrap the beads were not successful. This behavior indicates that the connection of the microspheres to the cytoskeleton had been reinforced. The reinforcement response was highly localized in an area of 8–12 μm^2 around the bead center. Microspheres coated with $\beta1$-integrin antibody did not exhibit a reinforcement response unless soluble FNIII7-10 fragments were added to the medium, indicating that integrin ligation to the ECM, and not integrin clustering alone, is required for cytoskeletal reinforcement.

The cytoskeletal reinforcement and development of the focal contact depend on locally applied force. In response to 1-μm beads coated with FNIII7-10 and held in place with a laser trap, both $\beta1$-integrins and vinculin were recruited to the site, but vinculin was not recruited without the application of force (Galbraith et al., 2002). In contrast, larger beads with a diameter of 6 μm induced recruitment of both $\beta1$-integrins and vinculin without force application. In both cases, the vinculin recruitment to bead contact areas correlated well with reinforcement. It is worthy to note that vitronectin-coated beads recruited only vinculin and underwent reinforcement when adhered to cells lacking c-Src. Thus, the mechanisms regulating the mechanoresponse are specific to individual adhesive ligands. In all cases, the vinculin accumulation was inhibited when the cells were pretreated with ML-7, a myosin light chain kinase inhibitor, or phenylarsine oxide, a tyrosine phosphatase inhibitor. This suggests that the response to externally applied force involves both tyrosine dephosphorylation events and the generation of intracellular force. One possibility for dephosphorylation is that the receptor-like protein tyrosine phosphatase α (RPTP-α) is key to the reinforcement response (von Wichert et al., 2003). Overexpression of RPTP-α enhanced adhesion and spreading, whereas recruitment of paxillin to microsphere adhesion sites was impaired in RPTP-α–null cells. Second, the idea that intracellular force generation serves to enhance the assembly of adhesion-associated structures is further supported by a study that measured force at individual adhesion sites and adhesive area (Balaban et al., 2001). This study indicated that cells exert 5.5 nN/μm^2 on each focal adhesion, independently of the individual adhesion site area. More importantly for this discussion, assembly of GFP-vinculin and stress per adhesion site were decreased after disruption of actomyosin contractility by treatment with 2,3-butanedione monoxime (BDM), and disassembly of vinculin and force relaxation occurred simultaneously. Thus, stress transmitted to adhesion sites regulates the biochemical interactions among adapter molecules, indicating that mechanotransduction can be triggered by strain or stress focusing at these sites.

Mechanisms explaining the transmission of these locally applied forces must involve the proteins that provide a direct link from integrins to the cytoskeleton, including talin, α-actinin, tensin, and filamin. In fibroblast-like cells deficient in talin-1, paxillin recruitment and cytoskeletal reinforcement near microspheres coated with FNIII7-10 were significantly reduced and were rescued by expressing wild-type talin (Giannone et al., 2003). Filamin A-deficient melanoma cells exhibited less reinforcement than filamin A-rescued cells, but the talin linkage to the cytoskeleton appeared to dominate the response. The role of talin in recruiting focal adhesion proteins depends on phosphatidylinositol phosphate kinase type Iγ (PIPKIγ), an enzyme that phosphorylates phosphatidylinositol 4-phosphate to form the second messenger phosphatidylinositol 4,5-bisphosphate (PIP_2). Talin binding activates PIPKIγ and serves to localize PIP_2 to focal adhesion sites in fibroblasts (Di Paolo et al., 2002). PIP_2 activates vinculin via a conformational change that allows it to bind talin and α-actinin to promote the formation of adhesion sites (Gilmore and Burridge, 1996). Thus, tension-dependent cytoskeletal reinforcement triggers a cascade of spatially localized events that regulate the transmission of force between the cytoskeleton and the ECM.

Probing the cell surface with microspheres induces the rapid formation of focal complex-like structures in response to a local force profile, but most physiological cell adhesion structures form at the interface with the ECM. Stretching cells for 2 min on a flexible substrate after allowing them to spread for 10 min induced the formation of focal adhesion sites in fibroblasts expressing talin-1, but talin-1–/– cells formed significantly fewer new adhesion sites (Giannone et al., 2003). However, allowing the cells to spread for 1 h before stretching the substrate abolished this difference in strain-induced focal adhesion formation between the two cell types. These data indicate that talin-1 is a critical component of early focal complex formation and reinforcement, but recruitment of additional adapter proteins and cytoskeleton-associated components contributes to this process at longer times after initial cell adhesion to the substrate. Thus, in analysis of mechanotransduction mechanisms, the adhesive state of the cell and the maturity of its adhesion structures must be considered. In fact, the adhesive structures that are probed in microsphere studies are nascent focal complexes, as they differ in composition from mature focal contacts.

Forces applied locally through the substrate can be transmitted to both nascent and mature adhesion sites. As discussed in Chapter 1, substrate rigidity guides cell migration rate and direction through a process of durotaxis. In addition, cells respond directly to local variations in tension applied by pulling or pushing with the blunt tip of a micropipet (Lo et al., 2000). Pulling the substrate with the microneedle served to increase the tension applied through the substrate to focal adhesions near the leading edge of migrating fibroblasts, and the cells increased their lamellipod spreading and migration toward the microneedle. Conversely, when the microneedle was pushed toward the leading edge of cells to decrease the substrate tension, the leading edge retracted, and lamellipodial extension began at the trailing edge. Cells began to migrate away from the needle. FAK–/– cells and cells expressing

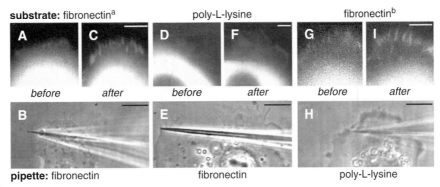

FIGURE 2.4 Force induces directional growth of adhesion structures through integrin–matrix interactions. (A–C) Dragging a microneedle coated with fibronectin across the surface of a cell on a fibronectin substrate induced formation of streaklike basal adhesions (labeled with GFP-vinculin) that were oriented in the direction of applied strain. (D–F) Adhesions did not form on a poly-L-lysine substrate, on which cells adhere nonspecifically by electrostatic interactions. (G–I) Nonspecific interactions via a poly-L-lysine–coated microneedle were sufficient to induce adhesion formation in cells on a fibronectin substrate. Bar = (C, F, I) 5 μm and (B. E. H) 10 μm. Reprinted, with permission, from Riveline et al. (2001).

FAK deficient in phosphorylation at tyrosine 397 could not distinguish between soft and rigid substrates (Wang et al., 2001). The proportion of activated Rac is decreased in FAK–/– cells, although the total protein expression level is not different from that of wild-type cells (Hsia et al., 2003). In serum-stimulated cells, GFP-FAK localizes to nascent focal complexes just behind the leading edge (in addition to existing focal adhesion sites), consistent with the location of Rac activation in extending lamellipodia. Furthermore, upregulation of Rac activity near these locations requires the association of FAK with paxillin (Ishibe et al., 2004). In contrast, the activity of Rho was upregulated in FAK–/– cells, and focal adhesion turnover rate was decreased (Ren et al., 2000). Dragging a fibronectin- or poly-L-lysine–coated microneedle across the apical cell surface promoted formation of streaklike basal adhesions that were oriented in the direction of applied strain (Fig. 2.4) (Riveline et al., 2001). Activation of mDia, a downstream effecter of Rho, was sufficient for the formation of these adhesions, which appeared similar to mature focal contacts. Taken together, these data implicate FAK in the balance between Rac activation associated with dynamic lamellipodium extension and Rho activation that mechanically stabilizes focal adhesions. As these are spatially distinct events in the cell, this transition would also be critical if similar mechanisms modulate the tactile exploration mechanism observed in response to local mechanical features in elastic substrates.

IV. CANDIDATE MECHANISMS FOR SENSING AND TRANSMITTING MECHANICAL CUES

The examples discussed above show that many of the proteins that constitute adhesion sites are also responsive to force transmitted through the structures they form. Although it is possible that these proteins serve as "mechanotransducers" that convert the mechanical energy of the applied force into their biochemical activity through conformation changes, mechanisms of cellular mechanotransduction are more accurately represented as complex systems of physical and biochemical interactions. The dynamics of these interactions are readouts of mechanotransmission through cell-matrix adhesions as well as inputs to feedback loops that modulate cell physiology. Thus, in addition to considering signaling networks associated with cell adhesion, modeling the cell as a complex mechanochemical system enhances understanding of endothelial adhesion behavior in the vascular wall.

A. Signaling Networks Associated with Mechanotransmission through Focal Adhesions

The response of the adhesion sites to mechanical stimuli goes well beyond simple changes in shape and localization. Shear stress induces the activation of many proteins localized to focal adhesions, including integrins, FAK, paxillin, c-Src, phosphatidylinositol 3-kinase, Akt kinase, Flk-1, and p130Cas. As a plethora of scaffolding and signaling proteins are directly and indirectly linked to integrins, their activation triggers a multitude of important signaling cascades. For example, integrin activation is intimately linked to the activity of the Rho family of small GTPases, which are critical in the assembly and dynamics of the cytoskeleton. RhoA promotes maturation of focal adhesion sites, which may provide a feedback signal that potentiates integrin activation by shear stress. Integrin activation is also required for shear stress-induced activation of extracellular signal-regulated kinase (ERK), p38, and c-jun N-terminal kinase (JNK), three of the mitogen-activated protein kinases (MAPK) that promote cell proliferation (Azuma et al., 2000). Two distinct pathways link integrin activation to ERK activity (Fig. 2.5).

In the first, autophosphorylation of FAK at Tyr-397 promotes its association with c-Src. The subsequent phosphorylation of paxillin and p130Cas recruits Crk and C3G and, ultimately, activates ERK via Rap1 and Raf. In a second pathway, shear stress activates Fyn, a protein that associates with both integrins and Shc at focal adhesion sites. Activated Fyn promotes the activation of Shc and subsequently the recruitment of Grb2 and Sos. The Shc–Grb2–Sos complex initiates ERK activation via Ras and Raf. Shear stress not only alters the activation state of molecules, but also affects their expression levels. The protein and mRNA levels of $\alpha 5$- and $\beta 1$-integrins are upregulated in cells exposed to shear stress, which enhances the adhesive capacity of the cells.

Cyclic strain of the substrate induces integrin-activated signaling networks that appear similar to those that respond to shear stress, including tyrosine phosphorylation of FAK and paxillin. Pyk2, a structural homolog to FAK, was also phosphorylated following cyclic strain, and this activation was downstream of Src phosphorylation and reactive oxygen species production (Cheng et al., 2002). In the intact aorta, FAK signaling was differentially regulated by steady or cyclic circumferential stretch (Lehoux et al., 2005). FAK phosphorylation after pulsatile pressure required integrin ligation and Src phosphorylation, whereas signaling downstream from FAK after a steady increase in pressure occurred independently of these events. Among these downstream signaling molecules are ERK, p38, and JNK, which parallel MAPK pathways that are also activated by shear stress (Azuma et al., 2000). Cyclic strain is not as potent as shear stress in activating p38 and ERK, but is equally robust in activating JNK.

In addition to activating integrins, strain also induces signaling activity from other structural proteins that are dynamically linked to focal adhesions.

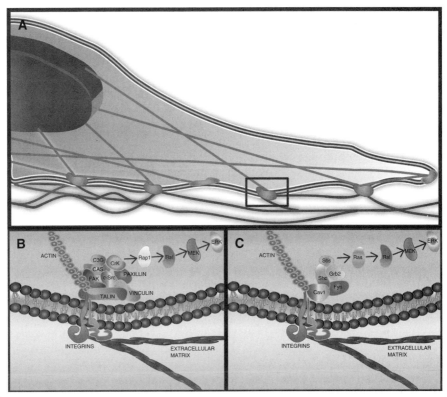

FIGURE 2.5 Adhesion sites induce MAPK signaling networks. (A) Transmission of tension from the actin cytoskeleton (red) to the ECM (purple) through focal adhesion sites (green) represents one mechanism of mechanotransduction. Complex interactions among the molecular components of a single adhesion site (black box) lead to ERK activation through at least two distinct signaling networks. (B) FAK activation serves as the critical initial step for assembly of myriad adapter and signaling proteins. (C) Assembly of the Shc–Grb2–Sos complex serves to activate Ras and Raf. (See Color Plate 3)

For example, cyclic strain enhanced the turnover rate and mRNA expression of zyxin, a crosslinking protein that usually assembles into actin stress fibers and focal adhesion sites (Cattaruzza et al., 2004). The N-terminus of zyxin contains the domain that mediates its assembly into focal adhesions, and the C-terminus contains a LIM domain, which is implicated in facilitating protein–protein and protein–DNA interactions. Interestingly, cyclic strain induced zyxin translocation to the nucleus, suggesting a role in transcriptional events. Indeed, inhibiting zyxin expression with antisense oligonucleotides abrogated the strain-induced upregulation of endothelin B receptor expression and enhanced the strain-induced expression of tenascin C. Endothelin B receptor is hypothesized to contribute to hypertension by inducing apoptosis in smooth muscle cells, and tenascin C is an ECM glycoprotein involved in vasculogenesis and wound healing. Although it is not known whether zyxin serves as a transcription factor directly, the presence of the C-terminal LIM domain suggests that zyxin could mediate transcription indirectly by facilitating protein–protein interactions within the nucleus. Thus, these data indicate a direct link between mechanotransmission through adhesion sites and regulation of gene expression involved in vascular wall remodeling.

Another possible mechanism linking mechanotransmission through adhesion sites to gene and protein expression involves tension-dependent recruitment of ribosomes and mRNA (Chicurel et al., 1998). Adhesion of microspheres coated with fibronectin induced localization of ribosomes and mRNA to clouds near the adhesion sites, and magnetic twisting of the microspheres to apply tension to the actin cytoskeleton augmented the frequency of ribosomal recruitment. Microspheres coated with anti-β1-integrin induced a more pronounced spatial localization than those coated with anti-αvβ3, and treatment with cytochalasin D or butanedione-2-monoxime (BDM) abrogated the response. Thus, mechanotransmission through a subset of focal adhesion complexes promotes not only structural stabilization of cytoskeleton–ECM connections, but also assembly of spatial microdomains that promote translation near sites of mechanical sensing. Such a hypothesis provides a mechanism for rapid post-transcriptional regulation consistent with cellular sensing of local mechanical cues in the ECM.

The passive mechanical properties of structural networks in the cytoplasm clearly contribute to force transmission to signaling components, but the relative role of active force-generating processes is less clear. Caveolin-1, the primary structural coat protein of cell surface caveolae, which are thought to be important for endothelial mechanotransduction, was rapidly and transiently phosphorylated after onset of 10 dyn/cm^2 shear stress (Radel and Rizzo, 2005). Src family kinases regulated the phosphorylation and recruitment of caveolin-1 near adhesion sites downstream of β1-integrin activation at those sites, where phosphorylated caveolin-1 was hypothesized to enable or enhance Src assembly and signaling through the Ras–ERK pathway. Interestingly, this interaction also modulates the association of the Src kinase inhibitor Csk with integrin-containing sites. Because downstream effects of Csk include regulation of shear-stimulated myosin light chain (MLC) phosphorylation, this adhesion site assembly is likely to affect cytoskeletal reorganization. Whether this mechanism also determines the level of myosin-mediated contractility and traction force generated by endothelial cells (see Chapter 1) has not yet been tested directly. However, in vascular smooth muscle cells, inhibition of MLC kinase prevented the polarized extension of lamellipodia that occurred after uniaxial stretch of the substrate (Katsumi et al., 2002). The polarized lamellipodial behavior corresponded to the spatial localization of Rac activity, suggesting that a relationship exists between small GTPase-activated actin polymerization and myosin-regulated force generation. Thus, it is likely that not only remodeling of cytoskeletal constitutive properties but also active contractile mechanisms are important in determining the forces generated against the ECM in response to an altered extracellular mechanical environment.

Mechanotransmission from the cytoplasm through focal adhesion sites to the ECM also directly affects force distribution and matrix structure outside of the cell. The fibronectin matrix is generally fibrillar in structure, and intracellular tension transmitted through focal adhesion sites may promote fibril assembly by stretching the fibronectin molecules. In fact, impairing cellular contractility by blocking Rho activity with C3 transferase abrogates fibronectin matrix assembly (Zhong et al., 1998). Even in the absence of cells, stretching a fibronectin-coated flexible substrate by 30% enhanced the binding of soluble fibronectin to the surface by greater than sevenfold, indicating that mechanical forces are transduced into increased biochemical affinity that promotes ECM assembly. The mechanism is likely to involve force-dependent alterations in the fibronectin molecule. In one estimate, a force on the order of piconewtons induces unfolding of fibronectin type III domains from an estimated length of 4 nm to 29 nm (Erickson, 1994). The type III domain is composed of seven β strands arranged in antiparallel fashion to form sheet structures. In steered molecular dynamics simulations (Krammer et al., 1999), the sheets unfolded sequentially so that the length of the type III domain increased significantly. In addition, excess stretch reduced the ability of fibronectin to bind to cells by transforming the integrin-binding RGD loop domain into a linear conformation. These predictions support the hypothesis that the fibronectin molecule itself can act as a mechanosensor.

A second force-dependent mechanism modulating ECM structure involves the activity of matrix metalloproteinases (MMPs) and tissue inhibitors of MMPs (TIMPs). These enzymes are responsible for the proteolysis of ECM components, thereby promoting reorganization of matrix structure. For example, cyclic stretch in the endothelium regulated expression levels of membrane type 1 MMP (Haas et al., 1999), and several studies indicate that MMP-2 is upregulated with the application of cyclic strain to endothelial cells on a time scale of hours (Sweeney et al., 2004; Wang et al., 2003). Both pro-MMP-2 and active MMP-2 levels were increased in cell culture medium and cell lysate after application of 5% cyclic strain. Inhibitors of MEK, ERK, and p38 MAPK all abrogated the strain-induced upregulation of pro-MMP-2 and MMP-2 release. One possible explanation is that substrate strain activates integrins in adhesion sites, leading to activation of MAPK networks that ultimately increase MMP expression. MMP-dependent matrix reorganization would then provide a level of feedback to the integrins by altering the adhesive ligands available to them.

B. A Systems Model of Mechanochemical Regulation in Endothelial Cells

In the search for the elusive "mechanotransducer," the adhesion interface contains numerous candidate molecules that likely participate in force transmission, including vinculin, the ERM (ezrin–radixin–moesin) family of proteins, talin, integrins, and pp60Src. Because these molecules exhibit distinct conformations that regulate their binding affinities, they are candidates to participate in a mechanically induced switching network that regulates structural dynamics and signaling (Fig. 2.6).

Although the physical arrangement of these molecules within adhesion sites remains unresolved, identification of components present in these molecular scaffolds suggests spatial and temporal coordination of interactions that trigger mechanosensory mechanisms. At the level of molecular structure, conformation or steric arrangement modulates not only chemical binding affinities, but also the ability to sustain tension. For example, tension acting on adhesions could generate molecular-scale slip to create physical gaps that energetically favor the recruitment of new molecules from the cytoplasm. Alternatively, force-induced rearrangements of adhesion-associated proteins would result in molecular sorting into microenvironments that alter both physical and chemical communication. Either of these processes must also include feedback that controls the subsequent decision to reinforce, maintain, or disassemble the adhesion structure.

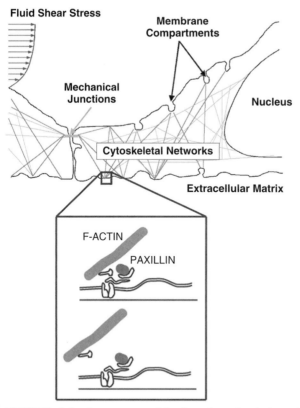

FIGURE 2.6 A systems model of mechanochemical regulation in endothelial cells. In response to fluid shear stress, cytoskeletal networks serve to integrate and transmit a complex array of signals that originate and act at locations such as specialized membrane compartments, intercellular junctions, adhesions to ECM, and the nucleus. Construction of accurate models with predictive capability requires measurement of switching network behavior involved in mechanotransmission, as might be indicated by physical separation of actin and paxillin in dual-wavelength high-resolution microscopy images.

A role for the cytoskeleton in biochemical sorting as part of such a feedback control loop remains largely unexplored. The cytoskeleton serves as a scaffold for numerous signaling molecules, including many enzymes whose activities are mechanosensitive. For example, microtubule disruption enhanced the flow-dependent elevation of nitric oxide production in endothelial cells (Knudsen and Frangos, 1997). In addition, stretching of detergent-treated cells enhanced the binding of paxillin, FAK, and the adapter protein p130Cas to the cytoskeleton. Analysis of proteomic interaction maps constructed from yeast two-hybrid screening revealed a strong correlation between cytoskeletal proteins and signaling networks relevant to mechanotransduction (Forgacs et al., 2004). This supports the hypothesis that the relationship between cytoskeletal and signaling proteins is indeed functional and not just physical. These studies have not yet considered the possibility that tethering and transport of signaling molecules by cytoskeleton-associated proteins contributes to coordinating the timing as well as the spatial localization of a complex array of signals. For example, Rho activity reaches a minimum only 5 min after onset of unidirectional laminar shear stress (Tzima et al., 2001), but upregulation of Rac activity peaks at 30 min

(Tzima et al., 2002). The mechanism that regulates the timing of these activities is unknown. The growing plus ends of microtubules are targeted to new focal complexes by the cytoplasmic linker protein CLIP-170. Application of tensile force enhances this process and promotes binding of filamin A and Rac1. Because the time needed to accomplish this assembly of Rac near the ruffling edge of the cell depends on microtubule dynamics, new focal complex assembly rate, and transport of several components to the cell edge, it is reasonable to hypothesize that this process is significantly slower than the inactivation of Rho, which is already located near these sites to stabilize actin filament bundles and adhesions. Elucidation of specific molecular mechanisms and the rate-limiting steps in such a system remains an unsolved problem.

Overall, the majority of these data indicate that interactions among the cytoskeleton and adhesion sites are the keys to integrative sensing. These cellular components play the primary role in deciphering the spatial code of physical and biochemical cues presented to the cell by the ECM. Coupling this hypothesis to new measurements of the dynamics of adhesion-activated signaling and adhesion–cytoskeleton interactions suggests a new model for mechanochemical regulation that represents intracellular events as a complex system of mechanochemical switches. Extracellular forces are redistributed by cytoskeletal components to intracellular locations that can trigger signaling events, while intracellular contractility and adhesion site nucleation simultaneously sample the mechanical microenvironment surrounding the cell. In response to either of these probing systems, spatial cues perceived by the cell to be functionally beneficial are reinforced by either biochemical signal amplification or structural remodeling, so that the cell adapts its structure and phenotype to support physiological (or pathological) functions. Thus, a complex array of inputs and feedback mechanisms based on cytoskeleton–adhesion–ECM interactions leads to outputs that the observer characterizes as phenotype. To understand this cellular phenotype control system, it is necessary to develop a multiscale systems approach that includes molecular-scale interactions among protein complexes, structural network interactions from the cytoplasm to the cell–matrix interface, and multicellular behaviors resulting from communication within the endothelial monolayer that determines tissue function, wound healing, and endothelial function on engineered biomaterials.

V. FUTURE DIRECTIONS

A complex systems model of mechanotransduction provides a route for formulating and testing hypotheses regarding fundamental mechanisms of cellular physiology and for answering several remaining questions. What serves as the "switchboard" to balance traffic flow among parallel mechanical and chemical signaling mechanisms? Can the evolution of this emergent network behavior be guided by engineering spatial cues into microenvironments that cells encounter during migration?

Answering these questions will be facilitated by a biomedical engineering approach that builds on novel biomaterials designs to control cell adhesion and phenotype. Advances in micrometer- to nanometer-scale fabrication techniques are rapidly showing potential for designing biocompatible surfaces that can probe cell–matrix interactions at the length scale of these multimolecular protein assemblies. Extension to three-dimensional material architectures will improve physiological function in engineered tissue constructs for vessel grafts and regenerative medicine. Thus, combining systems biology at the molecular and cellular levels with current technology at work in biomaterials and cell biology will significantly advance the state-of-the-art in vascular bioengineering.

ACKNOWLEDGMENTS

The authors thank Dr. A. D. Bershadsky and Dr. B. L. Langille for permission to reproduce figures. This work was supported in part by NIH Grant HL071958.

SUGGESTED READING

Chien, S., Li, S., and Shyy, Y. J. (1998). Effects of mechanical forces on signal transduction and gene expression in endothelial cells. *Hypertension* **31**: 162–169.
Davies, P. F. (1995). Flow-mediated endothelial mechanotransduction. *Physiol. Rev.* **75**: 519–560.
Gimbrone, M. A., Resnick, N., Nagel, T., Khachigian, L. M., Collins, T., and Topper, J. N. (1997). Hemodynamics, endothelial gene expression, and atherogenesis. *Ann. N. Y. Acad. Sci.* **811**: 1–10.
Key reviews of endothelial adaptation to hemodynamic shear stress.

Frangos, S. G., Knox, R., Yano, Y., Chen, E., Di Luozzo, G., Chen, A. H., and Sumpio, B. E. (2001). The integrin-mediated cyclic strain-induced signaling pathway in vascular endothelial cells. *Endothelium* **8**: 1–10.
Reviews integrin-activated signaling networks activated by substrate stretch.

Helmke, B. P. (2005). Molecular control of cytoskeletal mechanics by hemodynamic forces. *Physiology* **20**: 43–53.
Proposes molecular mechanisms that regulate cytoskeletal mechanics in response to changes in shear stress profile.

Janmey, P. A. (1998). The cytoskeleton and cell signaling: component localization and mechanical coupling. *Physiol. Rev.* **78**: 763–781.
Reviews spatial organization of signaling molecules and proposes that signaling networks function can be regulated by mechanical coupling through the cytoskeleton.

Ross, R. (1999). Atherosclerosis — an inflammatory disease. *N. Engl. J. Med.* **340**: 115–126.
Proposes that inflammatory mechanisms mediate atherosclerosis.

Shaul, P. W. (2002). Regulation of endothelial nitric oxide synthase: location, location, location. *Annu. Rev. Physiol.* **64**: 749–774.
Proposes that the spatial distribution of eNOS activity correlates with defects in vasodilatory activity observed in regions of atherosclerotic lesions.

Shyy, J. Y. J., and Chien, S. (2002). Role of integrins in endothelial mechanosensing of shear stress. *Circ. Res.* **91**: 769–775.
Reviews the shear stress–mediated activation and assembly of focal adhesion–associated signaling proteins.

REFERENCES

Ali, M. H., Pearlstein, D. P., Mathieu, C. E., and Schumacker, P. T. (2004). Mitochondrial requirement for endothelial responses to cyclic strain: Implications for mechanotransduction. *Am. J. Physiol.* **287**: L486–L496.

Azuma, N., Duzgun, S. A., Ikeda, M., Kito, H., Akasaka, N., Sasajima, T., and Sumpio, B. E. (2000). Endothelial cell response to different mechanical forces. *J. Vasc. Surg.* **32**: 789–794.

Balaban, N. Q., Schwarz, U. S., Riveline, D., Goichberg, P., Tzur, G., Sabanay, I., Mahalu, D., Safran, S., Bershadsky, A., Addadi, L., and Geiger, B. (2001). Force and focal adhesion assembly: A close relationship studied using elastic micropatterned substrates. *Nat. Cell Biol.* **3**: 466–472.

Bardy, N., Karillon, G. J., Merval, R., Samuel, J.-L., and Tedgui, A. (1995). Differential effects of pressure and flow on DNA and protein synthesis and on fibronectin expression by arteries in a novel organ culture system. *Circ. Res.* **77**; 684–694.

Caille, N., Tardy, Y., and Meister, J. J. (1998). Assessment of strain field in endothelial cells subjected to uniaxial deformation of their substrate. *Ann. Biomed. Eng.* **26**: 409–416.

Cattaruzza, M., Lattrich, C., and Hecker, M. (2004). Focal adhesion protein zyxin is a mechanosensitive modulator of gene expression in vascular smooth muscle cells. *Hypertension* **43**: 726–730.

Cheng, J.-J., Chao, Y.-J., and Wang, D. L. (2002). Cyclic strain activates redox-sensitive proline-rich tyrosine kinase 2 (PYK2) in endothelial cells. *J. Biol. Chem.* **277**: 48152–48157.

Chicurel, M. E., Singer, R. H., Meyer, C. J., and Ingber, D. E. (1998). Integrin binding and mechanical tension induce movement of mRNA and ribosomes to focal adhesions. *Nature* **392**: 730–733.

Chiquet, M., Matthisson, M., Koch, M., Tannheimer, M., and Chiquet-Ehrismann, R. (1996). Regulation of extracellular matrix synthesis by mechanical stress. *Biochem. Cell Biol.* **74**: 737–744.

Choquet, D., Felsenfeld, D. P., and Sheetz, M. P. (1997). Extracellular matrix rigidity causes strengthening of integrin–cytoskeleton linkages. *Cell* **88**: 39–48.

Davies, P. F., Robotewskyj, A., and Griem, M. L. (1994). Quantitative studies of endothelial cell adhesion: Directional remodeling of focal adhesion sites in response to flow forces. *J. Clin. Invest.* **93**: 2031–2038.

Dewey, C. F., Jr., Bussolari, S. R., Gimbrone, M. A., Jr., and Davies, P. F. (1981). The dynamic response of vascular endothelial cells to fluid shear stress. *J. Biomech. Eng.* **103**: 177–188.

Di Paolo, G., Pellegrini, L., Letinic, K., Cestra, G., Zoncu, R., Voronov, S., Chang, S., Guo, J., Wenk, M. R., and De Camilli, P. (2002). Recruitment and regulation of phosphatidylinositol phosphate kinase type 1γ by the FERM domain of talin. *Nature* **420**: 85–89.

Eckes, B., Dogic, D., Colucci-Guyon, E., Wang, N., Maniotis, A., Ingber, D., Merckling, A., Langa, F., Aumailley, M., Delouvee, A., *et al.* (1998). Impaired mechanical stability, migration and contractile capacity in vimentin-deficient fibroblasts. *J. Cell Sci.* **111**: 1879–1907.

Erickson, H. P. (1994). Reversible unfolding of fibronectin type III and immunoglobulin domains provides the structural basis for stretch and elasticity of titin and fibronectin. *Proc. Natl. Acad. Sci. USA* **91**: 10114–10118.

Fluck, M., Tunc-Civelek, V., and Chiquet, M. (2000). Rapid and reciprocal regulation of tenascin-C and tenascin-Y expression by loading of skeletal muscle. *J. Cell Sci.* **113**: 3583–3591.

Forgacs, G., Yook, S. H., Janmey, P. A., Jeong, H., and Burd, C. G. (2004). Role of the cytoskeleton in signaling networks. *J. Cell Sci.* **117**: 2769–2775.

Galbraith, C. G., Skalak, R., and Chien, S. (1998). Shear stress induces spatial reorganization of the endothelial cell cytoskeleton. *Cell Motil. Cytoskel.* **40**: 317–330.

Galbraith, C. G., Yamada, K. M., and Sheetz, M. P. (2002). The relationship between force and focal complex development. *J. Cell Sci.* **159**: 695–705.

Giannone, G., Jiang, G., Sutton, D. H., Critchley, D. R., and Sheetz, M. P. (2003). Talin1 is critical for force-dependent reinforcement of initial integrin-cytoskeleton bonds but not tyrosine kinase activation. *J. Cell Biol.* **163**: 409–419.

Gilmore, A. P., and Burridge, K. (1996). Regulation of vinculin binding to talin and actin by phosphatidyl-inositol-4–5-bisphosphate. *Nature* **381**: 531–535.

Girard, P. R., and Nerem, R. M. (1995). Shear stress modulates endothelial cell morphology and F-actin organization through the regulation of focal adhesion-associated proteins. *J. Cell. Physiol.* **163**: 179–193.

Gorfien, S. F., Howard, P. S., Myers, J. C., and Macarak, E. J. (1990). Cyclic biaxial strain of pulmonary artery endothelial cells causes an increase in cell layer-associated fibronectin. *Am. J. Respir. Cell Mol. Biol.* **3**: 421–429.

Gupte, A., and Frangos, J. A. (1990). Effects of flow on the synthesis and release of fibronectin by endothelial cells. *In Vitro Cell. Dev. Biol.* **26**: 57–60.

Haas, T. L., Stitelman, D., Davis, S. J., Apte, S. S., and Madri, J. A. (1999). Egr-1 mediates extracellular matrix-driven transcription of membrane type 1 matrix metalloproteinase in endothelium. *J. Biol. Chem.* **274**: 22679–22685.

Halliday, N. L., and Tomasek, J. J. (1995). Mechanical properties of the extracellular matrix influence fibronectin fibril assembly *in vitro. Exp. Cell Res.* **217**: 109–117.

Helmke, B. P., Goldman, R. D., and Davies, P. F. (2000). Rapid displacement of vimentin intermediate filaments in living endothelial cells exposed to flow. *Circ. Res.* **86**: 745–752.

Helmke, B. P., Rosen, A. B., and Davies, P. F. (2003). Mapping mechanical strain of an endogenous cytoskeletal network in living endothelial cells. *Biophys. J.* **84**: 2691–2699.

Helmke, B. P., Thakker, D. B., Goldman, R. D., and Davies, P. F. (2001). Spatiotemporal analysis of flow-induced intermediate filament displacement in living endothelial cells. *Biophys. J.* **80**: 184–194.

Homan, S. M., Martinez, R., Benware, A., and LaFlamme, S. E. (2002). Regulation of the association of $\alpha_6\beta_4$ with vimentin intermediate filaments in endothelial cells. *Exp. Cell Res.* **281**: 107–114.

Hsia, D. A., Mitra, S. K., Hauck, C. R., Streblow, D. N., Nelson, J. A., Ilic, D., Huang, S., Li, E., Nemerow, G. R., Leng, J., et al. (2003). Differential regulation of cell motility and invasion by FAK. *J. Cell Biol.* **160**: 753–767.

Ishibe, S., Joly, D., Liu, Z.-X., and Cantley, L. G. (2004). Paxillin serves as an ERK-regulated scaffold for coordinating FAK and Rac activation in epithelial morphogenesis. *Mol. Cell* **16**: 257–267.

Jalali, S., del Pozo, M. A., Chen, K.-D., Miao, H., Li, Y.-S., Schwartz, M. A., Shyy, J. Y.-J., and Chien, S. (2001). Integrin-mediated mechanotransduction requires its dynamic interaction with specific extracellular matrix (ECM) ligands. *Proc. Natl. Acad. Sci. USA* **98**: 1042–1046.

Janmey, P. A., Euteneuer, U., Traub, P., and Schliwa, M. (1991). Viscoelastic properties of vimentin compared with other filamentous biopolymer networks. *J. Cell Biol.* **113**: 155–160.

Janmey, P. A., Hvidt, S., Peetermans, J., Lamb, J., Ferry, J. D., and Stossel, T. P. (1988). Viscoelasticity of F-actin and F-actin/gelsolin complexes. *Biochemistry* **27**: 8218–8227.

Katsumi, A., Milanini, J., Kiosses, W. B., del Pozo, M. A., Kaunas, R., Chien, S., Hahn, K. M., and Schwartz, M. A. (2002). Effects of cell tension on the small GTPase Rac. *J. Cell Biol.* **158**: 153–164.

Knudsen, H. L., and Frangos, J. A. (1997). Role of cytoskeleton in shear stress–induced endothelial nitric oxide production. *Am. J. Physiol.* **273**: H347–H355.

Krammer, A., Lu, H., Isralewitz, B., Schulten, K., and Vogel, V. (1999). Forced unfolding of the fibronectin type III module reveals a tensile molecular recognition switch. *Proc. Natl. Acad. Sci. USA* **96**: 1351–1356.

Lee, J. S. H., Chang, M. I., Tseng, Y., and Wirtz, D. (2005). Cdc42 mediates nucleus movement and MTOC polarization in Swiss 3T3 fibroblasts under mechanical shear stress. *Mol. Biol. Cell* **16**: 871–880.

Lehoux, S., Esposito, B., Merval, R., and Tedgui, A. (2005). Differential regulation of vascular focal adhesion kinase by steady stretch and pulsatility. *Circulation* **111**: 643–649.

Leung, D. Y., Glagov, S., and Mathews, M. B. (1976). Cyclic stretching stimulates synthesis of matrix components by arterial smooth muscle cells *in vitro. Science* **191**: 415–477.

Levesque, M. J., and Nerem, R. M. (1985). The elongation and orientation of cultured endothelial cells in response to shear stress. *J. Biomech. Eng.* **107**: 341–347.

Li, S., Butler, P. J., Wang, Y., Hu, Y., Han, D. C., Usami, S., Guan, J.-L., and Chien, S. (2002). The role of the dynamics of focal adhesion kinase in the mechanotaxis of endothelial cells. *Proc. Natl. Acad. Sci. USA* **99**: 3546–3551.

Lo, C.-M., Wang, H.-B., Dembo, M., and Wang, Y.-l. (2000). Cell movement is guided by the rigidity of the substrate. *Biophys. J.* **79**: 144–152.

Malek, A. M., and Izumo, S. (1996). Mechanism of endothelial cell shape change and cytoskeletal remodeling in response to fluid shear stress. *J. Cell Sci.* **109**: 713–726.

Malek, A. M., Zhang, J., Jiang, J., Alper, S. L., and Izumo, S. (1999). Endothelin-1 gene suppression by shear stress: Pharmacological evaluation of the role of tyrosine kinase, intracellular calcium, cytoskeleton, and mechanosensitive channels. *J. Mol. Cell. Cardiol.* **31**: 387–399.

Mourgeon, E., Xu, J., Tanswell, A. K., Liu, M., and Post, M. (1999). Mechanical strain-induced posttranscriptional regulation of fibronectin production in fetal lung cells. *Am. J. Physiol.* **277**: L142–L149.

Nerem, R. M., Levesque, M. J., and Cornhill, J. F. (1981). Vascular endothelial morphology as an indicator of the pattern of blood flow. *J. Biomech. Eng.* **103**: 172–177.

Nicolas, A., Geiger, B., and Safran, S. A. (2004). Cell mechanosensitivity controls the anisotropy of focal adhesions. *Proc. Natl. Acad. Sci. USA* **101**: 12520–12525.

Noria, S., Xu, F., McCue, S., Jones, M., Gotlieb, A. I., and Langille, B. L. (2004). Assembly and reorientation of stress fibers drives morphological changes to endothelial cells exposed to shear stress. *Am. J. Pathol.* **164**: 1211–1223.

Qiu, Y., and Tarbell, J. M. (2000). Interaction between wall shear stress and circumferential strain affects endothelial cell biochemical production. *J. Vasc. Res.* **37**: 147–157.

Radel, C., and Rizzo, V. (2005). Integrin mechanotransduction stimulates caveolin-1 phosphorylation and recruitment of Csk to mediate actin reorganization. *Am. J. Physiol.* **288**: H936–H945.

Ren, X. D., Kiosses, W. B., Sieg, D. J., Otey, C. A., Schlaepfer, D. D., and Schwartz, M. A. (2000). Focal adhesion kinase suppresses Rho activity to promote focal adhesion turnover. *J. Cell Sci.* **113**: 3673–3678.

Riveline, D., Zamir, E., Balaban, N. Q., Schwarz, U. S., Ishizaki, T., Narumiya, S., Kam, Z., Geiger, B., and Bershadsky, A. D. (2001). Focal contacts as mechanosensors: Externally applied local mechanical force induces growth of focal contacts by an mDia1-dependent and ROCK-independent mechanism. *J. Cell Biol.* **153**: 1175–1186.

Sato, M., Ohshima, N., and Nerem, R. M. (1996). Viscoelastic properties of cultured porcine aortic endothelial cells exposed to shear stress. *J. Biomech.* **29**: 461–467.

Schiffers, P. M., Henrion, D., Boulanger, C. M., Colucci-Guyon, E., Langa-Vuves, F., van Essen, H., Fazzi, G. E., Levy, B. I., and De Mey, J. G. (2000). Altered flow-induced arterial remodeling in vimentin-deficient mice. *Arterioscler. Thromb. Vasc. Biol.* **20**: 611–616.

Shikata, Y., Rios, A., Kawkitinarong, K., DePaola, N., Garcia, J. G. N., and Birukov, K. G. (2005). Differential effects of shear stress and cyclic stretch on focal adhesion remodeling, site-specific FAK phosphorylation, and small GTPases in human lung endothelial cells. *Exp. Cell Res.* **304**: 40–49.

Sipkema, P., van der Linden, P. J. W., Westerhof, N., and Yin, F. C. P. (2003). Effect of cyclic axial stretch of rat arteries on endothelial cytoskeletal morphology and vascular reactivity. *J. Biomech.* **36**: 653–659.

Stamatas, G. N., and McIntire, L. V. (2001). Rapid flow-induced responses in endothelial cells. *Biotechnol. Prog.* **17**: 383–402.

Su, Y., Edwards-Bennett, S., Bubb, M. R., and Block, E. R. (2003). Regulation of endothelial nitric oxide synthase by the actin cytoskeleton. *Am. J. Physiol.* **284**: C1542–C1549.

Sugimoto, K., Yoshida, K., Fujii, S., Takemasa, T., Sago, H., and Yamashita, K. (1995). Heterogeneous responsiveness of the in situ rat vascular endothelial cells to mechanical stretching *in vitro. Eur. J. Cell Biol.* **68**: 70–77.

Sun, D., Huang, A., Sharma, S., Koller, A., and Kaley, G. (2001). Endothelial microtubule disruption blocks flow-dependent dilation of arterioles. *Am. J. Physiol.* **280**: H2087–H2093.

Svitkina, T. M., Verkhovsky, A. B., and Borisy, G. G. (1996). Plectin sidearms mediate interaction of intermediate filaments with microtubules and other components of the cytoskeleton. *J. Cell Biol.* **135**: 991–1007.

Sweeney, N. v. O., Cummins, P. M., Birney, Y. A., Redmond, E. M., and Cahill, P. A. (2004). Cyclic strain-induced endothelial MMP-2: role in vascular smooth muscle cell migration. *Biochem. Biophys. Res. Commun.* **320**: 325–333.

Takemasa, T., Sugimoto, K., and Yamashita, K. (1997). Amplitude-dependent stress fiber reorientation in early response to cyclic strain. *Exp. Cell Res.* **230**: 407–410.

Thoumine, O., Nerem, R. M., and Girard, P. R. (1995). Changes in organization and composition of the extracellular matrix underlying cultured endothelial cells exposed to laminar steady shear stress. *Lab. Invest.* **73**: 565–576.

Tsuruta, D., and Jones, J. C. R. (2003). The vimentin cytoskeleton regulates focal contact size and adhesion of endothelial cells subjected to shear stress. *J. Cell Sci.* **116**: 4977–4984.

Tzima, E., Del Pozo, M. A., Kiosses, W. B., Mohamed, S. A., Li, S., Chien, S., and Schwartz, M. A. (2002). Activation of Rac1 by shear stress in endothelial cells mediates both cytoskeletal reorganization and effects on gene expression. *EMBO J.* **21**: 6791–6800.

Tzima, E., Del Pozo, M. A., Shattil, S. J., Chien, S., and Schwartz, M. A. (2001). Activation of integrins in endothelial cells by fluid shear stress mediates Rho-dependent cytoskeletal alignment. *EMBO J.* **20**: 4639–4647.

Van Zanten, G. H., de Graaf, S., Slootweg, P. J., Heijnen, H. F., Connolly, T. M., de Groot, P. G., and Sixma, J. J. (1994). Increased platelet deposition on atherosclerotic coronary arteries. *J. Clin. Invest.* **93**: 615–632.

Von Wichert, G., Jiang, G., Kostic, A., De Vos, K., Sap, J., and Sheetz, M. P. (2003). RPTP-α acts as a transducer of mechanical force on α_V/β_3-integrin-cytoskeleton linkages. *J. Cell Biol.* **161**: 143–153.

Wang, B.-W., Chang, H., Lin, S., Kuan, P., and Shyu, K.-G. (2003). Induction of matrix metalloproteinases-14 and -2 by cyclical mechanical stretch is mediated by tumor necrosis factor-α in cultured human umbilical vein endothelial cells. *Cardiovasc. Res.* **59**: 460–469.

Wang, H.-B., Dembo, M., Hanks, S. K., and Wang, Y.-l. (2001). Focal adhesion kinase is involved in mechanosensing during fibroblast migration. *Proc. Natl. Acad. Sci. USA* **98**: 11295–11300.

Wang, N., Butler, J. P., and Ingber, D. E. (1993). Mechanotransduction across the cell surface and through the cytoskeleton. *Science* **260**: 1124–1127.

Wang, N., and Stamenovic, D. (2000). Contribution of intermediate filaments to cell stiffness, stiffening, and growth. *Am. J. Physiol.* **279**: C188-C194.

Weinbaum, S., Zhang, X., Han, Y., Vink, H., and Cowin, S. C. (2003). Mechanotransduction and flow across the endothelial glycocalyx. *Proc. Natl. Acad. Sci. USA* **100**: 7988–7995.

White, C. R., Haidekker, M., Bao, X., and Frangos, J. A. (2001). Temporal gradients in shear, but not spatial gradients, stimulate endothelial cell proliferation. *Circulation* **103**: 2508–2513.

Yano, Y., Geibel, J., and Sumpio, B. E. (1997). Cyclic strain induces reorganization of integrin alpha 5 beta 1 and alpha 2 beta 1 in human umbilical vein endothelial cells. *J. Cell. Biochem.* **64**: 505–513.

Zhong, C., Chrzanowska-Wodnicka, M., Brown, J., Shaub, A., Belkin, A. M., and Burridge, K. (1998). Rho-mediated contractility exposes a cryptic site in fibronectin and induces fibronectin matrix assembly. *J. Cell Biol.* **114**: 539–551.

3

USE OF HYDRODYNAMIC SHEAR STRESS TO ANALYZE CELL ADHESION

DAVID BOETTIGER

Department of Microbiology, University of Pennsylvania, Philadelphia, Pennsylvania, USA

Cell adhesion is central to understanding how cells interact and coordinate their behavior in multicellular organisms. It needs to be carefully regulated in development, homeostasis, and response to wounding and invasion. It is mediated by the specific binding of adhesion receptors to ligands expressed on the surface of adjacent cells or in the surrounding extracellular matrix. Because of the technical barriers to the analysis of adhesion receptor binding in the cell substrate contact zone, most of our data on the binding reaction and its control are based on the use of soluble systems. The objective of this chapter is to describe the progress made in bridging this technical gap using hydrodynamic shear stress methods. Methods are described in which the force developed by hydrodynamic shear is directly proportional to the number of adhesion receptor–ligand bonds that mediate the adhesion. The relationship can be used to provide a measure of the relative number of adhesive bonds and, if driven toward saturation, can be used to measure the effective K_D of adhesion receptor–ligand binding in the cell contact zone. This is the parameter that determines the proportion of receptors that are bound. The number of bound receptors determines both the adhesion strength and the strength of adhesion-mediated signals. Issues of bond strength, mechanisms of adhesion control, mechanisms of cell detachment under fluid shear stress, and cytoskeletal mechanics are considered in the adhesion context. This provides a model system in which to expand analysis of adhesion receptor binding in cell adhesion.

I. INTRODUCTION

The fundamental purpose of applying hydrodynamic shear to cells is to gain an understanding of cell adhesion. Over the past 50 years, it has become quite clear that cell adhesion is a function that is mediated by receptor–ligand binding (Pierschbacher and Ruoslahti, 1984). Although other elements in the cell substrate may modulate the process, mechanical adhesion appears to be the function of a limited set of cell surface receptors, which we call *adhesion receptors.* In this chapter, we focus on these receptor–ligand interactions in the context of cell adhesion. The problem is developing hydrodynamic shear methods in which there is a clear dose–response relationship between the number and strength of these adhesive bonds and the measured hydrodynamic parameter. If this can be accomplished, the relationship can be turned around, and the shear measurements can be used to measure the number and strength of adhesive bonds, providing the basis for understanding their biological regulation. In developing this approach, it is first necessary to construct a device for applying hydrodynamic shear. Here we compare the different approaches taken to determine those that display the required relationship. In refining the systems, knowledge of the hydrodynamic shear forces is important. This is most easily modeled in systems where there is laminar flow, and so the latter must be considered. Because it is our purpose to relate hydrodynamic shear to the properties of specific cell adhesion receptors, it is equally important to choose a cell system simple enough that one can relate the hydrodynamic measurements to the adhesive bonds. Hence, we began with K562 cells, which express a single cell surface integrin, $\alpha5\beta1$. This integrin cannot be activated or clustered in response to intracellular signals. Affinity is controlled by the addition of exogenous antibody to convert the receptor to active state. Once the principles of this system are understood, complexity can be increased by using cells in which these processes are controlled by intracellular signals. This is only the tip of the iceberg in terms of cell issues, but the point is that there are cellular parameters that need to be isolated one at a time and probed to determine their contribution and then to appropriately modify both the experimental and theoretical models.

The stated objective is to understand adhesive receptor–ligand interactions in the context of cell adhesion. The most intensively studied adhesion receptors are those of the integrin family (Hynes, 2002). Because they mediate principally the binding of cells to extracellular matrix (ECM), the models use a cell and a (purified) ligand rather than two cells. In addition, in the hematopoietic system, integrins are used as emergency adhesion molecules that exist in "off" (non–ligand-binding) and "on" (available for ligand binding) states. For example, $\alpha_{IIB}\beta_3$-integrin in platelets is required for effective platelet adhesion and clot formation in response to wounding of the blood vessel wall. However, the blood contains several ligands for $\alpha_{IIB}\beta_3$, at levels of hundreds of micrograms per milliliter. Thus, $\alpha_{IIB}\beta_3$-integrin needs to be shut down in the circulating platelet to avoid massive thrombosis (Shattil and Newman, 2004; Shattil et al., 1985). Obviously, this is a highly selected property in evolution. Analogous integrin models present in circulating leukocytes respond to parasitic invasion. This has led to model systems in which the regulation of integrin–ligand binding can be controlled by signaling reactions that can be triggered using soluble mediators and binding can be analyzed using soluble ligands and single-cell suspensions. This approach has led to major advances in our understanding of molecular mechanisms involved and to the development of structural models (Xiong et al., 2003; Hynes, 2002; Kim et al., 2003). However, the binding of soluble ligand to an adhesion receptor does not mediate adhesion. To what extent can results for soluble ligands be extended to actual cell adhesion? It is commonly assumed that the regulation of adhesion can be understood using these specialized integrin models and that similar regulatory circuits must apply generally to adhesion molecules.

To understand why these extrapolations of results into the realm of adhesion need to be questioned, one can begin with Bell's (1978) landmark theoretical article. This article explains the considerations in translating measurements of soluble binding parameters to the case of binding in the interface between two cells and in which both receptor and ligand are constrained to cell surfaces. At the time, the models used were based on antibody–antigen binding analyses because specific adhesion receptors had not yet been identified. With the identification and experimental analysis of specific receptors involved in cell adhesion, we only found that the system is even more complex than imagined by Bell. A major contribution to this increased complexity is the patterning of adhesion receptors on the cell surface, particularly when they are mediating adhesion and not on cells in suspension. This patterning is recognized in the formation of numerous adhesion junctions including focal adhesions, tight junctions, adherens junctions, desmosomes, and hemidesmosomes. It is unlikely that this encompasses the full complexity. The combination of these issues leads us to question the extension of the simple models that investigate soluble ligand binding to the case of cell adhesion. To actually analyze cell adhesion in terms of the receptors that mediate it, new approaches are required.

II. THE PROBLEM OF MEASURING CELL ADHESION

Numerous approaches have been developed for the analysis of cell adhesion. To understand the contribution that can be made using the hydrodynamic shear approaches developed later in the chapter, it is important to evaluate other approaches that have been developed and to ask what they can tell us about adhesion receptor–ligand binding.

A. Single Receptor–Ligand Bond Measurements

Although the objective is to measure cell adhesion in the context of the intact cell as a multiple-bonded state, all methods for the measurement of cell adhesion involve forces. In simple assays, these forces result from washing methods or shaking of cell cultures. In force-based approaches, the calibrated forces are used to measure cell detachment. We begin with the analysis of single receptor–ligand bonds because these measurements provide a yardstick to measure the forces applied relative to the expected bond strength.

To measure single receptor–ligand bonds it is necessary to isolate the single interaction and to develop tools that are sufficiently sensitive to detect the strength of the single bond. Two different tools have been used, laser tweezers and the atomic force microscope (AFM) (Litvinov et al., 2002; Zhang et al., 2004). Each has its strengths and limitations, but the general strategy for the experiment is similar. Measurements typically have used receptors *in situ* on a whole cell and a ligand attached to a surface. The objective is to bring the receptor and ligand into contact so that a bond can form and then to withdraw one partner to stress the bond until it ruptures. The force applied at the time of rupture is measured by the deflection of the AFM probe or the force exerted by the laser trap. The process is repeated under conditions that give a low probability of forming a bond in each cycle, but a large number of cycles are performed and multiple single-bond measurements have been made. The reason for the low binding probability conditions is to increase the probability that single bonds are measured. These are nonequilibrium measurements, and the actual bond strength measured is dependent on the loading rate at which the force is applied to the bond (Zhang et al., 2004). This dependency is predicted because the bond will sample a range of energy states over time that are predicted to exhibit differences in bond strength (Merkel et al., 1999). Slower pulling rates allow the sampling of more states and hence are more likely to include lower states. The sampling would be biased to the lowest state available. Because the sampling rates tend to be multiple times per second, the experimental loading rates are generally higher than would be expected in the organism.

The rate of change on bond strength as a function of loading rate is relatively small at the lower end of the measured range (Zhang et al., 2004). The loading rate dependency of bond strength suggests that rupture strength of adhesion receptor–ligand bonds depends on the conditions of the assay. The loading rates for the hydrodynamic shear experiments are low in comparison with those for the laser tweezer and AFM methods. Reported bond strengths for integrins, cadherins (cell–cell adhesion molecules in adherens junctions), and selectins are in the range of 50–100 pN per bond (Weisel et al., 2003). For the model used later with TS2/16 antibody–activated $\alpha 5\beta 1$-integrin on K562 cells, we measured forces of 50 pN per bond using laser tweezers, which is in agreement with the AFM measurements (Li et al., 2003); 50–100 pN appears to be a reasonable first approximation for adhesion receptor–ligand bonds.

B. Passive Methods

Passive adhesion assays are those that divide the cells into adherent and nonadherent populations using washing methods or shaking methods that involve low cell separation forces. Because the adhesive bonds are reasonably strong, the detachment forces that are present in these systems are similar to the strength of a single or few bonds at most. Therefore, these assays measure the initial binding assay but are not sensitive to variations in the number of bonds formed.

I. Aggregation Assays

Aggregation assays have been used to characterize cell–cell adhesion and have provided the basic approach to the identification of cell–cell adhesion receptors and their ligands. The assay was originally developed to study cell adhesion during embryonic development, and it has been extensively applied in the analysis of cadherins and Ig-CAMs. Single-cell suspensions of either specific cell populations or nonadherent cells like suspension L cells that have been transfected to express specific adhesion receptors are placed in flasks on a gyratory shaker, and the proportion of cells that join to form aggregates (or decrease in the single cells) is scored as a measure of aggregation (Roth, 1968). In general, the differences are very obvious and cells either aggregate or they do not. Cadherins and Ig-CAMs are commonly homophilic so that the expression of a single adhesion receptor serves to promote aggregation (Takeichi et al., 2000). Specificity is confirmed through the application of function–blocking antibodies.

2. Plate and Wash Assay

This assay is sometimes called the static adhesion assay. It is analogous to the aggregation assay, and the version described here came into general use with the identification of integrins (Pierschbacher and Ruoslahti, 1984). In this method, 96-well microtiter plates are coated with different concentrations of ECM ligand. Single-cell suspensions are plated in the wells, and after a specific time (10 min to several hours has been reported, with 20 min being optimal for most cells) the "nonadherent" cells are removed by washing, and the remaining cells quantified by counting cells or by using a dye-based assay with a microplate reader. The geometry of the small wells of the 96-well plate limits both the distribution and the level of shear stress involved in the washing procedure. There are no generally accepted standards for the washing step, and each laboratory adapts the method to the cell system used. In practice, this may not be important because the forces involved are small in comparison to the bond strength of the integrin–ligand or other receptor–ligand bonds. The value of this assay is its simplicity and sensitivity to receptor and ligand specificity. The actual meaning of the quantitative measurements has not been well defined.

C. Force-Dependent Methods

Force-dependent methods differ from passive assays in that they examine cells that are attached and, within that population, how much force is required for detachment.

I. Centrifugation

This method relies on centrifugal forces to detach cells from the substrate. The common application follows the plate and wash method in plating cells on ECM ligand-coated 96-well microtiter plates (Lotz et al., 1989). After plating of the cells, the wells are filled and the plates sealed. The sealed plates are first centrifuged right side up in a centrifuge equipped with a microplate carrier (usually 50 g for 5 min). This brings the cells into contact with the substrate and permits binding to occur. Then the plates are inverted and centrifuged again to detach the cells. Different protocols use different centrifuge speeds, but the practical limit is about 2000–3000 rpm, giving a force limit of about 500–600 g, although some protocols specify forces as low as 50 g for cell detachment. To calculate the force applied per cell it is necessary to measure the buoyant density of the cells. Cells that we commonly use, such as K562 erythroleukemia cells and HT1080 fibroblasts from a fibrosarcoma, have buoyant densities of 1.045 g/cm^3. According to the relationship $F = ma$, 50 g applies a force of 15 pN/cell, and 500 g, a force of 150 pN/cell. These forces are in the range of the forces required to break single integrin–ligand bonds. As expected from these calculations, the detachment profiles of adherent cells as a function of ligand density are very similar to those of plate and wash assays. These low detachment forces also raise concerns about nonspecific adhesion. Some cells can adhere nonspecifically to protein surfaces with a strength that requires up to 200 pN/cell for detachment, as measured by the spinning disk (Garcia et al., 1998). Thus, in its usual form, this assay is not significantly different from the plate and wash assays. It is theoretically possible to develop the centrifugation assay further using higher centrifugation forces that would detach cells, but this has not been well explored because of the high centrifugal forces required.

2. Micropipette

The micropipette method is basically a larger version of the single-bond method that is capable of assaying the breaking force of multiple bonds. It is certainly possible to develop sufficient force for cell detachment to occur. Analyses of cadherin (cell–cell) adhesion have been reported (Chu et al., 2004).

3. Hydrodynamic Shear

Two very different classes of adhesion assay are based on hydrodynamic shear. The more common type of assay is based on blood flow systems and typically uses hydrodynamic shear stress levels <15 dyn/cm^2 to mimic the flow in large arteries (Alon et al., 1997; Savage et al., 1998). These assays are run in parallel-plate chambers and follow the capture, rolling, and release of cells in flow. Integrins adhere only at very low shear stress ($<.05$ dyn/cm^2) (Alon et al., 1995). Selectin binding causes leukocytes to roll on the surface through the attachment and detachment of bonds (Alon et al., 1997). Shear stresses of 15 dyn/cm^2 generate detachment forces in the range of 500–1000 pN per cell on cells the size of leukocytes and, hence, are effective in measuring small numbers of engaged bonds.

The other application of hydrodynamic shear is in investigation of firm adhesion. This is the stable adhesion that is observed for most normally adherent cells in culture systems. In these systems hydrodynamic shear is used to detach adherent cells. The force required to cause this cell detachment provides a measure of cell adhesion. Because these cells adhere to the substrate through hundreds to thousands of adhesive bonds, the forces required for cell detachment are much higher. A fibroblast may express in the range of 2×10^5 α5β1-integrin receptors on the surface. Even if only 10% are bound and the average bond strength is 100 pN, about 4000 dyn/cm^2 would be required to detach all of the bonds at once.

III. APPLICATION OF HYDRODYNAMIC SHEAR TO FIRM ADHESION

Firm adhesion can be considered the steady-state condition for cells in tissues or adherent cells growing in tissue culture. It is maintained by multiple adhesion receptor–ligand bonds. The application of force to the cell may break the bonds in series by breaking one bond at a time in a peeling mechanism; or it may break bonds in parallel, where breaking of multiple bonds at the same time is required. The actual process could combine series and parallel breaking. Series breaking requires application of a small force for a long time, and parallel breaking requires application of a high force for a short time. The energy should be the same, but the hydrodynamic shear assay measures the peak force required.

A. Issues for Measurement of Firm Adhesion

1. Breaking the Weak Link

Application of force to the cell may detach the cell by breaking adhesive receptor–ligand bonds, or it may break the cell in other ways. It depends on what is the weak link in the chain that holds the cell on the substrate. The assumption in cell adhesion assays is that the cell adhesion receptor–ligand bond represents the weak link, and cell detachment occurs as a result of breaking this bond. The importance of this issue increases as the forces applied increase. Usually this has been left as an assumption. It has been shown that the application of force to an actin (substrate-bound)–integrin linkage results in reinforcement of the linkage (Choquet et al., 1997). Fibronectin-coated beads placed on the edge of a cell induce integrin binding, and the bound bead is translocated toward the center of the cell by actin filaments. If restrained with a laser trap, the bead escapes in a few seconds, and a large increase in trapping force is required to restrain the bead movement a second time. Hence the cell responds to the force and reinforces connections if allowed sufficient time. This suggests that the rate of force loading is an important issue. Fast loading is more likely to cause cell damage than slow loading.

Is there evidence to support the proposition that hydrodynamic shear causes cell detachment by breaking the adhesion receptor–ligand bonds? If the example of α5β1-integrin–mediated adhesion to fibronectin analyzed by the spinning disk is used, there are several approaches to this issue:

1. Breaking other linkages would be expected to leave α5β1-integrin and associated cytoplasmic proteins bound to the substrate. Analysis of the substrate after cell detachment using fluorescent phalloidin staining for actin or immunofluorescence staining for the focal adhesion proteins vinculin and paxillin reveals that no membrane/adhesion structure fragments are left by the detached cells.
2. The dose–response plot for detachment force versus number of bonds shows a direct linear relationship (see below), which is difficult to explain unless the breaking of this bond is the limiting event.
3. The α5β1-integrin–fibronectin bond can be strengthened by chemical crosslinking. This increases the measured detachment force twofold, implying that the α5β1-integrin–fibronectin was originally the weak link. Crosslinking in the absence of the α5β1-integrin–fibronectin bond produces no increase in detachment force, indicating that other crosslinked molecules contribute little.

Taken together these data strongly suggest that the α5β1-fibronectin bond is the weak link and hence the hydrodynamic shear assay measures this link.

2. Turbulence and Reynolds Numbers

Hydrodynamic shear measurements depend on the ability to predict the shear stress experienced by the cells. This becomes difficult if there is a transition from laminar to turbulent flow. This transition is governed by the ratio of inertial to viscous forces and is described in the Reynolds number,

$$\text{Re} = (\rho V D)/\mu, \tag{3.1}$$

where ρ is fluid density, V is volumetric flow, D is the characteristic length (the diameter of a pipe or the height of the parallel-plate chamber), and μ is dynamic viscosity. The Reynolds number at which there is a loss of laminar flow depends on the geometry of the chamber (size, height, shape), presence of fluid junctions, and roughness of the surface. It is dependent on both general geometry, which can be modeled, and actual construction, which is more difficult to model. For a parallel-plate system, Reynolds numbers <1000 have been considered laminar; for pipes, laminar flow is maintained at Reynolds numbers <2000. Because the Reynolds number increases with flow rate (V), it becomes increasingly important as the shear stress is increased by increasing the flow rate. In contrast, increasing the dynamic viscosity (μ) of the fluid increases the shear stress while reducing the Reynolds number. Given the uncertainties, it is wise to constrain the experimental system employed to laminar flow. One approach is described later for the spinning disk.

B. Devices for Hydrodynamic Shear

1. Parallel Plate

The most commonly used device is the parallel-plate system, in which two glass plates are separated by a gasket that defines the flow chamber. Medium is pumped through the chamber with a syringe pump. The apparatus is usually constructed so that it can be placed on an inverted microscope stage to view and quantify cell detachment. The critical dimension is chamber height, which is determined by gasket thickness, commonly 254 μm. Shear stress is determined with the formula

$$\tau = (6Q\mu)/(wh^2) \tag{3.2}$$

where τ is shear stress, μ is fluid dynamic viscosity, Q is volumetric flow, w is chamber width, and h is chamber height. A variation of this chamber, called the Hele–Shaw chamber, is also in use (Usami et al., 1993). This chamber is wedge-shaped and presents a shear stress gradient rather than a single shear stress. The shear stress is determined from the formula

$$\tau = (1 - d/l)(6Q\mu)/(wh^2) \tag{3.3}$$

where d is distance from the chamber entrance, and l is chamber length.

In hemodynamic applications to mimic arterial flow, these chambers are used with a maximal shear rate of 1500/s (Savage et al., 1998). For standard aqueous buffers, this is equivalent to 15 dyn/cm^2. Practical design considerations and the potential to induce turbulence have generally limited the shear stresses measured by these devices to <200 dyn/cm^2. In one report, shear stresses as high as 2500 dyn/cm^2 were used with a Hele–Shaw chamber, although there is no validation of the absence of turbulence (Moon et al., 2005).

2. Capillaries and Microfluidics

In principle, these devices are similar to parallel-plate systems, but provide potentially better flow control, particularly for the higher shear stress levels. In a microfluidic method, laser photolithography and contact molding are used to generate chambers (Lu et al., 2004). The chamber height is held to 25 μm to maximize shear stress and minimize the theoretical Reynolds number (see later text). By use of a series of chambers of varying width, a range of shear stresses can be applied. Shear stresses of 2000 dyn/cm^2 detached only 5% of the cells, whereas forces of 4000 dyn/cm^2 detached 95% of the cells. This apparent lack of linearity contrasts with other hydrodynamic shear systems described later. One wonders whether cell heights that must be about 30% of the total height of the flow path could induce fully three-dimensional flow at Reynolds numbers lower than expected from the theoretical calculations. In another system, capillaries of different diameters are used to generate a range of shear stresses (Nordon et al., 2004). Detachment forces up to 183 dyn/cm^2 were reported. These systems do offer promise, but are still at early stages of development and validation.

3. Radial Flow Chamber

In the radial flow chamber, the fluid is introduced in the center of the chamber and spreads out at a flow rate that decreases as distance from the center increases. Two different devices have been reported: one sits on a microscope stage and permits viewing of the detachment; the other consists of a plate and spacers that are inserted into a 60-mm tissue culture dish containing attached cells. The shear stress is given by the formula

$$\tau = (3Q\mu)/(\pi r h^2) \tag{3.4}$$

where r is radial position. This provides a shear stress gradient. To determine the critical shear stress, the area occupied by the cell was determined using stained cells at low magnification in a stereomicroscope (DiMilla et al., 1993). Data are reported that show that shear stresses up to 1000 dyn/cm (10 μdyn/μm^2) were used.

4. Spinning Disk

The spinning disk is based on the theoretical model of a disk spinning in an infinite fluid. As the disk spins, the edge experiences a higher wall shear stress because it is moving faster relative to the fluid. At the center of the disk, there is minimal shear stress as the cells are moving slowly with respect to the fluid. This produces a shear stress gradient that is described by the equation

$$\tau = 0.8r(\rho\mu\,\omega^3)^{1/2}, \tag{3.5}$$

where r is the distance from the center of rotation in centimeters, ρ is the density of the fluid in grams per cubic centimeter, μ is dynamic viscosity in Poise, and ω is rotation speed in radians per second. This provides a linear shear gradient from the center to the edge (Garcia et al., 1998).

IV. APPLICATION OF THE SPINNING DISK: PRACTICAL CONSIDERATIONS

A. Design Issues

1. Laminar Flow in a Spinning Disk Device

Early versions of the spinning disk device illustrated how easy it is to generate turbulence and vortexes with this type of device. Hence, significant effort was directed to developing designs that would control turbulence and provide the required laminar flow at the surface. Because we did not know the level of shear stress that would be required, over the course of experimentation, we developed multiple models, two of which are in current use. Model A (Fig. 3.1A) is better for lower shear stress ranges. The 4-in.-diameter chamber contains four vertical baffles and a top. It can be used for speeds up to 2400 rpm to measure critical shear stresses of 200 dyn/cm^2. This model uses a belt drive instead of the direct drive used in an earlier model to move the motor out from under the chamber and to make a simpler system for alignment. Some leakage through the lip seal gaskets that seal the shaft entry into the chamber is inevitable. This design allows a 25-mm coverslip to be attached to the disk and seeding of cells on this coverslip while it is in place. Thus, manipulations that could remove cells from the surface are minimized. Hence, with proper care, this design can be used to measure shear stresses for cell detachment down to about 5 dyn/cm^2. In the K562 cell model, this can represent nonspecific adhesion levels.

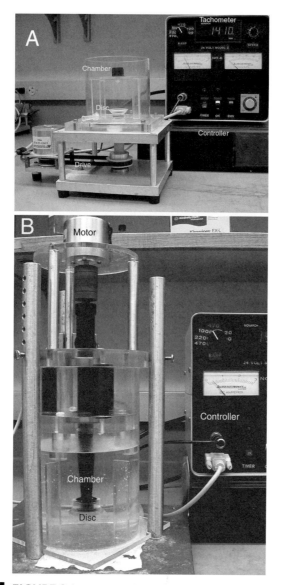

FIGURE 3.1 Spinning disk device. (A) The spinning chamber and disk are shown on the left and the controller unit at the right. Cells are plated directly on coverslips attached to the disk holder. The chamber is filled with buffer and spun at predetermined acceleration and maximum rotation speed and time. (B) Modified version for higher-shear-stress applications. The cells are first seeded on coverslips in dishes on the bench top, and then the disk is attached to the holder and the system in immersed in buffer upside down.

For higher shear stresses, we have adopted a different design in which a direct drive is used but the disk is inverted in the spinning chamber (Model B, Fig. 1B). This design is more convenient because it does not require emptying of the chamber after each spin to remove the disk. A 5-in.-diameter chamber with a more centrally positioned spinning platform and similar baffling strategy is used. In standard aqueous buffers, this device can be used up to 5000 rpm and measures critical shear stresses up to 800 dyn/cm^2. Addition of

5% dextran allows rotational speeds of at least 7000 rpm and can measure shear stresses up to 3500 dyn/cm^2. This has been more than sufficient to detach a variety of cells including long-term spread cells, with the exception of activated, spread platelets.

2. Validation of Laminar Flow

To ensure laminar flow at the surface of the spinning disk, each of the designs was analyzed under standard experimental conditions using a fero/fericyanide electrochemical reaction. Oxidation occurs at the electrode surface and is the rate-limiting step in the reaction. As the voltage is increased, the current increases to a plateau where surface transport becomes limiting. An electrode is placed on the spinning disk. As the speed of rotation increases, the flow across the surface increases and the limiting transport rate increases. Therefore, the plateau value for the current (maximum current) increases proportionately to speed as long as laminar flow is maintained. The transition to turbulence causes acceleration of the transport rate and the rate of increase in current. In Fig. 3.2, plateau current values are plotted as a function of disk rotational speed, showing that laminar flow was maintained at the surface up to at least 7000 dyn/cm^2 using 5% dextran and the model B device. Similar data were previously published for the model A device (Garcia et al., 1997). These data also show that 5% dextran behaves as a Newtonian fluid over this shear stress range.

3. Controlling Shear Stress and the Rate of Shear Loading

The DC motors that drive the spinning disk device are powered by a specifically designed controller to provide maximum consistency between

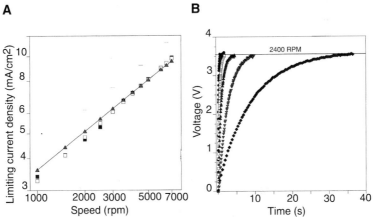

FIGURE 3.2 Characterization of spinning disk. (A) Test for turbulence in buffer containing 5% dextran (500,000 MW) to increase dynamic viscosity (μ). A platinum electrode was attached to the spinning platform. The buffer incorporated a ferro/fericyanide electrochemical reaction that is limited by the transport reaction at the platinum electrode. The experimental points ( ) represent the maximum (plateau) current as a function of rotational speed. The linear relationship demonstrates laminar flow under the set of conditions used. The line represents the theoretical relationship; ■, experimental points; □, fit; ▲, $P = 0.5$. (B) The rate of increase in speed was determined from the voltage up to a maximum of 2400 rpm using the model A device. Introduction of capacitors was used to dampen the acceleration rate. ●, 100 μF; □, 220 μF; ▲, 470 μF; ▼, 1500 μF; ◆, 4700 μF.

runs (Fig. 3.1A). The controller adjusts the maximum speed of rotation based on an integral tachometer. The time of the spin is determined by a timer; typically 5-min spins are used. The duration of spinning has essentially no effect on measurement of the mean cell detachment force, suggesting that the system is at a quasi-equilibrium condition. The controller also provides a means for controlling the rate of acceleration so that the shear stress loading rate can be controlled. This was based on the idea of reinforcement described earlier. In practice, at the highest acceleration rates, more than 10 s is needed to reach speed, and this appeared to provide a sufficiently slow loading rate; further decreases in loading rate had little effect on the measurements. Our standard loading rate appears to detach cells by separating the integrin–ligand bonds.

B. Quantification of Mean Cell Detachment Forces

The problem of translating the raw data from shear stress analyses to mean cell detachment forces is similar for all hydrodynamic shear devices. The spinning disk provides a particularly advantageous format because it represents a shear stress gradient that can be adjusted by varying the rotational speed to generate a full-range detachment profile from 100% attached in the center to 0% attached (or background) at the edge. The advantage of the full profile is that it provides a more reliable curve fit. Because cells in a population will exhibit a range of cell sizes and levels of surface adhesion receptor density, it is expected that there would be a range of detachment forces that would be normally distributed. Thus, plot of the cells that resist detachment as a function of shear stress would fit a probabilistic sigmoid curve. The disks are fixed with ethanol, stained with the DNA-binding dye ethidium homodimer, and counted using a mechanical stage that scans and counts sixty-one $10\times$ objective fields, four fields per radial position, 15 different radial positions, plus the center. This amounts to about 12% of the total disk surface and represents 20,000 cell fates on a typical disk. These cell counts are normalized to the cell count at the center, and the normalized counts are plotted as a function of the calculated shear stress at that position. The plot is curve fit using the formula (Fig. 3.3)

$$f = 1/(1 + \exp[b(\tau - \tau_{50})]) \qquad (3.6)$$

where f is the normalized cell density, τ is the shear stress at that density, τ_{50} is the shear stress at the inflection point, and b is the slope at the inflection point. The fit is used to obtain the value for τ_{50}, which represents the mean shear stress required to detach a cell in the population assayed. Analogous measurements could be made using other devices to apply shear stress. Increasing flow rates have been used for the parallel plate system, but this has the disadvantage that the measurements are not all taken at the same time and adhesion strength may vary with time.

C. Experimental Models

The ability to use these measurements to address biological problems requires cells that are simple enough (minimal variables) to take advantage of the quantitative measurements, but retain the critical biological properties. Many previous analyses that used hydrodynamic shear to measure adhesive receptor–ligand interactions employed beads and antigen–antibody reaction or interaction between the Fc region of immunoglobulin and Fc receptors as surrogates for cells. Instead, we focus initially on a very simple cell model.

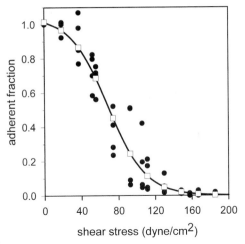

FIGURE 3.3 Detachment profile. K562 cells activated with TS2/16 monoclonal antibody were plated on a fibronectin-coated glass coverslip (160 ng/cm^2). The cells are allowed 15 min to adhere; the disk is spun for 5 min, removed, fixed in ethanol, and stained with the fluorescent DNA dye ethidium homodimer. Sixty-one fields were counted at 15 different radial positions (4 each) plus the center and normalized to center counts. The wall shear stress was calculated (see text), and the proportion of cells remaining was plotted as a function of shear stress (●) and fit to a sigmoid *curve* (○)(see text).

I. K562 Cells

Integrins are widely expressed cell adhesion receptors for ECM. Most cell types express multiple integrins, and often multiple integrins can bind to a common ligand (Sonnenberg, 1993). In addition, integrin–ligand binding is subject to regulation by intracellular signals (Hynes, 2002). Hence, the initial analysis required a system that expressed a single integrin on the surface and one in which regulation of ligand binding by that single integrin could be either known specifically or experimentally controlled. K562 cells grow in suspension and maintain a rounded shape when plated on an adherent substrate, which makes them a relatively simple hydrodynamic model. They express a single integrin on the cell surface, α5β1-integrin. And the binding of that α5β1-integrin to fibronectin is dependent on activation by reaction with either Mn^{2+} or an activating monoclonal antibody. There is no known intracellular signaling that affects α5β1-integrin binding in these cells (Garcia et al., 1998). This provides an ideal system in which to validate the basic parameters of hydrodynamic analysis of cell adhesion. K562 cells can also serve as recipients for expression of other integrin receptors following transfection (Garcia et al., 1998; Kawaguchi and Hemler, 1993). This expands the range of integrin types that can be assayed and provides a platform for mutational analyses.

2. HT1080 Cells

The central questions in the analysis of cell adhesion are how adhesion receptors are regulated for their function in both cell adhesion and cell signaling, and how the adhesion receptors function to send signals from the extracellular ligand to intracellular signaling pathways. These issues cannot be addressed using the K562 system. HT1080 cells express multiple integrins in their cell surface (α5β1, α3β1, α6β1, αvβ3, α2β1), binding of these integrins to ligands occurs without adding factors, the binding is regulated over time, and ligand binding stimulates intracellular signals. The system can be simplified by using fibronectin as the ligand. The cells express both $\alpha 5\beta 1$ and $\alpha v\beta 3$ that can bind fibronectin; however, binding is reduced to background levels by blocking only $\alpha 5\beta 1$ and blocking $\alpha v\beta 3$ has no significant effect. It appears that $\alpha v\beta 3$ is effectively inactive under these conditions. From the perspective of hydrodynamic shear, HT1080 cells do spread over time in culture. The initial phases of this spreading are reasonably symmetric, giving a morphology that resembles a hemispherical cap. Hence, it fits reasonably well to relatively simple morphological models.

3. Receptor Density

Although the number of adhesive bonds is expected to be dependent on the cell surface density of the receptor, quantitative binding analyses are often reported without consideration of receptor density. This may reflect the predominant idea that it is the number of active integrins and not the total number of integrins on the cell surface that is important. This presumes that the cell surface integrins represent a mixed population of active and inactive forms. It is true that there will be bound and free adhesion receptors. What is less clear is that there is an activation step that precedes binding in nonhematopoietic cells.

The number of cell surface receptors per cell can be determined by saturation binding of soluble ligand (Faull et al., 1993; Garcia et al., 1998). This method has been used to quantify the number of $\alpha 5\beta 1$-integrins per K562 cell. The value has been stable across laboratories and time in culture and hence, it can be used as a standard to determine the number of $\alpha 5\beta 1$-integrins on the surface of other cells by FACS. This gives the number of $\alpha 5\beta 1$ per cell. To calculate receptor surface density requires measures of the cell surface area. A first approximation is obtained from the geometric shape. Measurements of total surface area have been made by swelling cells in suspension (Dustin et al., 1997). This gives a total area of about twice the surface of the suspended cell modeled as a simple sphere. If there are issues of integrin activation state, one can compare experiments done using Mg^{2+} as the divalent cation with experiments using the activating cation Mn^{2+} (Masumoto and Hemler, 1993).

4. Ligand Density

Ligand density is also critical to determining the number of bound integrins. Measurement of ligand density has two components. The first is the level of ligand that has adsorbed on the substrate. Extracellular matrix ligands like fibronectin and collagen are long proteins that adsorb like a plate of spaghetti. Such ligands require a minimal length association with the surface to bind, which becomes less probable as the concentration increases. At some point there is a change in the slope of the adsorption profile, indicating a phase change from binding to substrate to binding to already adsorbed protein. Above this concentration, not only is there a shift in the proportionality between concentration in solution and adsorbed density, but there is a qualitative change in that the adsorbed protein is bound to protein rather than the inert substrate. Hence, it is not valid to compare directly the adhesion of cells in different phases of the adsorption isotherm. The second component is the availability of the adsorbed ligand for receptor binding. Part of this issue is geometric in that due to the random nature of the adsorption process, only a portion of the ligands have the receptor binding face exposed. Also, adsorption onto a surface can result in partial denaturation of the protein,

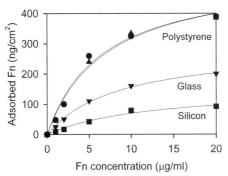

FIGURE 3.4 Adsorption of fibronectin. ^{125}I-Labeled fibronectin was adsorbed to bacterial culture dishes (●, untreated polystyrene); tissue culture–treated polystyrene (▲); glass (▼); or silicon wafers (■). The charge differences in the polystyrene surfaces had little effect. The glass and the silicon are chemically similar, but the silicon surface is smooth at the nanometer level, the glass is rougher, and the polystyrene is rougher than the glass. The apparent correlation is with surface roughness. The difference in adsorption between the glass and polystyrene has little effect on adhesion, whereas the silicon has about a twofold reduction in adhesion reflecting the reduced adsorption. Surface roughness appears to affect both the level of adsorption (more surface, more adsorption) and presentation (there is a limit before the fibronectin piles up and increased fibronectin is not accompanied by increased numbers of available binding sites).

which could have subsequent effects on ligand binding. The alternative approach is to control both the chemical environment and the orientation of the ligand. For example, we used the 13G12 monoclonal antibody, which binds to the cell binding domain of fibronectin but on the opposite face of the molecule (Akiyama et al., 1995). 13G12 was bound to a nitrocellulose surface and the available binding sites could be titrated by soluble antigen (fibronectin fragments). Now the cell binding domain fragments could be added and would have a relatively uniform presentation in a defined chemical environment and minimal denaturation.

The most common expression of ligand levels is the concentration used for coating. Common ECM proteins reach surface saturation at coating concentrations of 10 μg/ml. The actual amount adsorbed appears to depend primarily on surface roughness. For example, adsorption of fibronectin to smooth silicon surfaces gives a saturation of 80 ng/cm^2, a rougher coverslip glass surface adsorbs about 160 ng/cm^2, and polystyrene surfaces used for tissue culture plastics adsorb about 355 ng/cm^2 (Fig. 3.4).

V. ANALYSIS OF BIOLOGICAL PROBLEMS IN CELL ADHESION

The objective of the analysis in this chapter is to address the issue of firm adhesion of cells. It should be clear from the preceding sections that firm adhesion is mediated by multiple adhesion receptor–ligand bonds, and it takes place on a backdrop of adhesion receptors that are both patterned on the surface and dynamically linked to the cytoskeleton. These properties are only maintained in the intact cell. In this discussion, α5β1-integrin is used as the model system. Here, we keep the hydrodynamic properties of the cells constant by using the K562 model system and address issues of comparing different cell types and stages of cell spreading in the next section.

A. Adhesion Strength (τ_{50}) Is Proportional to the Number of Integrin–Ligand Bonds

In the preceding section, the issue of using hydrodynamics to measure the mean shear stress required for cell detachment was considered. In this section, this measurement (τ_{50}, or adhesion strength) is used to determine the number of adhesive bonds in the cell–substrate space. The analysis is based on the Law of Mass Action and the assumption is that it applies equally to bonds formed in the cell–substrate interface.

1. The Simplified Cell Model

To test the hypothesis that τ_{50} is proportional to the number of adhesive bonds, the following relationship was developed based on the mechanical conditions and the formula for chemical equilibrium:

$$\tau_{50} = G\Psi N_L N_R + \lambda. \tag{3.7}$$

Here, G is the geometric function that converts shear stress to force per cell, Ψ is the bond factor, N_L is ligand density, N_R is receptor density, and λ is the nonspecific binding contribution. This relationship was tested using K562 cells in which $\alpha 5 \beta 1$-integrin is activated by the TS2/16 monoclonal antibody. Cells are allowed to adhere for 15 min to glass coverslips coated with graded densities of fibronectin and blocked with bovine serum albumin to reduce nonspecific binding of cells directly to the glass. A linear relationship was obtained (Fig. 3.5A), which showed that doubling the number of bonds doubled the shear stress required for cell detachment. The data in this graph represent the compilation of 50 experiments performed over more than 1 year, demonstrating the reproducibility of the measurements. The linear regression has a P value of $<10^{-8}$. To vary N_R, the K562 cells were transfected with expression constructs for $\alpha 5$-integrin, and clones were selected that had different levels of $\alpha 5 \beta 1$ expression. The cells already express excess $\beta 1$-integrin, which is degraded in the absence of the $\alpha 5$ partner, so it was not necessary to transfect in more $\beta 1$. The maximum level of increased expression was about sixfold. This high level of overexpression is unusual for integrins and probably indicates that growth of the cells in suspension does not involve $\alpha 5 \beta 1$ and, hence, there is minimal negative selection. Figure 3.5B illustrates the linear relationship between N_R and τ_{50}. When the data from Figs. 3.5A and B are combined, the product $N_R N_L$ also exhibits a linear relationship with τ_{50}, which appears to extend over two orders of magnitude (Fig. 3.5C). Thus, the hypothetical formula appears to be valid. This result raised three critical questions:

1. Why is there no apparent saturation of the binding at high ligand densities? The simplest answer to the saturation issue is that ligand binding curves are very close to linear when the ligand density is $<<K_D$ for the reaction. Although the ligand densities used may saturate the surface, they are not necessarily sufficient to saturate the receptor. However, this would mean that the proportion of $\alpha 5 \beta 1$ that bound to fibronectin under these conditions must be a small proportion of the total, although, based on the small volume of the cell–substrate space and the published values for K_D, it was expected that the majority of the integrin receptors would be bound.

2. The bonds on the side facing the flow are expected to have higher loads than those on the opposite side, leading to detachment by a peeling

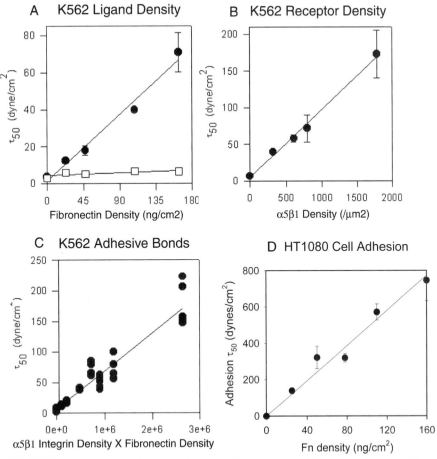

FIGURE 3.5 Detachment force as a function of bond number. (A) τ_{50} was determined from detachment profiles as in Fig. 3.3 for K562 cells plated on different densities of fibronectin, both with (●) and without (□) activation by TS2/16. (B) τ_{50} was determined for K562 cells expressing different levels of $\alpha5\beta1$ on the surface (expt. ●). (C) τ_{50} was determined for K562 cells for different combinations of fibronectin density and surface $\alpha5\beta1$ levels plotted as the product of $\alpha5\beta1$ and fibronectin densities (expt. ●). (D) HT1080 fibroblasts were plated, allowed to spread for 1 h (partially spread cells), and analyzed in the spinning disk to determine τ_{50} as a function of fibronectin density (expt. ●). This adhesion is fully $\alpha5\beta1$ integrin-dependent. (A)–(C) are reproduced from Garcia, A.J., Huber, F., and Boettiger, D. (1998) *Journal of Biological Chemistry* **273**: 10988–10993.

mechanism. This would not give the linear relationship observed. So, why is it linear? Is it clear that the forces cannot be redistributed by the cell to equalize the forces on the bonds? Does the relationship really distinguish peeling models from nonpeeling models?

3. It was expected that integrin clustering or localization in focal adhesions would contribute to the binding process and, hence, there should be a second–order dependency on receptor density. Again, why was the relationship linear? First, although $\alpha5\beta1$ binding to substrate does induce integrin clustering and formation of focal contacts in many cell types, it does not lead to clustering on K562 cells (data not shown). However, this does not appear to be the full answer.

2. More Realistic Cell Systems

Although the K562 system results demonstrate that hydrodynamic shear analysis can be used to analyze the number of bonds in the adhesive interface, the K562 cell system is artificial in that it uses preactivated $\alpha5\beta1$ that is insensitive to cellular controls. To test whether the analysis could be applied to more complex systems, HT1080 cells were analyzed. In Fig. 3.5D the plot of τ_{50} as a function of fibronectin density reproduces the linearity observed for the K562 cells. This plot represents HT1080 cells that were incubated for 60 min before assay, by which stage the localization of integrin into focal contacts has well progressed. Hence, the linearity is not simply a reflection of the absence of integrin clustering in the K562 cell model. The other difference is that much higher τ_{50} values were required for cell detachment. However, because of the differences in cell size and shape, it is difficult to make direct comparisons. This issue is addressed later.

Theoretical objections notwithstanding, the data show a clear relationship between the number of bonds expected on the basis of the Law of Mass Action and the measurement of τ_{50}, for the simple K562 cell model in which integrins are uniformly activated by the addition of activating monoclonal antibody. The approximate linear relationship was maintained over two orders of magnitude. This relationship demonstrates that τ_{50} can be used as a measure of the number of bonds. It is not limited to $\alpha5\beta1$-integrin but has been extended to $\alpha v\beta3$ and $\alpha6\beta1$. This analysis can be widely applied to adhesion receptors even outside the integrin family. The limitation has been the availability of model systems in which adhesion could be limited to specific receptor–ligand reactions and purified, well-characterized ligand substrates could be produced.

B. Separating the Number of Bonds from Bond Strength

Up to this point, a means of measuring the *relative* number of adhesive bonds in homologous systems has been developed. To address the important biological questions, it is necessary to be able to generalize the measurements and to compare the measurements made on different systems. Because force is used to break the adhesive bonds, the τ_{50} that is measured contains a component that includes bond strength (Ψ in Eq. 3.7). The total detachment force would be the product of the number of bonds and the average bond strength. The problem is to separate these two parameters.

The experiments on $\alpha5\beta1$-integrin binding to fibronectin described earlier yielded linear relationships between the number of bonds and τ_{50} over the full measurable range. It was suggested that this relationship could reflect an effective K_D that was higher than the fibronectin surface saturation density. This experimental limitation led to the search for other systems that would not have the same limitation. Thus, adhesion receptor–ligand pairs with lower K_D values were sought. Fractalkine is the ligand for the seven-transmembrane G-protein–coupled chemokine receptor CX3CR1 (Bazan et al., 1997). Fractalkine has a cytoplasmic and transmembrane domain linked to an extracellular mucin-rich stalk and topped with the receptor binding domain. It can mediate adhesion of cells to a substrate that resists hydrodynamic shear (Haskell et al., 2000). For the analysis, a closely related receptor–ligand pair was included in which the receptor was CXCR1 and the ligand was interleukin (IL)-8. IL-8 is homologous to the receptor binding domain of fractalkine, but it is expressed only as a soluble ligand, and was used to replace the fractalkine receptor binding domain (Haskell et al., 2000). The K_D values of the soluble ligands for CXCR1 and CX3CR1 were 10^{-9} and 3×10^{-9} M, respectively, two orders of magnitude lower than that for $\alpha 5\beta 1$-fibronectin. In Fig. 3.6, measurements of τ_{50} are plotted as a function of ligand density. Both plots show that the reactions do saturate as predicted and that they can be fit to the predicted hyperbolic form. This is a critical result, because it demonstrates that saturation binding can be observed and that the $\alpha 5\beta 1$-fibronectin data were indeed linear because they were at a much less effective K_D for the binding reaction in the contact zone. (It answers question 1 in Section V.A.1.)

With the expected hyperbolic saturation curve, it is possible to use curve fitting to extract parameters for both the effective K_D (half-maximal binding) and the saturation asymptote. For CXCR1, the effective $K_D = 873/\mu m^2$; and $\tau_{max} = 7$ pN/μm^2 (Fig. 6C). For CX3CR1, the effective $K_D = 2488/\mu m^2$; and $\tau_{max} = 22$ pN/μm^2 (Fig. 3.6B). Note that the term *effective* K_D is used rather than K_D. K_D is defined by solution binding analyses and is related to bond energy. The effective K_D for the reaction in the cell–substrate interface is affected by the additional elements of this more complex environment; hence, effective K_D reflects this composite reaction. The accuracy of effective K_D determination is limited by two factors. The first is the accuracy of the determination of ligand

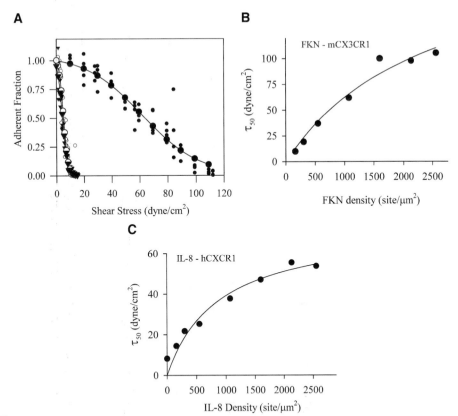

FIGURE 3.6 Analysis of saturation equilibrium binding for chemokine receptors. (A) Adhesion profiles for CX3CR1 binding to fractalkine (●, data and fit) at 1000/μm^2 and control (no fractalkine, only immunoglobin (○) or IL-8 substrate (▼)). (B) Plot of τ_{50} versus fractalkine density (●) and curve fit to ligand binding $\tau_{50} =$ constant $N_L/ (K_D + N_L)$, where N_L is ligand density and K_D is the effective K_D measured from curve fit. (C) Same as (B) for CXCR1–IL-8. Reproduced from Lee, F-H, Haskell, C., Charo, I.F., and Boettiger, D. (2004) *Biochemistry* **43**: 7179–7186, with permission.

density. For surface-adsorbed ligands, this is not only the adsorbed density, but is influenced by orientation and denaturation. (See earlier discussion of ligand density.) To control these factors in the CXCR1 and CX3CR1 experiments, fractalkine was modified to replace the transmembrane and cytoplasmic domains with a 6-his tag. The 6-his tag was presented by anti-6-his antibody adsorbed to the surface. In this way, it is possible to ensure both a uniform presentation and an accurate measure of the density. The second factor is having experimental points distributed both above and below the effective K_D value to constrain the curve fit. This was accomplished by using binding reactions of higher affinity (reduced K_D). With the effective K_D, the ligand density, and the receptor density, the number of bonds can be calculated.

What is the meaning of the asymptotic values (τ_{max})? In the solution binding equation, this would represent the total number of receptors available for binding. In the transformation of the binding saturation equations to the spinning disk, two additional terms were introduced (Eq. 7). The geometric factor (G) included the cell shape and provides the conversion of shear stress to the applied force per cell. The bond factor (Ψ) represents bond strength and the number of bonds that detach in parallel when adhesion fails. Given that the level of expression of CXCR1 on transfected 300.19 cells was 2×10^5/cell and the expression level of CX3CR1 on transfected 300.19 cells was 10^5/cell, there would be twice as many bonds for about one-third the force (τ_{max}) for the CXCR1–IL-8 bonds. The cells were the same except for receptor expression. Hence, the shape and proportion of bonds that break at failure should be the same. Thus, CX3CR1–fractalkine bonds are about sixfold stronger than CXCR1–IL-8 bonds. This is not the expected result because the bond with the higher affinity, based on solution binding, and hence the higher bond energy (ΔG), is the weaker bond under tension (Bell, 1978; Dembo et al., 1988; Kuo and Lauffenburger, 1993). Conversion of these values to bond strengths depends on assumptions and models for cell shape and attachment failure mechanisms discussed later.

C. Analysis of Adhesion-Mediated Signaling

Receptor-mediated signaling reactions generate intracellular signals, the magnitude of which is proportional to the level of ligand bound within certain limits. As the reaction proceeds down a signaling pathway to generate a binary response, it is necessary to sharpen the dose–response curve by increasing the Hill coefficient (Ferrell, 1998). This leads to two questions: (1) Does adhesion-mediated signaling exhibit the expected dose-response relationship (like growth factor-stimulated signals)? (2) Can this process be used to identify the proximal elements of the signaling pathway?

To analyze these issues, HT1080 cells were serum starved to reduce the signaling background, plated for 1 h on different densities of fibronectin, and assayed with the spinning disk to determine the relative number of bonds and the phosphorylation of focal adhesion kinase (FAK) to measure the downstream signaling. Figures 3.7A and B; illustrate the linear relationship between the level of bound $\alpha 5\beta 1$ and the level of FAK Y397, the direct dose–response relationship between ligand binding and signaling. Because the plots of $\alpha 5\beta 1$ binding and FAK phosphorylation as a function of ligand density have the same slope, this would suggest that stimulation of FAK phosphorylation is an early event in the signaling cascade.

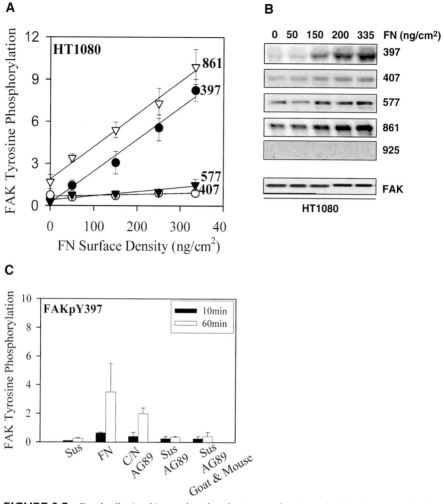

FIGURE 3.7 Focal adhesion kinase phosphorylation as a function of α5β1–fibronectin bonds. (A) Specific FAK phosphorylation (phosphorylated FAK/total FAK) as a function of fibronectin density. (B) Western blot using phospho-specific antibodies to the phosphorylation sites in FAK Y397, Y407, Y577, and Y861, and with anti-FAK for standard. (C) HT1080 cells in suspension (sus), plated on fibronectin (FN); plated with AG89 monoclonal antibody to β1-integrin plated on nitrocellulose (binds the AG89); AG89 in suspension, and AG89 plus anti-mouse antibody to increase clustering. Y397 requires α5β1 attached to substrate, but it does not matter whether the ligand is fibronectin or antibody. This works with all anti-α5β1 antibodies tested (4). Reproduced from Shi, Q., and Boettiger, D. (2003) *Mol. Biol. Cell* **14**: 4306–4315, with permission.

What happens to the mechanical elements of the signal? The binding of adhesion receptors to ligands involves not only ligand binding, but also the effect of the receptor's becoming tethered to the substrate through this receptor–ligand bond. The binding of soluble fibronectin to α5β1 on HT1080 cells in suspension did not stimulate FAK Y397 phosphorylation, but it did stimulate the phosphorylation of Y861, when primary and secondary antibodies were used to induce α5β1 clustering (Fig. 3.7C). Although these complexes attached to α5β1-integrin did not stimulate the phosphorylation of FAK Y397 for cells in suspension, attachment of this complex to a surface was sufficient to stimulate FAK Y397 phosphorylation (Fig. 3.7C).

It did not matter whether α5β1 was attached through fibronectin, activating monoclonal antibody, or adhesion-inhibiting monoclonal antibody. Each stimulated a dose-dependent increase in Y397 phosphorylation. Thus, there may be multiple modes of integrin-dependent signaling, clustering of integrins, ligand occupancy, and tethering of integrin to the substrate. The last mode allows the application of mechanical stress and hence is sensitive to mechanical cues.

Does the binding of α5β1 to substrate-associated fibronectin always induce signaling (as measured by FAK Y397 phosphorylation)? We found two cases when the level of bound α5β1 measured by the spinning disk does not translate into the phosphorylation of FAK Y397: when the fibronectin is plated on uncharged polystyrene surfaces, and when the fibronectin is adsorbed to positively charged surfaces like poly-L-lysine (Miller and Boettiger, 2003). It is not yet evident how these discrepancies can be reconciled with either the mechanical signaling models or the bidirectional allosteric models of integrin signaling. There may be additional layers in the process to uncover.

These analyses demonstrate that integrin-dependent signaling is proportional to the level of integrin binding, but that binding may not be sufficient and additional elements may be required for downstream signaling. Because the signaling is dependent on both ligand binding and downstream issues, it is essential to have the ligand binding data to draw conclusions about the contributions to signaling. For example, there is an increase in FAK phosphorylation that occurs with cell spreading and integrin clustering. So, one might conclude that integrin clustering causes FAK signaling. However, if integrin clustering also increases the number of integrin–ligand bonds, then the increased signaling could be caused by increased receptor binding and not be dependent on clustering.

VI. CALCULATION AND INTERPRETATION

In the preceding sections, the approach to measuring cell detachment forces and the application of these measurements to biological problems in cell adhesion have been presented. The use of hydrodynamic shear methods for analysis of the number of adhesive bonds that form in the cell–substrate interface has been demonstrated to be feasible. In the generation of models for cell adhesion, it is necessary to convert shear stress to applied force and to consider the meaning of the total cell detachment forces in terms of receptor–ligand bonds. Understanding these issues also provides a basis for the comparison of the cell binding reactions during cell spreading or between different cell types.

A. Converting Shear Stress to Forces on Cells

To compare quantitative measurements between different cell types or different cell lines and to use the quantitative data in the development of mathematical models, it is necessary to convert the τ_{50} measurement into the force applied per cell. Unfortunately, cell shape is difficult to model, so simplifications are required. Choosing an approximation depends on the shape of the cell during detachment in hydrodynamic flow. By 15 min of spreading for HT1080 cells, the edges can be seen beginning to protrude beyond the projected cell sphere (Fig. 3.8).

HT1080 Cell Spreading

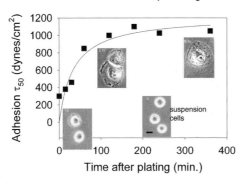

FIGURE 3.8 HT1080 cell spreading. The curve indicates the increase in projected cell area with time. Photographs show cells at 15, 60, and 360 min at appropriate parts of the curve, and suspended cells. Note the initial spreading, where cells begin to spread beyond the projected spherical cell area at 15 min.

When a spherical cell from suspension is plated on a ligand-coated surface, there is an initial phase of isotropic spreading, much like a drop wetting a surface (Giannone et al., 2004). Observations using interference reflection microscopy (IRM), which provides a view of the cell–substrate distance, show that the cell–substrate interface has a fairly uniform separation at 5 min that has an area similar to the projected area of the cell. It is not like a hard ball placed on a surface.

Based on the morphological observations and IRM data, we have developed a simple modification of the spherical cap model. Because of the inaccuracies involved in microscopic measurements of cell height, we assume that the cell maintains a constant volume. This seems a reasonable assumption considering that cells maintain a constant volume through multiple cycles of trypsinization and replating during tissue culture. The projected cell area is used to calculate an average radius, and the suspended cell diameter is used to calculate cell volume. Then height can be determined from the formula for the volume of a spherical cap.

$$V = \pi/6(r^2 + h^2)h, \tag{3.8}$$

where r is the radius and h is the height These parameters are used in the Truskey and Proulx model for spherical caps (Truskey and Proulx, 1993).

$$F_D = 4.33\pi r^2 \tau \qquad \text{Calculation of the drag force} \tag{3.9}$$

$$T = 2.44\pi r^2 h\tau \qquad \text{Calculation of torque on the cell} \tag{3.10}$$

$$F_y = 0.75\pi/a(T + F_D h) \qquad \text{Correction for force in the } y \text{ dimension} \tag{3.11}$$

$$F_T = (F_D^2 + F_y^2)^{1/2} \qquad \text{Vector addition to determine net detachment force} \tag{3.12}$$

F_D is the drag, r is the radius for the projected cell, τ is the shear stress applied, h is the height of the cell, a is the radius of the contact area, and F_T is the total force applied to the cell. Application of these corrections to the HT1080 cell spreading model is illustrated in Fig. 3.8. The cells spread relatively uniformly during the short term. Later, the shape becomes elongated as the ECM is remodeled. This uniform spreading suggests that the cap model is reasonable. As cells spread, diameter increases and height decreases. For the purpose of applying hydrodynamic shear using the spinning disk, the earliest time point we have measured is 5 min. By this time, the IRM analysis shows that the projected cell radius and the radius of the contact area are approximately equal. Hence we have made the approximation that $a = r$. F_T can then be expressed as shear stress (τ) times a conversion factor. For the spreading of HT1080 cells from 300 μm^2 (which is equivalent to the projected area of the cell in suspension) to 1200 μm^2 (which represents a steady-state condition that is maintained for several hours), the conversion factor varies by <10%. The reasons for this unexpectedly low variation are, first, that the contact area and the projected area are equal and, second, that the volume remains constant. The cell is not a hard geometrical object, and during the application of shear stress to fibroblasts, like HT1080 cells, the nucleus is displaced in the direction of flow, pulling a distorted cytoplasm behind. Under these conditions, cell spreading would not be expected to have a large effect on the detachment force, in agreement with the variation in the Truskey and Proulx model presented earlier. The displacement of the nucleus up from the surface as well as in the direction of flow suggests that the calculated model is likely to be an underestimate of the applied detachment force per cell. Additional models are under development.

B. Converting Force per Cell to Force per Bond

The conversion of force per cell to force per bond depends on the distribution of stress to the bonds. Although the earliest models did not account for differences in the stress applied to the bonds (Hammer and Lauffenburger, 1987), this has been a significant issue in subsequent models (Kuo et al., 1997; Ward et al., 1994). The models are based on the cell as a hard sphere. If hydrodynamic forces are applied to the sphere, it tends to roll (torque). If the sphere is attached to the surface by bonds, the bonds on the side toward the flow would be stretched and those on the opposite side would be compressed. Therefore, the bonds would not be equally stressed, and the ones on the side toward the flow would be the first to break as the flow rate (shear stress) is increased. This is the extreme model. Analysis of the effect of hydrodynamic shear in the detachment of single *Dictyostelium* cells has been investigated by interference contrast reflection microscopy, which reveals a decreasing contact area beginning at the edge toward the flow as shear stress is applied (Decave et al., 2002). This fills the basic prediction of the peeling model based on the unequal loading of bonds. In another model, one end of an elongated muscle myotube was aspirated into a wide-bore pipet and allowed to attach. The hydrodynamic shear was applied by flow down the pipet against the tip of the cell (Griffin et al., 2004). In this case the cells detach by a peeling mechanism. Thus, both models and experimental analyses have favored the argument that the cells must detach by a peeling mechanism.

These models, however, are stacked in favor of the peeling mechanism, and it is likely that cell detachment depends on the loading rate, the forces involved, and how the force is applied to the cell. If the tip of an elongated cell is pulled on, how else can it detach except by a peeling mechanism? The case of *Dictyostelium* involves the detachment of a cell that is not adherent to the substrate through specific bonds as it crawls along, and indeed, the forces involved in the detachment are much less than those involved in cell detachment as described earlier. When fluid is directed at a spread cell, it is clear that the edge toward the flow will receive more shear stress than the opposite edge. There is also flow across the top of the cell to create a drag that is distributed over the surface, raising the total force with less effect on differential stresses. The dogmatic view that the only way the cell can detach under these conditions is by peeling assumes the cell is a passive element. However, the evidence that cells respond to forces is indisputable. Cells respond to the application of force by reinforcing the stressed connections (Choquet et al., 1997). Stress induces the accumulation of vinculin at sites of adhesive stress which is also thought to increase the adhesive contact (Riveline et al., 2001). In effect, the cell responses act like a *tensegrity* system, redistributing and equalizing the effects of the force (see Chapter 4). These responses occur in the time frame of a second to a few seconds and, hence, are rapid enough to compensate for the applied force as long as the loading rate does not exceed the cell's ability to respond. All the data we have collected with spinning disk analyses cannot eliminate the possibility that peeling plays a role in the cell detachment, but the analysis does not favor the peeling model.

It is important to point out that whatever the mechanism of cell detachment turns out to be, it does not affect the data analysis and application in the determinations of the relative number of receptor–ligand bonds and the effective K_D. It does affect the application of hydrodynamic shear stress analysis to measurements of bond strength.

VII. DISCUSSION

Hydrodynamic shear methods have been used to analyze the functions of the hemopoietic organs that function in the fluid environment of the blood. In contrast, this chapter has focused on the development of hydrodynamic shear as a tool for the analysis of cells that are normally adherent and reside in solid tissues. It was demonstrated that the hydrodynamic shear stresses that must be applied in this analysis are one to two orders of magnitude higher than those experienced by cells in the blood circulation. This level of hydrodynamic shear is clearly nonphysiological. However, cells in the body do experience stresses from other sources. The ability of cells to resist stress has been selected through evolution and balanced both by the cost of maintaining the ability to resist and by the demands of other functions. It is remarkable that the cells can withstand these high shear stresses and still only break at the adhesion receptor–ligand bond. It is worth considering what underlies this property. The spinning disk, as it has been developed here, has a reasonably slow loading rate, and this loading rate is in the range of known cell responses to applied stress. High loading rates may result in fractures that break apart the cell.

The high shear stresses are unexpected from another perspective. They show that the detachment of these normally adherent cell types from a substrate requires high peak forces, and this can be explained only by the detachment of many bonds in parallel. Take the simple calculation for HT1080 cells. The cells express 2×10^5 α5β1-integrins on the surface, and when fully spread, 10^5 integrins could be in the cell–substrate interface. If 10% of them are bound, then there would be 10^4 adhesive bonds. The force required to detach this cell measured by the spinning disk is 6×10^5 pN, or 60 pN per bond (assuming that they ALL detach in parallel). This is very close to the bond strengths of 50–100 pN measured with the single-bond methods. Of course, the 10% figure could be an underestimate, but the equilibrium binding data suggest that it is not a large underestimate. This suggests that the peeling mechanism of detachment (serial bond breaking) contributes relatively little to the detachment process, which does not seem to make sense if the cell is regarded as a passive element. Thus, one can presume that the calculation is wrong, or that the cell is alive and responds to the application of force by redistribution of reinforcements, like a tensegrity structure (see Chapter 4).

The high shear stress required for the detachment of spread cells also creates design problems for hydrodynamic shear devices. The high flow rates required increase the tendency of turbulent flow to develop. For our spinning disk device, laminar flow has been confirmed experimentally for our experimental conditions. Because it has been more extensively analyzed, it provides a model for both development of new hydrodynamic methods and investigation of biological control. There is a linear relationship between the number of bonds and the shear stress required to detach the cell. This was applied over the full available experimental range and was limited primarily by the surface ligand density and the effective K_D. This relationship has held for over 8 years of experiments and 14,000 disks analyzed for several cell types and receptor types. In addition, the results have been confirmed by chemical crosslinking, which depends on different assay principles (Shi and Boettiger, 2003). Given the weight of these data, it is difficult to ascribe the results to artifact. Hence, these data provide a benchmark by which other hydrodynamic shear approaches can be measured. Although a number of other devices have been reported to deliver hydrodynamic shear to cells, none have been extensively developed for the measurement of adhesive bond numbers. The relationship between the number of bonds and critical shear has been the subject of little investigation. Systems for which these relationships are not simple or cannot be worked out are of limited use because the spinning disk analysis demonstrates that the relationship does exist.

The problem of measuring the number of adhesion receptor–ligand bonds has been difficult to address. Although several issues concerning hydrodynamic shear methods have been resolved, there are some outstanding issues.

1. It is difficult to distinguish the contribution of the number of receptors versus the bond strength of the bond to total adhesion strength. The single-bond analyses have found single-bond strengths for specific receptor–ligand bonds (Weisel et al., 2003). The analysis only measured bonds at the breaking point. It is possible that whatever states the bonds were in prior to stress, stress to the breaking point brings them all to the same final state.

2. It is difficult to determine the number (proportion) of receptor–ligand bonds that need to break in parallel for detachment to occur. It would greatly simplify the analysis if all bonds broke at the same time.

3. Equilibrium binding analyses are limited by ligand density. Hence, the precision of determination of effective K_D is limited. Some methods are suggested for increasing the effective ligand density: using smaller ligand fragments and providing for more efficient presentation. More technical innovation could be applied to this issue.

VIII. FUTURE PROSPECTS

Most of the data available on the binding of integrin receptors to their ligands is based on the use of soluble components. The analysis presented in this chapter implies that these data cannot be directly extrapolated to the binding parameters of integrins and ligands in cell adhesion. When a cell is placed on a ligand-coated surface, the cell adheres through integrin receptors; these integrins also redistribute into clusters, bind to a multitude of cytoplasmic proteins, either directly or through adapters, connect to the cytoskeleton, and initiate a variety of intracellular signaling cascades (Hynes, 2002). Moreover, the transmembrane information appears to be bidirectional, with cell signals affecting ligand binding and ligand binding affecting cell signals. Although the evidence for inside-out signals is strong for platelets and leukocytes, for which prevention of binding and coagulation is required, the evidence in adherent cells is weak. In fact, it is quite difficult to measure, unless the relative number of bound receptors can be measured. For the outside-in arm of the signaling, it is important to know the effect of the protocol used on the level of binding. Therefore, it is important to apply this analysis to the problems of regulation of integrin binding and regulation of integrin-mediated signaling.

To date this analysis of receptor–ligand binding has been applied to only a few integrins. The integrins represent a large family with overlapping specificities. A quantitative understanding of how the binding occurs and is regulated in the context of adhesion may provide clues to how the different integrins are adapted to different adhesive functions or to function in different adhesive contacts. There are many examples of cases in which multiple integrins for the same ligand are present on the cell surface but only one of them is used for adhesion. It is important to extend these analyses to other adhesion receptors including cell–cell receptors.

The ability to measure receptor–ligand bonds was acquired only recently and as yet is not widely used. It is a critical element that needs to be incorporated into the analyses of adhesion regulation and adhesion-mediated signaling.

SUGGESTED READING

Bell, G.I. (1978). Models for the specific adhesion of cells to cells. *Science* **200:** 618–627.

A model is developed to analyze adhesion between cell-bound receptors and cell-bound ligands. This is the starting point for most mathematical models of cell adhesion.

Choquet, D., Felsenfield, D.P., and Sheetz, M.P. (1997). Extracellular matrix rigidity causes strengthening of integrin–cytoskeletal linkages. *Cell* **88:** 39–48.

This article demonstrates that application of force to integrins results in strengthening of the cytoskeletal connections.

DiMilla, P.A., Stone, J.A., Quinn, J.A., Albelda, S.M., and Lauffenburger, D.A. (1993). Maximal migration of human smooth muscle cells on fibronectin and type IV collagen occurs at an intermediate attachment strength. *J. Cell. Biol.* **122:** 729–737.

This article represents the first attempt to measure adhesion of fibroblasts using a quantitative hydrodynamic shear approach. It develops the radial flow plate method.

Hynes, R.O. (2002). Integrins: bidirectional, allosteric signaling machines. *Cell* **110**: 673–687.

This review describes the current bidirectional allosteric model for integrin regulation and the experimental basis of the model. It provides a good introduction to the regulation of adhesion receptor binding.

Garcia, A.J., Ducheyne, P., and Boettiger, D. (1997). Quantification of cell adhesion using a spinning disc device and application to surface-reactive materials. *Biomaterials* **18**: 1091–1098.

This paper describes the construction and initial analysis of the spinning disk device for the analysis of cell adhesion.

Garcia, A.J., Huber, F., and Boettiger, D. (1998). Force required to break alpha5beta1 integrin–fibronectin bonds in intact adherent cells is sensitive to integrin activation state. *J. Biol. Chem.* **273**: 10988–10993.

This article decribes the application of the spinning disk device to a simple cell model and develops some of the validation procedures.

Giannone, G., Dubin-Thaler, B.J., Dobereiner, H.G., Kieffer, N., Bresnick, A.R., and Sheetz, M.P. (2004). Periodic lamellipodial contractions correlate with rearward actin waves. *Cell* **116**: 431–443.

This article provides a new view of the processes of cell spreading and describes the mechanical and signaling issues involved.

Shi, Q., and Boettiger, D. (2003). A novel mode for integrin-mediated signaling: Tethering is required for phosphorylation of FAK Y397. *Mol. Biol. Cell* **14**: 4306–4315.

This article applies the spinning disk method to spread cells and demonstrates a direct relationship between the number of integrin–ligand bonds and the strength of adhesion receptor signaling as measured in specific cellular phosphorylation reactions.

Weisel, J.W., Shuman, H., and Litvinov, R.I. (2003). Protein–protein unbinding induced by force: Single-molecule studies. *Curr. Opin. Struct. Biol.* **13**: 227–235.

This review provides an excellent simple overview of the single bond strength analysis and provides a survey of bond strengths for different bonds.

REFERENCES

Akiyama, S. K., Aota, S., and Yamada, K. M. (1995). Function and receptor specificity of a minimal 20 kilodalton cell adhesive fragment of fibronectin. *Cell Adhes. Commun.* 3: 13–25.

Alon, R., Chen, S., Puri, K. D., Finger, E. B., and Springer, T. A. (1997). The kinetics of L-selectin tethers and the mechanics of selectin-mediated rolling. *J. Cell Biol.* 138: 1169–1180.

Alon, R., Kassner, P. D., Carr, M. W., Finger, E. B., Hemler, M. E., and Springer, T. A. (1995). The integrin VLA-4 supports tethering and rolling in flow on VCAM-1. *J. Cell Biol.* 128: 1243–1253.

Bazan, J. F., Bacon, K. B., Hardiman, G., Wang, W., Soo, K., Rossi, D., Greaves, D. R., Zlotnik, A., and Schall, T.J. (1997). A new class of membrane-bound chemokine with a CX3C motif. *Nature* 385: 640–644.

Bell,G. I. (1978). Models for the specific adhesion of cells to cells. *Science* 200: 618–627.

Choquet, D., Felsenfield, D. P., and Sheetz, M. P. (1997). Extracellular matrix rigidity causes strengthening of integrin-cytoskeletal linkages. *Cell* **88**: 39–48.

Chu, Y. S., Thomas, W. A., Eder, O., Pincet, F., Perez, E., Thiery, J. P., and Dufour, S. (2004). Force measurements in E-cadherin-mediated cell doublets reveal rapid adhesion strengthened by actin cytoskeleton remodeling through Rac and Cdc42. *J. Cell Biol.* **167**: 1183–1194.

Decave, E., Garrivier, D., Brechet, Y., Bruckert, F., and Fourcade, B. (2002). Peeling process in living cell movement under shear flow. *Phys. Rev. Lett.* **89**: 108101–1 *to* 108101–4.

Dembo, M., Torney, D. C., Saxman, K., and Hammer, D. (1988). The reaction-limited kinetics of membrane-to-surface adhesion and detachment. *Proc R Soc London Ser. B* **234**: 55–83.

DiMilla, P. A., Stone, J. A., Quinn, J. A., Albelda, S. M., and Lauffenburger, D. A. (1993). Maximal migration of human smooth muscle cells on fibronectin and type IV collagen occurs at an intermediate attachment strength. *J. Cell Biol.* **122**: 729–737.

Dustin, M. L., Golan, D. E., Zhu, D. M., Miller, J. M., Meier, W., Davies, E. A., and van der Merwe, P. A. (1997). Low affinity interaction of human or rat T cell adhesion molecule CD2 with its ligand aligns adhering membranes to achieve high physiological affinity. *J. Biol. Chem.* **272**: 30889–30898.

Faull, R. J., Kovach, N. L., Harlan, J. M., and Ginsberg, M. H. (1993). Affinity modulation of integrin alpha 5 beta 1: Regulation of the functional response by soluble fibronectin. *J. Cell Biol.* **121**: 155–162.

Ferrell, J. E., Jr. (1998). How regulated protein translocation can produce switch-like responses. [review] [37 refs]. *Trends. Biochem. Sci.* **23**: 461–465.

Garcia, A. J., Ducheyne, P., and Boettiger, D. (1997). Quantification of cell adhesion using a spinning disc device and application to surface-reactive materials. *Biomaterials* **18**: 1091–1098.

Garcia, A.J., Huber, F., and Boettiger, D. (1998). Force required to break alpha5beta1 integrin–fibronectin bonds in intact adherent cells is sensitive to integrin activation state. *J. Biol. Chem.* **273**: 10988–10993.

Giannone, G., Dubin-Thaler, B. J., Dobereiner, H.G., Kieffer, N., Bresnick, A.R., and Sheetz, M. P. (2004). Periodic lamellipodial contractions correlate with rearward actin waves. *Cell* **116**: 431–443.

Griffin, M. A., Sen, S., Sweeney, H. L., and Discher, D. E. (2004). Adhesion–contractile balance in myocyte differentiation. *J. Cell Sci.* **117**: 5855–5863.

Hammer, D. A., and Lauffenburger, D. A. (1987). A dynamical model for receptor-mediated cell adhesion to surfaces. *Biophys. J.* **52**: 475–487.

Haskell, C. A., Cleary, M. D., and Charo, I.F. (2000). Unique role of the chemokine domain of fractalkine in cell capture: Kinetics of receptor dissociation correlate with cell adhesion. *J Biol Chem.* **275**: 34183–34189.

Hynes, R. O. (2002). Integrins: Bidirectional, allosteric signaling machines. *Cell* **110**: 673–687.

Kawaguchi, S., and Hemler, M. E. (1993). Role of the alpha subunit cytoplasmic domain in regulation of adhesive activity mediated by the integrin VLA-2. *J. Biol. Chem.* **268**: 16279–16285.

Kim, M., Carman, C. V., and Springer, T. A. (2003). Bidirectional transmembrane signaling by cytoplasmic domain separation in integrins. *Science* **301**: 1720–1725.

Kuo, S. C., Hammer, D. A., and Lauffenburger, D. A. (1997). Simulation of detachment of specifically bound particles from surfaces by shear flow. *Biophys. J.* **73**: 517–531.

Kuo, S. C., and Lauffenburger, D. A. (1993). Relationship between receptor/ligand binding affinity and adhesion strength. *Biophys. J.* **65**: 2191–2200.

Li, F., Redick, S. D., Erickson, H. P., and Moy, V. T. (2003). Force measurements of the alpha(5)beta(1) integrin–fibronectin interaction. *Biophys. J.* **84**: 1252–1262.

Litvinov, R. I., Shuman, H., Bennett, J.S., and Weisel, J. W. (2002). Binding strength and activation state of single fibrinogen-integrin pairs on living cells. *Proc. Natl. Acad. Sci. USA.* **99**: 7426–7431.

Lotz, M. M., Burdsal, C. A., Erickson, H. P., and McClay, D. R. (1989). Cell adhesion to fibronectin and tenascin: Quantitative measurements of initial binding and subsequent strengthening response. *J. Cell Biol.* **109**: 1795–1805.

Lu, H., Koo, L. Y., Wang, W. M., Lauffenburger, D. A., Griffith, L. G., and Jensen, K. F. (2004). Microfluidic shear devices for quantitative analysis of cell adhesion. *Anal. Chem.* **76**: 5257–5264.

Masumoto, A. and Hemler, M. E. (1993). Multiple activation states of VLA-4: Mechanistic differences between adhesion to CS1/fibronectin and to vascular cell adhesion molecule-1. *J. Biol. Chem.* **268**: 228–234.

Merkel, R., Nassoy, P., Leung, A., Ritchie, K., and Evans, E. (1999). Energy landscapes of receptor–ligand bonds explored with dynamic force spectroscopy [comment]. *Nature* **397**: 50–53.

Miller, T., and Boettiger, D. (2003). Control of intracellular signaling by modulation of fibronectin Conformation at the cell–materials interface. *Langmuir* **19**: 1730–1737.

Moon, J.J., Matsumoto, M., Patel, S., Lee, L., Guan, J.L., and Li, S. (2005). Role of cell surface heparan sulfate proteoglycans in endothelial cell migration and mechanotransduction. *J. Cell. Physiol.* **203**: 166–176.

Nordon, R. E., Shu, A., Camacho, F., and Milthorpe, B. K. (2004). Hollow-fiber assay for ligand-mediated cell adhesion. *Cytometry A* **57**: 39–44.

Pierschbacher, M.D. and Ruoslahti, E. (1984). Cell attachment activity of fibronectin can be duplicated by small synthetic fragments of the molecule. *Nature* **309**: 30–33.

Riveline, D., Zamir, E., Balaban, N. Q., Schwarz, U. S., Ishizaki, T., Narumiya, S., Kam, Z., Geiger, B., and Bershadsky, A. D. (2001). Focal contacts as mechanosensors: Externally applied local mechanical force induces growth of focal contacts by an mDia1-dependent and ROCK-independent mechanism. *J. Cell Biol.* **153**: 1175–1186.

Roth, S. (1968). Studies on intercellular adhesive selectivity. *Dev. Biol.* **18**: 602–631.

Savage, B., Almus-Jacobs, F., and Ruggeri, Z. M. (1998). Specific synergy of multiple substrate–receptor interactions in platelet thrombus formation under flow. *Cell* **94**: 657–666.

Shattil, S. J., Hoxie, J. A., Cunningham, M., and Brass, L. F. (1985). Changes in the platelet membrane glycoprotein IIb–IIIa complex during platelet activation. *J. Biol. Chem.* **260**: 11107–11114.

Shattil, S. J., and Newman, P. J. (2004). Integrins: Dynamic scaffolds for adhesion and signaling in platelets. *Blood* **104**: 1606–1615.

Shi, Q., and Boettiger, D. (2003). A novel mode for integrin-mediated signaling: Tethering is required for phosphorylation of FAK Y397. *Mol. Biol. Cell* **14**: 4306–4315.

Sonnenberg, A. (1993). Integrins and their ligands. *Curr. Top. Microbiol. Immunol.* **184**: 7–35.

Takeichi, M., Nakagawa, S., Aono, S., Usui, T., and Uemura, T. (2000). Patterning of cell assemblies regulated by adhesion receptors of the cadherin superfamily. *Philos. Trans. R. Soc. London B* **355**: 885–890.

Truskey, G. A., and Proulx, T. L. (1993). Relationship between 3T3 cell spreading and the strength of adhesion on glass and silane surfaces. *Biomaterials* **14**: 243–254.

Usami, S., Chen, H. H., Zhao, Y., Chien, S., and Skalak, R. (1993). Design and construction of a linear shear stress flow chamber. *Ann. Biomed. Eng.* **21**: 77–83.

Ward, M. D., Dembo, M., and Hammer, D. A. (1994). Kinetics of cell detachment: Peeling of discrete receptor clusters. *Biophys. J* **67**: 2522–2534.

Weisel, J.W., Shuman, H., and Litvinov, R. I. (2003). Protein–protein unbinding induced by force: Single-molecule studies. *Curr. Opin. Struct. Biol.* **13**: 227–235.

Xiong, J. P., Stehle, T., Goodman, S. L., and Arnaout, M. A. (2003). New insights into the structural basis of integrin activation. *Blood* **102**: 1155–1159.

Zhang, X., Chen, A., De Leon, D., Li, H., Noiri, E., Moy, V. T., and Goligorsky, M. S. (2004). Atomic force microscopy measurement of leukocyte–endothelial interaction. *Am. J. Physiol. Heart Circ. Physiol.* **286**: H359–H367.

4

CELLULAR TENSEGRITY MODELS AND CELL–SUBSTRATE INTERACTIONS

DIMITRIJE STAMENOVIĆ, NING WANG, and DONALD E. INGBER

Department of Biomedical Engineering, Boston University, Physiology Program, Harvard School of Public Health, Departments of Pathology and Surgery, Children's Hospital, and Harvard Medical School, Boston, Massachusetts, USA

Mammalian cells control their shape and function by altering their mechanical properties through structural rearrangements and changes in molecular biochemistry at the nanometer scale. Yet, little is known about the molecular and biophysical basis of cell mechanics. Most of the existing engineering models of cells are *ad hoc* descriptions based on measurements obtained under particular experimental conditions, and these continuum models usually ignore contributions of subcellular structures and molecular components. More than 20 years ago, we introduced an alternative model of the cell based on tensegrity architecture which proposes that isometric tension in the cytoskeleton is critical for cell shape stability. Key to this model is the concept that this stabilizing tensile "prestress" results from a complementary force balance between multiple, discrete, molecular support elements, including microfilaments, intermediate filaments and microtubules in the cytoskeleton, as well as external adhesions to extracellular matrix and to neighboring cells. In this chapter, we review progress in the area of cellular tensegrity, including development of theoretical formulations of the tensegrity model that have led to multiple *a priori* predictions relating to both static and dynamic cell mechanical behaviors that have now been confirmed in experimental studies with living cells. We describe how the cytoskeleton and extracellular matrix form a single, tensionally integrated structural system as predicted by tensegrity, and how distinct molecular biopolymers (e.g., microfilaments vs microtubules) may bear either tensile or compressive loads inside the cell. The tensegrity model is also compared and contrasted with other models of cell mechanics. Taken together, these recent theoretical and experimental studies confirm that the cellular tensegrity model is a useful model because it provides a mechanism to link mechanics to structure at the molecular level, in addition to helping to explain how mechanical signals are transduced into biochemical responses within living cells and tissues.

I. INTRODUCTION

Mechanotransduction, the cellular response to mechanical stress, is governed by the cytoskeleton (CSK), a molecular network composed of different types of biopolymers (microfilaments, microtubules, and intermediate filaments) that mechanically stabilizes the cell and actively generates contractile forces. To carry out certain behaviors (e.g., crawling, spreading, division, invasion), cells must modify their CSK to become highly deformable and almost fluid-like, whereas to maintain their structural integrity when mechanically stressed, the CSK must behave like an elastic solid. These responses are governed by the passive material properties of the CSK, as well as stress-induced changes in biochemistry that modify cytoskeletal structure. Although the mechanical properties of cells govern their form and function and, when abnormal, lead to a wide range of diseases (Ingber, 2003b), little is known about the linked biophysical and biochemical mechanisms by which cells control their deformability.

Because CSK filaments can chemically depolymerize and repolymerize, it was assumed in the past that cells alter their mechanical properties via sol–gel transitions (cf. Stossel, 1993). However, cells can change shape from round to fully spread without altering the total amount of CSK microfilament or microtubule polymers in the cell (Mooney et al., 1995). In fact, individual bundles of microfilaments (stress fibers), intermediate filaments, and microtubules often remain structurally intact for periods of minutes to hours, even though individual subcomponents may bind and unbind to filaments on the outer surface of the bundle. Only in small regions of the cell that undergo rapid extension and retraction, such as within ruffles at the leading edge of actively migrating cells, are rapid waves of new cytoskeletal filament polymerization observed.

Instead, studies with cultured cells suggest that adherent cells control their mechanical behavior by altering the level of "prestress" borne by their CSK (Hubmayr et al., 1996; Pourati et al., 1998; Rosenblatt et al., 2004; Stamenović et al., 2002a, 2004; Wang and Ingber, 1994; Wang et al., 2001, 2002). Prestress here refers to the preexisting tensile stress that exists in the CSK prior to application of an external load. This prestress results from the action of tensional forces that are generated in contractile actin microfilaments (active prestress), transmitted over intermediate filaments, and resisted by adhesive tethers to extracellular matrix (ECM), known as focal adhesions (FAs), to neighboring cells through cell–cell junctions, and by other CSK filaments (e.g., microtubules) inside the cell. The prestress is also generated passively by mechanical distension of the cell.

These observations are consistent with the idea that the CSK is organized as a tensegrity structure; this mechanical, stress-supported network maintains its structural stability through the agency of tensile prestress (Ingber, 1993, 2003a; Stamenović et al., 1996; Volokh et al., 2000). Additional experiments show that changes in CSK prestress can alter cell static mechanical properties (e.g., stiffness) (Hubmayr et al., 1996; Pourati et al., 1998; Wang and Ingber, 1994; Wang et al., 2001, 2002), as well as dynamic cell rheological behaviors (Fabry et al., 2001; Rosenblatt et al. 2004; Stamenović et al., 2002a, 2004). Changes in this cellular force balance may alter cell deformability by promoting structural rearrangements within the CSK and by altering the dynamic mechanical behavior of individual CSK filaments. Shifts in force between discrete internal and external load-bearing elements in the CSK and ECM also may promote biochemical changes, such as remodeling of FAs that physically couple the CSK to the ECM (Burridge et al., 1988) and alterations in CSK filament assembly that are critical to the cellular adaptation to stress, as well as to cell migration and growth (Davies, 1995; Ingber, 1997).

Many previous sophisticated attempts to develop a comprehensive model of cell rheology have typically focused on examining only specific molecular mechanisms or individual types of cytoskeletal components, while disregarding the architectural organization of the whole cytoskeletal network. This is not at all to say that molecular details are unimportant, but only that the controlling physics occurs at a higher level of structural organization. Specifically, our past results suggest that cell mechanical behaviors are determined at the level of integrative lattice properties and are governed by the prestress borne by the CSK. In addition, the tensegrity model also provides a mechanism to link these integrative system-level properties to changes in forces transferred between distinct load-bearing elements at the molecular level, which is not possible using existing models of cell mechanics.

In this chapter, we first review the key role that cell shape and mechanics play in control of cell function, and how cell shape stability is controlled through cell–substrate interactions. We then describe how the original conceptual model of cellular tensegrity has been translated into mathematical terms over the past decade, and how this model can be used as a framework to describe both steady-state (static) cell mechanical behavior and cell rheological behaviors under dynamic loading. In the process, we explain how molecular-scale shifts in force between discrete internal and external load-bearing elements in the CSK and ECM adhesive substrate control the mechanical behavior of the whole cell.

II. COUPLING BETWEEN CELL SHAPE AND FUNCTION

Mechanical distortion of cell shape can impact many cell biological behaviors, including motility, contractility, growth, differentiation, and apoptosis (Chen et al., 1997; Folkman and Moscona, 1978; Parker et al., 2002; Polte et al., 2004; Singhvi et al., 1994). Cell shape can be altered by distorting flexible culture substrates on which cells adhere (cf. Harris et al., 1980; Pelham and Wang, 1997; Pourati et al., 1998; Rosenblatt et al., 2004), directly manipulating the surface membrane with adhesive particles (cf. Wang et al., 1993; Bausch et al., 1998) or pipettes (cf. Maniotis et al., 1997), applying local distortion by suction (cf. Sato et al., 1990) or indentation (cf. Petersen et al., 1982; Shroff et al., 1995), altering ECM coating densities on otherwise nonadhesive dishes (Wang and Ingber, 1994), or culturing cells on adhesive islands of defined shape and size on the micrometer scale created with microfabrication techniques (cf. Singhvi et al., 1994). For example, cells can be switched between growth, differentiation, and apoptosis (programmed cell death) by plating cells on different-sized adhesive islands coated with the same density of ECM protein (Chen et al., 1997). Endothelial cells that spread on large islands proliferate; cells on intermediate-sized islands differentiate; and cells on the smallest islands that fully prevent distortion die, even though all cells are bathed in medium containing a saturating amount of soluble growth factors. Similar results have been observed with a variety of cell types (Singhvi et al., 1994; Chen et al., 1997).

Mechanical distortion of cells produces these changes in cell function by inducing restructuring of the CSK and thereby impacting cellular biochemistry and gene expression through largely unknown mechanisms (Flusberg et al., 2001; Huang et al., 1998; Ingber, 1997; Mammoto et al., 2004; Numaguchi et al., 2004). Motile cells also can sense the mechanical stiffness of their ECM substrate, and preferentially move toward areas of greater rigidity (Pelham and Wang, 1997). The changes in cell extension and movement that make cell migration possible depend on the ability of the substrate to resist cell tractional forces and thereby alter the cellular force balance (Parker et al., 2002; Sheetz et al., 1998), which, in turn, alters cellular biochemistry to further strengthen the cell's ability to resist the applied load (Sheetz et al., 1998; Wang et al., 1993). Even at the cytoplasmic and nuclear levels, a key role for mechanical distention is evident in the control of subcellular structure and function (cf. Ingber, 1993, Maniotis et al., 1997).

These observations suggest that regulation of many vital cellular behaviors can be influenced by cell–substrate interactions that alter cell spreading and the mechanical distending stress borne by the CSK. An effective model of the cell must therefore allow us to relate cytoskeletal mechanics to biochemistry at the molecular level, and to translate this description into quantitative, mathematical terms. The former permits us to define how specific molecular components contribute to cell behaviors. The latter allows us to develop computational approaches to address levels of complexity and multicomponent interactions that exist in cells but cannot be described or explained by current approaches.

III. THE CELLULAR TENSEGRITY MODEL

The idea that deformability of the CSK network may govern mechanotransduction and thereby regulate cellular functions was conceptualized for the first time in the 1980s, when one of us (Ingber) proposed the tensegrity model of the CSK (Ingber et al., 1981). As many signaling molecules are physically linked to the CSK, deformation of the CSK caused by externally applied forces to the cell should produce molecular distortion. This, in turn, should affect chemical signaling in the cell through alterations in molecular kinetic and thermodynamic parameters (Ingber, 1997). Mechanical signals, therefore, may be integrated with other environmental signals and transduced into biochemical responses through force-dependent changes in CSK architecture. Tensegrity architecture may therefore provide a mechanism by which cells can focus mechanical signals on molecular transducers, as well as orchestrate and tune cellular response.

Tensegrity architecture (cf. Fuller, 1961) describes a class of discrete network structures that maintain their structural integrity because of prestress in their cable-like structural members (Fig. 4.1).

Ordinary elastic materials (e.g., rubber, polymers, metals) by contrast, require no such prestress. The central mechanism by which these structures develop restoring stress in the presence of external loading is primarily by geometrical rearrangement of their pretensed members. The greater the

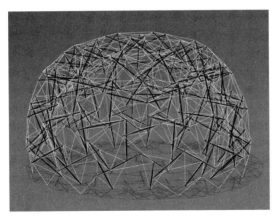

FIGURE 4.1 A typical cable-and-strut tensegrity structure, Dome Image ©1999 Bob Burkhardt. In this structure, tension in the cables (white lines) is partly balanced by the compression of the struts (thick black lines) and partly by the attachments to the substrate. Reprinted, with permission, from Burkhardt (2004).

pretension carried by these members, the less geometrical rearrangements they undergo under an applied load and, thus, the less deformable (more rigid) the structure. A hallmark property that stems from this feature is that structural rigidity (stiffness) of the network is proportional to the level of prestress it supports (cf. Stamenović and Ingber, 2002). As distinct from other stress-supported structures falling within the class, in tensegrity architecture a part of the prestress in the cable network is balanced by compression of internal elements (i.e., elements that are contained within the self-stabilizing network) that are called struts. Another portion of the prestress is resisted by the anchoring forces at points of the structure's attachment to the external world; this is a fundamental feature of hierarchical tensegrity structures (Fuller, 1961; Ingber, 1993, 2003a).

According to the cellular tensegrity model (cf. Ingber, 1993, 2003a), actin filaments and intermediate filaments are envisioned as tensile elements, whereas microtubules and thick crosslinked actin bundles (e.g., within filopodia) act as compression elements. In addition to these compression elements within the CSK, the cell's tethers to the ECM (which is physically connected to the CSK via binding to transmembrane integrin receptors, and is critical to cell shape stability) also balance a portion of the prestress. One difference from static tensegrity sculptures or buildings is that in the cell, the elements that constitute the struts and cables are often composed of bundles of multiple smaller molecular filaments (e.g., stress fibers, intermediate filaments), and individual components within the structure may undergo dynamic assembly or disassembly. However, the principles that govern how the mechanical stability of the entire structure is determined through balancing forces between its component elements are the same. Moreover, the overall architecture of the structural systems maintains the same underlying pattern of force distributions, even though individual subcomponents may come and go.

Two key features of the cellular tensegrity model are that: (1) the prestress carried by the actin microfilament network and intermediate filaments confers shape stability to the cell, and (2) this prestress is partly balanced by CSK-based microtubules and partly by the ECM (cf. Ingber, 2003a). Thus, a disturbance of this complementary force balance would cause load transfer between these distinct load-bearing systems that would, in turn, affect cell deformability and alter stress-sensitive biochemical activities at the molecular level.

A unique property of tensegrity structures is that a mechanical stress may be transferred over long distances within the tensionally linked structural network, a phenomenon referred to as "action at a distance" (Ingber, 1993; 2003a). This is based on the property that because tensegrity structures resist externally applied loads by geometrically rearranging their structural members, a local disturbance can result in a global rearrangement of the structural lattice that may be manifested at points distal from the point of an applied load. This is quite different from continuum models, in which local disturbances produce only local responses, which dissipate inversely with the distance from the point of load application. In complex multimodular or hierarchical tensegrity structures, this action at a distance may not always be observable because the distance the force is transferred depends on the stiffness of the individual support elements and the strength of coupling between different structural modules (Ingber, 2003a). For example, the force may be dissipated locally if the structural members at sites of force application are highly flexible, or it could fade away at points distal from the point of load application if links between smaller and larger modules in the structural hierarchy are weakly connected.

Physical and mathematical tensegrity models of the cell have been successful in describing various mechanical properties of living cells. These behaviors include prestress-induced stiffening (McGarry and Prendergast, 2004; Stamenović et al., 1996; Volokh et al., 2000), strain hardening (McGarry and Prendergast, 2004; Stamenović et al., 1996; Thoumine et al., 1995; Wang et al., 1993; Wendling et al., 1999), the effect of fluid shear on endothelial cell shape (Thoumine et al., 1995), the effect of cell distention on nuclear distortion (Ingber, 1993), the effect of microtubule buckling on cellular mechanics (Coughlin and Stamenović, 1997; Volokh et al., 2000), the effect of cell spreading on cell deformability (Coughlin and Stamenović, 1998; McGarry and Prendergast, 2004), the contribution of intermediate filaments to cell deformability (Wang and Stamenović, 2000), as well as viscoelastic transient and oscillatory behaviors of cultured cells (Cañadas et al., 2002; Sultan et al., 2004). All of these results were obtained using tensegrity models with idealized geometries and a limited number of cables and struts.

Although no other model of cellular mechanics could account for so many disparate features exhibited by living cells, the simplicity of the past cellular tensegrity models has made critics of the tensegrity idea skeptical about the results obtained from these models. For example, some critics have suggested that cutting one or two cables or struts in a simple (e.g., 30-element) tensegrity model would cause their total collapse and dysfunction, whereas in the cell, disruption of actin filaments or microtubules would compromise but not entirely abolish cell shape stability (Petersen et al., 1982; Sato et al., 1990; Stamenović et al., 2002a; Wang et al., 1993). Although disruption of a single cable can cause collapse of some simple tensegrity structures (e.g., a three-strut, nine-cable model), it does not cause complete structural failure in a multimodular or hierarchical tensegrity structure; only the local module is lost. This behavior is similar to the way a person loses function of the foot when the Achilles tendon is torn, yet still retains function of the remainder of the hierarchical musculoskeletal system. Disrupting individual stress fibers with laser irradiation similarly results in retraction of portions of the cell, localized between surrounding ECM adhesions, that help stabilize the remainder of the cell (unpublished data).

Importantly, the simple 30-element tensegrity structure that was used in early theoretical formulations embodies most of the key mechanical properties of more complex multimodular and hierarchical tensegrities that more closely mimic the localized responses of living cells, and more closely resemble how cells are structured at the molecular level (Ingber, 2003a). In addition, we recently modeled the CSK as an affine network that embodies basic principles of cellular tensegrity, namely, prestress and the three-way force balance between the tension-bearing and compression-bearing components of the CSK and the ECM (Stamenović, 2005). However, the oversimplification of CSK geometry of the earlier tensegrity models that was a source of many controversies is avoided in the affine tensegrity model, which does not depend on any specific architecture. Although the affine approach uses methods of continuum mechanics, it still provides insight into how mechanics of the microstructure and force balances between different types of discrete load-bearing elements relate to the overall mechanical behavior of the network. Importantly, this affine approximation of the tensegrity model also predicts behaviors that are exhibited by living cells, including their ability to shift compressive loads between internal microtubules and ECM adhesions (Stamenović, 2005).

Another criticism of past tensegrity models is their static elastic nature because the CSK of living cells is a nonelastic and dynamic network. The cellular tensegrity model also has been criticized as being purely mechanical, in that it does not have the ability to account for biochemical remodeling events that likely also influence distortion-dependent behaviors over longer time scales, such as cell movement and growth.

Although these criticisms may have sometimes been justified when addressing individual aspects of past simplified tensegrity models, they have not disproved the hallmark of the tensegrity idea, namely, that: (1) CSK prestress confers shape stability to the cell, and (2) the interconnectedness of the CSK and its physical linkage to the ECM facilitates load shifts between various molecular components that are vital for control of cell mechanics, shape, and function. These two features are the underlying principles of previous cellular tensegrity models (McGarry and Prendergarst, 2004; Stamenović, 2005; Sultan et al., 2004; Wendling et al., 1999), notwithstanding how simple those models are, and they are the key reasons why those models yield results that are consistent with various experimental observations in living cells.

The other important feature of the tensegrity model is that it is a discrete network structure, rather than a mechanical continuum. This provides a mechanism for cells to channel mechanical stresses along discrete molecular load-bearing elements throughout the cell, and to concentrate these forces on key sites, such as FA sites and the nuclear membrane, where mechanochemical transduction responses required for dynamic cell adaptation to stress take place. Individual cytoskeletal filaments, such as microtubules, also may dynamically assemble or disassemble in response to changes in compressive forces imposed on their ends through a tensegrity force balance These combined features may explain the tensegrity model's past successes at predicting various cell static and dynamic mechanical behaviors over time scales that we know involve both biochemical and mechanical responses, even though the model is a purely mechanical system. Although it is not known at present exactly how this works, we speculate that tensegrity provides both a mechanical response mechanism and a mechanism to focus forces on biochemical transduction components.

IV. EXPERIMENTAL EVIDENCE CONSISTENT WITH THE CELLULAR TENSEGRITY MODEL

For decades, the controversy surrounding the cellular tensegrity model was attributable primarily to the lack of quantitative experimental data that could rigorously test *a priori* predictions of this model (Ingber et al., 2000; Jenkins 2003). During the first few years following publication of the cellular tensegrity model, experimental data that were used in support of this model were circumstantial or, at best, qualitative. For example, the observation that actin filaments appear curved (cf. Mackintosh et al., 1995) and microtubules straight (cf. Mizushima-Sugano, 1983) when analyzed in isolation *in vitro*, whereas the opposite is true in living cells, was interpreted to indicate that microfilaments carry tension whereas microtubules are bent and under compression (Ingber 1993). The findings that contracting cells cause wrinkling of malleable substrates (Harris et al., 1980) and that cells rapidly retract when their attachments to the substrate are severed by trypsin (Sims et al., 1992) or when they are cut by a microneedle (Pourati et al., 1998) have been interpreted as evidence of mechanical interactions between the CSK and the ECM, and that the CSK is under preexisting tension that is partially balanced by the anchoring forces at focal adhesions (Ingber, 1993).

Alternative mechanisms, however, also have been raised to explain these observations. For example, Heidemann et al. (1999) viewed the curved shape of microtubules as an evidence of their high flexibility and almost fluid-like behavior. This would imply that microtubules cannot carry appreciable compressive forces, contrary to the prediction of the tensegrity model. Second, the fact that the cells retract rapidly after they detach from the substrate or after they are cut could be interpreted as an active response to cell injury, rather than a result of disturbed force balance between the CSK and the ECM. Yet, biophysical analysis of microtubules has revealed that they have a persistence length on the order of millimeters, and that they rarely, if ever, display the degree of curvature observed in living cells when studied *in vitro* (cf. Gittes et al., 1993). The latter suggestion that detachment is an active injury response is also contradicted by the finding that membrane-permeabilized cells undergo similar coordinated cell, cytoskeletal, and nuclear retraction in the presence of ATP, and that this response can be inhibited by preventing actomyosin-based rigor complex formation in the CSK (Sims et al., 1992).

Critics of the cellular tensegrity model have used the lack of firm experimental evidence for the compression-supporting role of microtubules as a basis for dismissing the model as a viable description of cellular mechanics (Heidemann et al., 1999; Ingber et al., 2000, Jenkins, 2003). For many years, this issue has completely overshadowed a much more important feature of the model, namely, the central role of prestress as a determinant of cytoskeletal shape stability. Nevertheless, new developments in mechanical cytometry techniques (Fabry et al., 2001; Hu et al., 2003; Pelham and Wang, 1997) and new theoretical approaches (Stamenović, 2005; Stamenović et al., 2002b) made it possible to rigorously test some key features and *a priori* predictions of the cellular tensegrity model and to clear some of the confusion that has surrounded it. In the sections that follow, we review major accomplishments related to the testing and validation of this model.

A. Prestress-Induced Stiffening

An *a priori* prediction of all prestressed structures is that their stiffness increases in a nearly direct proportion with prestress. Therefore, failure of cells to conform to this prestress-induced stiffening behavior would once and for all rule out the cellular tensegrity model. A number of experiments in various cell types have confirmed that living mammalian cells display prestress-induced stiffening. For example, it has been shown that mechanical (Cai et al., 1998; Pourati et al., 1998; Wang and Ingber, 1994), pharmacological (Hubmayr et al., 1996), and genetic (Cai et al., 1998) modulations of CSK prestress are paralleled by changes in cell stiffness. Developments of the traction cytometry technique made it possible to quantitatively measure various indices of cytoskeletal prestress (Balaban et al., 2001; Dembo and Wang, 1999; Pelham and Wang, 1997; Wang et al., 2002). When these data were compared with data obtained from measurements of cell stiffness, it was found that cell stiffness (G) increases linearly with increasing contractile stress (P) in cultured human airway smooth muscle cells whose contractility was altered by graded doses of contractile and relaxant agonists (Stamenović et al., 2002a, 2004; Wang et al., 2001, 2002) (Fig. 4.2).

The G-versus-P relationship predicted from the affine tensegrity model, $G \approx 1.2P$ (Stamenović, 2005), provides a close quantitative correspondence to these experimental data (Fig. 4.2). Although this association between cell stiffness and contractile stress does not preclude other interpretations, it is a hallmark of tensegrity structures that secure shape stability through the agency of the prestress. Other possible interpretations of this finding are discussed later.

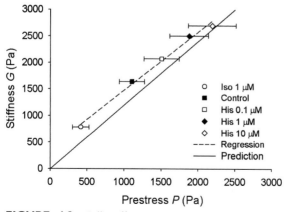

FIGURE 4.2 Cell stiffness (G) increases linearly with increasing cytoskeletal contractile prestress (P). Measurements were made in cultured human airway smooth muscle cells, the contractility of which was modulated by graded doses of histamine (constrictor) and isoproterenol (relaxant). Stiffness was measured using the magnetic cytometry technique, and prestress was measured by the traction cytometry technique (Wang et al., 2002). The slope of the regression line is ~1.1 (dashed line). The affine tensegrity model (Stamenović, 2005) predicts a slope of ~1.2 (solid line).

It has been shown that in addition to generating contractile force, pharmacological contractile agonists also may induce actin filament polymerization (Mehta and Gunst, 1999). Thus, the observed stiffening in response to contractile agonists could be nothing more than the result of actin polymerization. However, An et al. (2002) showed that agonist-induced actin polymerization in smooth muscle cells accounts for only a portion of the observed stiffening, whereas most of the stiffening response is associated with contractile force generation. Another potential mechanism that could explain the G-versus-P relationship in Fig. 4.2 is the effect of crossbridge recruitment. It is known from studies of isolated smooth muscle strips in uniaxial extension that both muscle stiffness and muscle force are directly proportional to the number of attached crossbridges. Thus, the association between cell stiffness and prestress could reflect nothing more than the effect of changes in the number of attached crossbridges in response to pharmacological stimulation. However, results from a theoretical model of myosin crossbridge kinetics (Mijailovich et al., 2000) conflict with this possibility. Results of both experiments and theoretical analysis predict an oscillatory response qualitatively different from that measured in airway smooth muscle cells (Fabry et al., 2001). Thus, the kinetics of crossbridges cannot explain all aspects of CSK mechanics.

B. Action at a Distance

To investigate whether the CSK can exhibit action at a distance, we performed experiments in which the tip of a glass micropipette that was coated with fibronectin and bound to integrin receptors on the surface of living endothelial cells was pulled laterally using a micromanipulator (Maniotis et al., 1997). If the CSK is organized as a discrete tensegrity structure and integrins are physically linked to the CSK, then pulling on these surface membrane receptors should produce an observable deformation at points distant from the site of load application, deep inside the cell. We observed that the nuclear border moved along the line of applied pulling force and that nucleoli inside the nucleus exhibited molecular realignment along the applied tension field lines, as measured by birefringence microscopy; these findings are direct demonstrations of action at a distance. Additional evidence for long-distance force transfer through the CSK was provided in our recent study using intracellular stress tomography (Hu et al., 2003). This technique enabled us to observe stress distributions within the CSK in response to a shear disturbance applied locally to cell surface integrin receptors by oscillatory twisting of receptor-bound magnetic beads. Islands of stress concentrations were found at distances $>20~\mu m$ from the point of application of shear loading (Fig. 4.3a) using this method. Interestingly, when the actin lattice was disrupted using cytochalasin D, the action-at-a-distance effect disappeared (Fig. 4.3b), suggesting that connectivity of the actin network is essential for transmission of mechanical signals throughout the CSK. In addition, long-distance force transfer was suppressed by dissipating cytoskeletal prestress in cells expressing caldesmon. The action at a distance effect also has been observed by other groups in neurons (Ingber et al., 2000) and in response to application of apical fluid shear stress to endothelial cells (Helmke et al., 2003).

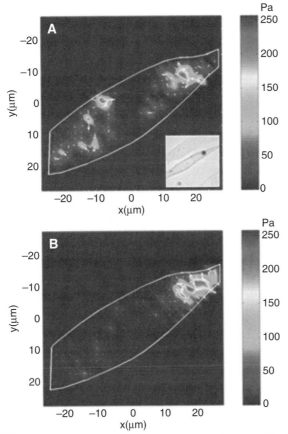

FIGURE 4.3 Evidence of action at a distance affecting living cells. Stress maps in living human airway smooth muscle cells were obtained using the intracellular stress tomography technique (Hu et al., 2003). Stress was applied to the cell by twisting a ferromagnetic bead bound to integrin receptors on the apical cell surface. The bead position is shown on the phase-contrast image of the cell (A, inset); the black dot on the image is the bead. The white arrows indicate the direction of the stress field; the color map represents the magnitude of stress. Stress does not decay inversely with increasing distance from the bead center. Appreciable islands of stress concentration could be seen >20 μm from the bead (A), consistent with the action-at-a-distance effect. After disruption of the actin network by cytochalasin D (1 μg/ml), the stress concentration could be detected only in the local vicinity of the bead (B), demonstrating that connectivity of the actin CSK lattice plays an important role in intracellular stress distribution. Reprinted, with permission, from Hu et al. (2003). (See Color Plate 4)

On the other hand, Heideman et al. (1999) failed to observe this phenomenon in living fibroblasts. These investigators applied various mechanical stresses to cells using a micromanipulator to pull on a glass micropipette coated with the ECM molecule laminin, which should bind to cell surface integrin receptors. They found that such disturbances produced only local deformations. However, the authors did not confirm that binding of these pipettes induces formation of Fas at the point of force application. Focal adhesion formation is critical for efficient mechanical transfer between integrins and the internal cytoskeletal lattice (Wang et al., 1993; Matthews et al., 2004). Thus, their results remain controversial. In a recent review, however, Heidemann and Wirtz (2004) pointed out that both long-distance force transfer and short-distance force transfer can coexist in living cells.

C. Microtubules as Compression Elements

Microscopic visualization of green fluorescent protein (GFP)-labeled microtubules of living cells shows that microtubules buckle as they oppose contraction of the actin network (Kaech et al., 1996; Wang et al., 2001; Waterman-Storer and Salmon, 1997). It was not known, however, whether the compression that causes this buckling could balance a substantial fraction of the contractile prestress. To investigate this possibility, we measured changes in traction and energy associated with buckling of microtubules using the traction cytometry technique (Stamenović et al., 2002b; Wang et al., 2002). The prediction of the tensegrity model is that the cytoskeletal prestress is balanced partly by compression of microtubules and partly by the tractional stress at the cell–ECM interface (Stamenović, 2005; Stamenović et al., 2002b). Consequently, disruption of microtubules should result in a shift of the compression stress and associated energy stored from microtubules to the flexible ECM substrate. The transferred energy was obtained as work done by traction forces during contraction.

In highly stimulated and spread human airway smooth muscle cells, we found that the traction increased, on average, by 13% following disruption of microtubules using colchicine; this result suggests that microtubules balance ~13% of the prestress, whereas the remaining portion of the prestress is balanced by the substrate. The work of traction increases on average by ~30% relative to the state before disruption, and equals 0.13 pJ (Stamenović et al., 2002b).

This result was then used in an energetic analysis. We assumed, based on the model of Brodland and Gordon (1990), that CSK-based microtubules are slender elastic rods laterally supported by intermediate filaments. Using the postbuckling equilibrium theory of Euler struts, we calculated that the energy stored during buckling of microtubules is ~0.18 pJ, which is close to the measured value of ~0.13 pJ (Stamenović et al., 2002b). This is further evidence in support of the idea that microtubules are intracellular compression-bearing elements. Potential concerns are that disruption of microtubules may activate myosin light-chain phosphorylation (Kolodney and Elson, 1995) or could cause a release of intracellular calcium (Paul et al., 2000). Thus, the observed increase in traction and work of traction following disruption of microtubules could be due entirely to chemical mechanisms rather than mechanical load transfer. These concerns were alleviated by studies that confirmed that microtubule disruption results in an increase in traction in cells optimally stimulated with contractile agonists, that is, even when the level of myosin light-chain phosphorylation and the level of calcium do not change (Stamenović et al., 2002b; Wang et al., 2001). An increase in traction in response to disruption of microtubules has also been observed in different cell types, by other investigators, but has not been quantified (Danowski, 1989; Kolodney and Wysolmerski, 1992).

Importantly, we recently demonstrated that the contribution of microtubules to balancing the prestress and to the energy budget of the cell depends on the extent of cell spreading on the ECM (Hu et al., 2004), as specifically predicted by the tensegrity model. Using the traction cytometry technique, we found that traction and the energy associated with buckling of microtubules increase with decreasing cell spreading in airway smooth muscle cells. For example, as the cell projected area in contact with ECM decreased from 1800 to 500 μm^2, the percentage change in traction following disruption of microtubules increased from a few percent to 80%. On the basis of these findings, we calculated, using a method described in our previous study (Stamenović et al., 2002b), that in poorly spread cells, microtubules could balance up to 45% of the prestress, whereas in highly spread cells, their contribution to balancing prestress is negligible (Fig. 4.4).

Because in their natural habitat cells seldom exhibit highly spread forms, the aforementioned results suggest that *in vivo*, the contribution of microtubules in balancing the prestress can be very significant.

Using the affine tensegrity model, we predicted the effect of microtubules on cell stiffness at different states of spreading from the data in Fig. 4.4. We found that in highly spread cells stiffness slightly increases in response to disruption of microtubules, whereas in poorly spread cells the stiffness decreases with disruption of microtubules (Stamenović, 2005). This finding is quantitatively consistent with experimental data from cultured human airway smooth muscle cells that show that disruption of microtubules by cholchicine causes a 10% increase in cell stiffness (Stamenović et al., 2002a), whereas the model predicts an 8% increase (Stamenović, 2005). Only the tensegrity model of the cell could explain this type of complementary force balance between the contractile actin CSK, microtubules, and ECM adhesions in a prestressed cell and its effect on cell deformability.

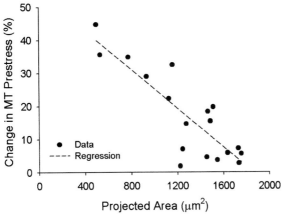

FIGURE 4.4 The part of the cytoskeletal contractile prestress balanced by microtubules (MTs) decreases with increasing cell spreading, as predicted by the cellular tensegrity model. Measurements were made in human airway smooth muscle cells micropatterned on adhesive circular islands of different surface areas (500–1800 μm^2) (Hu et al., 2004). Prestress balanced by microtubules was calculated from an increase in traction from controls following disruption of microtubules by colchicine (1 μM) measured by traction force microscopy (Stamenović et al., 2002b). Throughout these measurements, the total contractile prestress was maintained constant by use of saturating doses of contractile agonist (histamine 10 μM), as confirmed by biochemical analysis of myosin light-chain phosphorylation and intracellular calcium (Wang et al., 2001). The data are expressed as percentage change from controls; each symbol corresponds to one cell (dashed line indicates linear regression).

D. The Role of Intermediate Filaments

According to the cellular tensegrity model, CSK-based intermediate filaments also carry prestress and link the nucleus to the cell surface and the CSK (Ingber, 1993, 2003a). There is experimental evidence that supports this view. First, vimentin-deficient fibroblasts exhibit reduced contractility and reduced traction on the substrate in comparison to wild-type cells (Eckes et al., 1998). Second, the intermediate filament network alone is sufficient to transfer mechanical load from cell surface to the nucleus in cells in which the actin and microtubule networks are chemically disrupted (Maniotis et al., 1997). Taken together, these observations suggest that intermediate filaments play a role in transferring the contractile prestress to the substrate and in long-distance load transfer within the CSK. Both are key features of the cellular tensegrity model. In addition, inhibition of intermediate filaments causes a decrease in cell stiffness (Eckes et al., 1998; Stamenović et al., 2002a; Wang and Stamenović, 2000; Wang et al., 1993), as well as cytoplasmic tearing in response to high applied strains (Eckes et al., 1998; Maniotis et al., 1997). In fact, it appears that the contribution of intermediate filaments to the cell's ability to resist shape distortion is substantial only at relatively large strains (Wang and Stamenović, 2000). Another role of intermediate filaments is suggested by Brodland and Gordon (1990). According to these authors, intermediate filaments provide a lateral stabilizing support to microtubules as they buckle due to compression by opposing contractile forces transmitted by the actin CSK. This description is consistent with experimental data (Stamenović et al., 2002b).

V. OTHER MODELS OF CELLULAR MECHANICS

There are different structural models of the CSK in the literature, most notably models based on a cortical network (Discher et al., 1998), open cell foam (Satcher and Dewey, 1996), tensed cable network (Coughlin and Stamenović, 1999), and percolation (Forgacs, 1995). These models have been successful at explaining some aspects of cellular mechanics, but lack the ability to describe many other mechanical behaviors that are important for cell function. In particular, the open cell foam model and the percolation model do not account for the effect of cytoskeletal prestress on cell deformability, and none of these models takes into account the contribution of microtubules to cell mechanical behaviors. The cortical network model, the open cell foam model, and the percolation model also ignore the contribution of the ECM to cellular mechanics, and the cortical network model cannot explain the observed transmission of mechanical signals from the cell surface to the nucleus as well as to basal FAs ("action at a distance") (Hu et al., 2003; Maniotis et al., 1997; Wang et al., 2001).

On the other hand, all of these features (and many others) can be explained by models that depict the CSK as a prestressed tensegrity structure (cf. Ingber, 1993, 2003a). Moreover, none of the other models provides a mechanism to explain how mechanical stresses applied to the cell surface results in force-dependent changes in biochemistry at discrete sites inside the cell (e.g., FAs, nuclear membrane, microtubules), whereas tensegrity can (Ingber, 1997). Thus, we believe that the cellular tensegrity model represents a good platform for further research on the cell structure–function relationship.

It is important to clarify that the cortical membrane model and the tensed cable network model also fall into the category of stress-supported structures. In fact, according to a mathematical definition that is based on considerations of structural stability (cf. Connelly and Back, 1998), all stress-supported structures are tensegrity structures. They differ from each other only in the manner by which they balance prestress. However, in the structural mechanics literature, this difference is used to make a distinction between various types of stress-supported structures. Consequently, tensed cable nets and tensegrity architecture are considered two distinct types of stress-supported structures. In a real structure, however, a cable net cannot remain tensed unless it is balanced by some other element that cannot be compressed, and the overall shape stability of this whole couple system is determined by the level of prestress in the system.

VI. CONTROL OF DYNAMIC CELL RHEOLOGICAL BEHAVIORS BY PRESTRESS

In the preceding sections we showed how tensegrity-based models could account for the steady-state elastic behavior of cells. However, cells are known to exhibit time- and rate of deformation-dependent viscoelastic behavior (cf. Bausch et al., 1998; Laurent et al., 2002; Petersen et al., 1982; Sato et al., 1990; Shroff et al., 1995; Stamenović et al., 2002a; Yamada et al., 2000). Because in their natural habitat, cells are often exposed to dynamic loads (e.g., pulsatile blood flow, periodic stretching of lung ECM), their viscoelastic properties are important determinants of their mechanical behavior. Given that the tensegrity-based models have provided a good description of the elastic behavior of adherent cells, it is of considerable interest to investigate whether these models can be extended to describe cell viscoelastic behavior.

Rheological measurements of different cell types with various techniques have established that the dynamic stiffness of the CSK of adherent cells increases with the frequency (ω) of the imposed deformation as a weak power law, ω^α, where α is a number that takes a value between 0 and 1 (Fabry et al., 2001; Yamada et al., 2000). In the limit of $\alpha \to 0$, the stiffness becomes independent of ω, indicating elastic solid behavior, whereas in the limit $\alpha \to 1$, it becomes proportional to ω, indicating viscous Newtonian fluid behavior. For $0 < \alpha < 1$, the rheological behavior is neither elastic nor viscous, but rather viscoelastic. Thus, the power law exponent α is an index of the cell transition between fluid-like and solid-like behavior.

Importantly, we recently discovered that there is a unique inverse relationship between α and prestress P of the CSK (Stamenović et al., 2004). Because α is an index of cell rheological behavior, this association between α and P indicates that P may control the transition between the fluid-like and solid-like behavior of cells. The question remains: Is this apparent link between α and P facilitated by a tensegrity-based mechanism?

The actin network of the CSK is a major determinant of the mechanical behavior of whole living cells (cf. Petersen et al., 1982; Pourati et al., 1998; Sato et al., 1990; Stamenović et al., 2002a; Wang et al., 1993). Studies of rheology of actin polymer networks *in vitro* have provided some insight into mechanisms that govern cell rheology (cf. Janmey et al., 1990). However, actin networks reconstituted with myosin II dispersed in them (Humphrey et al., 2002) fail to mimic the mechanical behaviors of living cells. In such reconstituted systems, activation of dispersed myosin by ATP causes fluidization of the entangled actin polymer gel and a dramatic reduction in network stiffness (Humphrey et al., 2002). This result is opposite in sign from results obtained in living cells, which stiffen when molecular motors are activated in response to stimulation by contractile agonists (Stamenović et al., 2004; Wang et al., 2001, 2002).

A possible explanation for this discrepancy between the *in vitro* gel experiments and the living cell experiments is that myosin activation does not increase contractile prestress within polymerized actin gels because there are no sites for mechanical anchorage that could counterbalance this stress. Instead, the force generated by the molecular motors pushes the actin filaments through the network, thereby enhancing their longitudinal motion and causing an apparent increase in gel fluidity (Humphrey et al., 2002). On the other hand, the tensegrity model predicts that the contractile forces generated by molecular motors are transmitted by the actin network and intermediate filaments to microtubules and, ultimately, via FAs to the ECM, which counterbalance these forces and thereby prestress the CSK (Ingber, 1993, 2003a; Stamenović et al., 2002b). This would explain why contractile stress enhances stability rather than induces fluidity of the CSK within intact living cells.

Other theories have been used to describe the transition between solid-like and fluid-like cell behavior, most notably, models based on soft glass rheology (Fabry et al., 2001) or gel–sol transitions (i.e., polymerization–depolymerization) of the CSK network (cf. Janmey et al., 1990; Stossel, 1993). Both theories can account for the observed power law behavior of living cells, and the gel–sol transition theory also has been implicated in cell locomotion (cf. Stossel, 1993). Recent findings, however, indicate that agonist-induced actin polymerization is not sufficient to explain the observed stiffening response of smooth muscle cells, and that myosin activation is also needed to account for this response (An et al., 2002). Moreover, cells can undergo extensive changes in shape from round to fully spread, and vice versa (e.g., during trypsinization), without any change in total microfilament or microtubule polymer in their CSK (Mooney et al., 1995).

The soft glass rheology theory is an energy activation theory (cf. Sollich, 1998), but its physical origins have not yet been identified, although there is a possibility that in smooth muscle cells this energy activation may be ATP-dependent (Gunst and Fredberg, 2003). Most importantly, neither soft glass rheology theory nor gel–sol transition theory explicitly accounts for the effect of cytoskeletal prestress on cell rheological behavior. In contrast, the basic mechanisms of the tensegrity model can potentially explain the dependence of cell viscoelastic properties on prestress (Sultan et al., 2004). However, the current embodiments of the tensegrity model cannot completely explain the power law behavior observed in cells. Origins of this power law behavior are still not well understood. It appears that a part of this behavior could be tied to the dynamics of polymers of the CSK (thermal fluctuations, etc.), which, in turn, may be influenced by the prestress (cf. Stamenović et al., 2004).

VII. CONCLUSION AND FUTURE DIRECTIONS

Numerous experimental studies support the notion that living cells use a tensegrity mechanism to sense, respond, and adapt to changes in their mechanical environment, including stresses applied at the cell–ECM interface.

The tensegrity model identifies mechanical prestress borne by the CSK as a key determinant of shape stability within living cells and tissues. It also shows how mechanical interactions between the CSK and ECM come into play in the control of various cellular functions. This does not at all preclude the numerous chemically and genetically mediated mechanisms (e.g., CSK remodeling, actomyosin motor kinetics, phosphorylation, crosslinking) that are known to regulate CSK filament assembly and force generation. Rather, it elucidates a higher level of organization in which these events normally function and are regulated. The tensegrity model therefore appears, at present, to be the most robust among existing models of cellular mechanics. Nevertheless, the model needs further improvements and refinements, and most importantly it needs to be linked to biochemical functions of the cell in a quantitative manner. Therefore, future work should focus on the relationship between the cytoskeletal prestress, cell mechanical behaviors, and the underlying microstructural mechanisms, as well as on how shifts in forces between different molecular load-bearing elements in the cell impact biochemical remodeling within FAs and the CSK.

The cellular tensegrity model provides a good theoretical framework for such research because it provides a way to channel mechanical forces in distinct patterns, to shift them between different load-bearing elements in the CSK and ECM, and to focus them on particular sites where biochemical remodeling may take place. If successful, this approach may show the extent to which prestress plays a unifying role in terms of both determining cell rheological behavior and orchestrating mechanical and chemical responses within living cells. Moreover, it will elucidate potential mechanisms that link cell rheology to the mechanical prestress of the CSK, from the level of molecular dynamics and biochemical remodeling events, to the level of whole-cell mechanics. In addition, the tensegrity model also may reveal how cytoskeletal structure, prestress, and the ECM come into play in the control of cellular information processing at the whole-cell level (Ingber, 2003b).

ACKNOWLEDGMENTS

We thank all of the students, fellows, and technicians in our laboratories who contributed to the development and testing of the tensegrity model, as well as related studies on cell–substrate interactions. This work was supported by grants from NIH and NASA.

SUGGESTED READING

Alenghat, F., and Ingber, D. E. (2002) Mechanotransduction: All signals point to cytoskeleton, matrix, and integrins. *Science's STKE,* www.stke.org/cgi/content/full/OC_sigtrans;2002/119/pe5.

This review describes recent experimental evidence that supports key roles for the cytoskeleton, extracellular matrix, and transmembrane integrin receptors that mechanically couple these structures in the process of cellular mechanotransduction.

Connelly, R., and Back, A. (1998) Mathematics and tensegrity. *Am. Scientist* **86:** 142–151.

This review article describes how mathematicians define and describe tensegrity structures.

Fuller, B. (1961) Tensegrity. *Portfolio Artnews Annu.* **4**: 112–127.

This description of the tensegrity building system is written by the person who defined the concept and coined the term *tensegrity.*

Ingber, D. E. (1993) The riddle of morphogenesis: A question of solution chemistry or molecular cell engineering? *Cell* **75**: 1249–1252.

This article introduces the fundamental role of structural scaffolds and solid-phase biochemistry in living cells, and discusses the critical importance of combining approaches from engineering, physics, computer science, and biology to fully understand complex cell behaviors.

Ingber, D. E. (1997) Tensegrity: The architectural basis of cellular mechano-transduction. *Annu. Rev. Physiol.* **59**: 575–599.

This review describes how use of tensegrity architecture by cells may impact the mechanism of cellular mechanotransduction — the process by which cells sense mechanical signals and convert them into changes in intracellular biochemistry.

Ingber, D. E. (1998) The architecture of life. *Sci. Am.* **278**: 48–57.

This easy-to-read article describes the origin of the theory of cellular tensegrity, as well as how this design principle may be used on smaller and larger scales in the hierarchy of life.

Ingber, D. E. (2003) Cellular tensegrity revisited: I. Cell structure and hierarchical systems biology. *J. Cell Sci.* **116**: 1157–1173.

Ingber, D. E. (2003) Tensegrity: II. How structural networks influence cellular information-processing networks. *J. Cell Sci.* **116**: 1397–1408.

These two articles provide the most in-depth and up-to-date review of the field of cellular tensegrity, as well as its implications for cell and developmental control.

Snelson, K. (1996) Snelson on the tensegrity invention. *Int. J. Space Struct.* **11**: 43–48.

This article by the sculptor who constructed the first prestressed tensegrity structures describes various kinds of tensegrity structures and their applications.

REFERENCES

An, S.S., Laudadio, R. E., Lai, J., Rogers, R. A., and Fredberg, J. J. (2002) Stiffness changes in cultured airway smooth muscle cells. *Am. J. Physiol. Cell Physiol.* **283**: C792–C801.

Balaban, N. Q., Schwarz, U. S., Riveline, D., Goichberg, P., Tzur, G., Sabanay, I., Mahalu, D., Safran, S., Bershadskym A., Addadi, L., and Geiger, B. (2001) Force and focal adhesion assembly: A close relationship studied using elastic micropatterned substrates. *Nat. Cell Biol.* **3**: 466–472.

Bausch, A., Ziemann, F., Boulbitch, A. A., Jacobson, K., and Sackmann, E. (1998) Local measurements of viscoelastic parameters of adherent cell surfaces by magnetic bead microrheometry. *Biophys. J.* **75**: 2038–2049.

Brodland, G. W., and Gordon, R. (1990) Intermediate filaments may prevent buckling of compressively loaded microtubules. *ASME J. Biomech. Eng.* **112**: 319–321.

Burkhardt, R. (2004). A technology for designing tensegrity domes and spheres. http://www.channel1.com/users/bobwb/prospect/prospect.htm.

Burridge, K., Fath, K., Kelly, T., Nucko, G., and Turner, C. (1988) Focal adhesions: Transmembrane junctions between the extracellular matrix and cytoskeleton. *Annu. Rev. Cell Biol.* **4**: 487–252.

Cai, S., Pestic-Dragovich, L., O'Donnell, M. E., Wang, N., Ingber, D. E., Elson, E., and Lanorelle, P. (1998) Regulation of cytoskeletal mechanics and cell growth by myosin light chain phosphorylation. *Am. J. Physiol. Cell Physiol.* **275:** C1349–1356.

Cañadas, P., Laurent, V. M., Oddou, C., Isabey, D., and Wendling, S. (2002) A cellular tensegrity model to analyze the structural viscoelasticity of the cytoskeleton. *J. Theor. Biol.* **218:** 155–173.

Chen, C. S., Mrksich, M., Huang, S., Whitesides, G. M., and Ingber, D. E. (1997) Geometric control of cell life and death. *Science* **276:** 1425–1428.

Coughlin, M. F., and Stamenović, D. (1997) A tensegrity structure with buckling compression elements: application to cell mechanics. *ASME J. Appl. Mech.* **64:** 480–486.

Coughlin, M. F., and Stamenović, D. (1998) A tensegrity model of the cytoskeleton in spread and round cells. *ASME J. Biomech. Eng.* **120:** 770–777.

Coughlin, M. F., and Stamenović, D. (2003) A prestressed cable network model of adherent cell cytoskeleton. *Biophys. J.* **84:** 1328–1336.

Davies, P. F. (1995) Flow-mediated endothelial mechanotransduction. *Physiol. Rev.* **75:** 519–560.

Dembo, M., and Wang, Y.L. (1999) Stress at the cell-to-substrate interface during locomotion of fibroblasts. *Biophys. J.* **76:** 2307–2316.

Discher, D. E., Boal, D. H., and Boey, S. K. (1998) Simulations of the erythrocyte cytoskeleton at large deformation: II. Micropipette aspiration. *Biophys. J.* **75:** 1584–1597.

Eckes, B., Dogic, D., Colucci-Guyon, E., Wang, N., Maniotis, A., Ingber, D., Merckling, A., Langa, F., Aumailley, M., Delouvée, A., Koteliansky, V. Babinet, C., and Krieg, T. (1998) Impaired mechanical stability, migration and contractile capacity in vimentin-deficient fibroblasts. *J. Cell Sci.* **111:** 1897–1907.

Fabry, B., Maksym, G. N., Butler, J. P., Glogauer, M., Navajas, D., and Fredberg, J. J. (2001) Scaling the microrheology of living cells. *Phys. Rev. Lett.* **87:** 148102(1–4).

Flusberg, D. A., Numaguchi, Y., and Ingber, D. E. (2001) Cooperative control of Akt phosphorylation and apoptosis by cytoskeletal microfilaments and microtubules. *Mol. Biol. Cell* **12:** 3087–3094.

Folkman, J., and Moscona, A. (1978) Role of cell shape in growth and control. *Nature* **273:** 345–349.

Forgacs, G. (1995) Of possible role of cytoskeletal filamentous networks in intracellular signaling: An approach based on percolation. *J. Cell Sci.* **108:** 2131–2143.

Gittes, F., Mickey, B., Nettleton, J., and Howard, J. (1993) Flexural rigidity of microtubules and actin filaments measured from thermal fluctuations in shape. *J. Cell Biol.* **120:** 923–934.

Harris, A. K., Wild, P., and Stopak, D. (1980) Silicon rubber substrata: A new wrinkle in the study of cell locomotion. *Science* **208:** 177–179.

Heidemann, S.R., and Wirtz, D. (2004) Towards a regional approach to cell mechanics. *Trends Cell Biol.* **14:** 160–166.

Heidemann, S. R., Kaech, S., Buxbaum, R. E., and Matus, A. (1999) Direct observations of the mechanical behavior of the cytoskeleton in living fibroblasts. *J. Cell Biol.* **145:** 109–122.

Helmke, B. P., Rosen, A. B., and Davies, P. F. (2003) Mapping mechanical strain of an endogenous cytoskeletal network in living endothelial cells. *Biophys. J.* **84:** 2691–2699.

Hu, S., Chen, J., Fabry, B., Namaguchi, Y., Gouldstone, A., Ingber, D. E., Fredberg, J. J., Butler, J. P., and Wang, N. (2003) Intracellular stress tomography reveals stress and structural anisotropy in the cytoskeleton of living cells. *Am. J. Physiol. Cell Physiol.* **285:** C1082–C1090.

Hu, S., Chen, J., and Wang, N. (2004) Cell spreading controls balance of prestress by microtubules and extracellular matrix. *Front. Biosci.* **9:** 2177–2182.

Huang, S., Chen, C. S., and Ingber, D. E. (1998) Control of cyclin D1, p27[Kip1] and cell cycle progression in human capillary endothelial cells by cell shape and cytoskeletal tension. *Mol. Biol. Cell* **9:** 3179–3193.

Hubmayr, R. D., Shore, S. A., Fredberg, J. J., Planus, E., Panettieri, Jr, R. A., Moller, W., Heyder, J., and Wang, N. (1996) Pharmacological activation changes stiffness of cultured human airway smooth muscle cells. *Am. J. Physiol. Cell Physiol.* **271:** C1660–C1668.

Humphrey, D., Duggan, C., Saha, D., Smith, D., and Käs, J. (2002). Active fluidization of polymer networks through molecular motors. *Nature* **416:** 413–416.

Ingber, D. E. (1993) Cellular tensegrity: Defining new rules of biological design that govern the cytoskeleton. *J. Cell Sci.* **104:** 613–627.

Ingber, D. E. (1997) Tensegrity: The architectural basis of cellular mechanotransduction. *Annu. Rev. Physiol.* **59:** 575–599.

Ingber, D. E. (2003a) Cellular tensegrity revisited: I. Cell structure and hierarchical systems biology. *J. Cell Sci.* **116:** 1157–1173.

Ingber, D. E. (2003b) Tensegrity: II. How structural networks influence cellular information-processing networks. *J. Cell Sci.* **116:** 1397–1408.

Ingber, D. E., Heidemann, S. R., Lamoroux, P., and Buxbaum, R. E. (2000) Opposing views on tensegrity as a structural framework for understanding cell mechanics. *J. Appl. Physiol.* **89:** 1663–1670.

Ingber, D. E., Madri, J. A., and Jameison, J. D. (1981) Role of basal lamina in the neoplastic disorganization of tissue architecture. *Proc. Natl. Acad. Sci. USA* **78:** 3901–3905.

Janmey, P. A., Hvidt, S., Lamb, J., and Stossel, T. P. (1990) Resemblance of actin-binding protein actin gels of covalently crosslinked networks. *Nature* **345:** 89–92.

Jenkins, S. (2003) Does tensegrity make the machine work? *Scientist* **17:** 26–27.

Kaech, S., Ludin, B., and Matus, A. (1996) Cytoskeletal plasticity in cells expressing neuronal microtubule-expressing proteins. *Neuron* **17:** 1189–1199.

Kolodney, M. S., and Wysolmerski, R. B. (1992) Isometric contraction by fibroblasts and endothelial cells in tissue culture: A quantitative study. *J. Cell Biol.* **117:** 73–82.

Kolodney, M. S., and Elson, E. L. (1995) Contraction due to microtubule disruption is associated with increased phosphorylation of myosin regulatory light chain. *Proc. Natl. Acad. Sci. USA* **92:** 10252–10256.

Laurent, V. M., Henon, S., Planus, E., Fodil, R., Balland, M., Isabey, D., and Gallet, F. (2002) Assessment of mechanical properties of adherent living cells by bead micromanipulation: Comparison of magnetic twisting cytometry vs optical tweezers. *ASME J. Biomech. Eng.* **124:** 408–421.

MacKintosh, F. C., Käs, J., and Janmey, P. A. (1995) Elasticity of semiflexible biopolymer networks. *Phys. Rev. Lett.* **75:** 4425–4428.

Mammoto, A., Huang, S., Moore, K., Oh, P., and Ingber, D. E. (2004) Role of RhoA, mDia, and ROCK in cell shape-dependent control of the Skp2-p27(kip1) pathway and the G(1)/S transition. *J. Biol. Chem.* **279:** 26323–26330.

Maniotis, A. J., Chen, C. S., and Ingber, D. E. (1997) Demonstration of mechanical connectivity between integrins, cytoskeletal filaments, and nucleoplasm that stabilize nuclear structure. *Proc. Natl. Acad. Sci. USA* **94:** 849–854.

Matthews, B. D., Overby, D. R., Alenghat, F. J., Karavitis, J., Numaguchi, Y., Allen, P. G., and Ingber, D. E. (2004) Mechanical properties of individual focal adhesions probed with a magnetic microneedle. *Biochem. Biophys. Res. Commun.* **313:** 758–764.

McGarry, J. G., and Prendergast, P. J. (2004) A three-dimensional finite element model of an adherent eukaryotic cell. *Eur. Cells Mater.* **7:** 27–34.

Mehta, D., and Gunst, S. J. (1999) Actin polymerization stimulated by contractile activation regulates force development in canine tracheal smooth muscle. *J. Physiol.* (London) **519:** 829–840.

Mijailovich, S. M., Butler, J. P., and Fredberg, J. J. (2000) Perturbed equilibrium of myosin binding in airway smooth muscle: Bond-length distributions, mechanics, and ATP metabolism. *Biophys. J.* **79:** 2667–2681.

Mooney, D., Langer, R., and Ingber, D. E. (1995) Cytoskeletal filament assembly and the control of cell shape and function by extracellular matrix. *J. Cell Sci.* **108:** 2311–2320.

Numaguchi, Y., Huang, S., Polte, T. R., Eichler, G. S., Wang, N., and Ingber, D. E. (2003) Caldesmon-dependent switching between capillary endothelial cell growth and apoptosis through modulation of cell shape and contractility. *Angiogenesis* **6:** 55–64.

Parker, K. K., Brock, A. L., Brangwynne, C., Mannix, R. J., Wang, N., Ostuni, E., Geisse, N., Adams, J. C., Whitesides, G. M., and Ingber, D. E. (2002) Directional control of lamellipodia extension by constraining cell shape and orienting cell tractional forces. *FASEB J.* **16:** 1195–1204.

Paul, R. J., Bowman, P., and Kolodney, M. S. (2000) Effects of microtubule disruption on force, velocity, stiffness and [Ca^{2+}]$_i$ in porcine coronary arteries. *Am. J. Physiol. Heart Circ. Physiol.* **279:** H2493–H2501.

Pelham, R. J., and Wang, Y. L. (1997) Cell locomotion and focal adhesions are regulated by substrate flexibility. *Proc. Natl. Acad. Sci. USA* **94:** 13661–13665.

Petersen, N. O., McConnaugey, W. B., and Elson, E. L. (1982) Dependence of locally measured cellular deformability on position on the cell, temperature, and cytochalasin B. *Proc. Natl. Acad. Sci. USA* **79:** 5327–5331.

Polte, T. R., Eichler, G. S., Wang, N., and Ingber, D. E. (2004) Extracellular matrix controls myosin light chain phosphorylation and cell contractility through modulation of cell shape and cytoskeletal prestress. *Am. J. Physiol. Cell Physiol.* **286:** C518–C528.

Pourati, J., Maniotis, A., Spiegel, D., Schaffer, J. L., Butler, J. P., Fredberg, J. J., Ingber, D. E., Stamenović, D., and Wang, N. (1998) Is cytoskeletal tension a major determinant of cell deformability in adherent endothelial cells? *Am. J. Physiol. Cell Physiol.* **274:** C1283–C1289.

Rosenblatt, N., Hu, S., Chen, J., Wang, N., and Stamenović, D. (2004) Distending stress of the cytoskeleton is a key determinant of cell rheological behavior. *Biochem. Biophys. Res. Commun.* **321**: 617–622.

Satcher, R. L., Jr., and Dewey, C. F. Jr. (1996) Theoretical estimates of mechanical properties of the endothelial cell cytoskeleton. *Biophys. J.* **71**: 109–118.

Sato, M., Theret, D. P., Wheeler, L. T., Ohshima, N., and Nerem, R. M. (1990) Application of the micropipette technique to the measurements of cultured porcine aortic endothelial cell viscoelastic properties. *ASME J. Biomech. Eng.* **112**: 263–268.

Sheetz, M. P., Felsenfeld, D. P., and Galbraith, C. G. (1998) Cell migration: Regulation of force on extracellular-matrix–integrin complexes. *Trends Cell Biol.* **8**: 51–54.

Shroff, S. G., Saner, D. R., and Lal, R. (1995) Dynamic micromechanical properties of cultured rat arterial myocytes measured by atomic force microscopy. *Am. J. Physiol. Cell Physiol.* **269**: C286–C292.

Sims, J. R., Karp, S., Ingber, D. E. (1992) Altering the cellular mechanical force balance results in integrated changes in cell, cytoskeletal and nuclear shape. *J. Cell Sci.* **103**: 1215–1222.

Singhvi, R., Kumar, A., Lopez, G. P., Stephanopoulos, G. N., Wang, D.I.C., Whitesides, G. M., and Ingber, D. E. (1994) Engineering cell shape and function. *Science* **264**: 696–698.

Sollich, P. (1998) Rheological constitutive equation for a model of soft glassy materials. *Phys. Rev. Lett. E* **58**: 738–759.

Stamenović, D. (2005) Microtubules may harden or soften cells, depending on the extent of cell distension. *J. Biomech.*, **38**: 1728–1732.

Stamenović, D., Fredberg, J. J., Wang, N., Butler, J. P., and Ingber, D. E. (1996) A microstructural approach to cytoskeletal mechanics based on tensegrity. *J. Theor. Biol.* **181**: 125–136.

Stamenović, D., Liang, Z., Chen, J., and Wang, N. (2002a) The effect of cytoskeletal prestress on the mechanical impedance of cultured airway smooth muscle cells. *J. Appl. Physiol.* **92**: 1443–1450.

Stamenović, D., Mijailovich, S. M., Tolíc-Nørrelykke, I. M., Chen, J., and Wang, N. (2002b) Cell prestress: II. Contribution of microtubules. *Am. J. Physiol. Cell Physiol.* **282**: C617–C624.

Stamenović, D., Suki, B., Fabry, B., Wang, N., and Fredberg, J. J. (2004) Rheology of airway smooth muscle cells is associated with cytoskeletal contractile stress. *J. Appl. Physiol.* **96**: 1600–1605.

Stossel, T. P. (1993) On the crawling of animal cells. *Science* **260**: 1086–1094.

Sultan, C., Stamenović, D., and Ingber, D. E. (2004) A computational tensegrity model predicts dynamic rheological behaviors in living cells. *Ann. Biomed. Eng.* **32**: 520–530.

Thoumine, O., Ziegler, T., Girard, P. R., and Nerem, R.M. (1995) Elongation of confluent endothelial cells in culture: The importance of fields of force in the associated alterations of their cytoskeletal structure. *Exp. Cell Res.* **219**: 427–441.

Venier, P., Maggs, A. C., Carlier, M. -F., and Pantaloni, D. (1994) Analysis of microtubule rigidity using hydrodynamic flow and thermal fluctuations. *J. Biol. Chem.* **269**: 13353–13360.

Volokh, K. Y., Vilnay, O., and Belsky, M. (2000), Tensegrity architecture explains linear stiffening and predicts softening of living cells. *J. Biomech.* **33**: 1543–1549.

Wang, N., and Ingber, D. E. (1994) Control of the cytoskeletal mechanics by extracellular matrix, cell shape, and mechanical tension. *Biophys. J.* **66**: 2181–2189.

Wang, N., and Stamenović, D. (2000) Contribution of intermediate filaments to cell stiffness, stiffening and growth. *Am. J. Physiol. Cell Physiol.* **279**: C188–C194.

Wang, N., Butler, J. P., and Ingber, D. E. (1993) Mechanotransduction across cell surface and through the cytoskeleton. *Science* **26**: 1124–1127.

Wang, N., Naruse, K., Stamenović, D., Fredberg, J. J., Mijailovich, S. M., Tolić-Nørrelykke, I. M., Polte, T., Mannix, R., and Ingber, D. E. (2001) Mechanical behavior in living cells consistent with the tensegrity model. *Proc. Natl. Acad. Sci. USA* **98**: 7765–7770.

Wang, N., Tolić-Nørrelykke, I. M., Chen, J., Mijailovich, S. M., Butler, J. P., Fredberg, J. J., and Stamenović, D. (2002) Cell prestress: I. Stiffness and prestress are closely associated in adherent contractile cells. *Am. J. Physiol. Cell Physiol.* **282**: C606–C616.

Waterman-Storer, C. M., and Salmon, E. D. (1997) Actomyosin-based retrograde flow of microtubules in lamella of migrating epithelial cells influences microtubule dynamic instability and turnover and is associated with microtubule breakage and treadmilling. *J. Cell Biol.* **139**: 417–434.

Wendling, S., Oddou, C., and Isabey, D. (1999) Stiffening response of a cellular tensegrity model. *J. Theor. Biol.* **196**: 309–325.

Yamada, S., Wirtz, D., and Kuo, S.C. (2000) Mechanics of living cells measured by laser tracking microrheology. *Biophys. J.* **78**: 1736–1747.

II

PHYSICAL CHEMISTRY
OF THE CELL SURFACE

5

BOND FORMATION DURING CELL COMPRESSION

ELENA LOMAKINA and RICHARD E. WAUGH

Department of Pharmacology and Physiology and Department of Biomedical Engineering, University of Rochester, Rochester, New York, USA

In this chapter we examine the role that mechanical forces play in the formation of adhesive contacts between leukocytes (principally neutrophils) and an adhesive substrate. After a brief review of the principal molecules involved and their roles in cell recruitment to the endothelium, we summarize fundamental aspects of the kinetics of bond formation at a membrane interface. Mechanisms by which force may enhance the adhesive process cross molecular, microscopic, and macroscopic length scales. At the molecular level, forward rates of reaction may be affected by repulsive interactions between cell surfaces that prevent molecules from interacting. Increasing impingement stress can overcome these repulsive interactions and significantly increase bond formation rates. On a macroscopic level, increasing impingement force increases the area of a contact region, creating increased opportunities for bond formation. At intermediate length scales, microvilli protruding from the cell surface may limit the regions within the macroscopic contact zone where membranes are in molecularly close contact, and mechanical forces can alter the microtopography of the surface, increasing the percentage of area within the macroscopic contact zone where bonds can form. Quantitative examples showing the effects of force through these various mechanisms are presented, demonstrating the substantial effects that force can have on adhesion.

I. INTRODUCTION

The topic of bond formation under compression has its most direct relevance to adhesive interactions that occur in the circulation. In the complex hydrodynamic environment of the vasculature, circulating cells are constantly colliding with each other and with the vessel wall. Although one might imagine other circumstances in which compressive forces may play a role in the formation of adhesive contacts, in most cases outside the circulation, cells actively deform and modify their shape as adhesive interactions occur. Therefore, even though many of the principles discussed in this chapter may have relevance in other contexts, we have chosen to examine the effects of compressive forces on bond formation in the particular context of interactions between leukocytes and endothelium.

The migration of leukocytes into the tissues is a central event in inflammation and is critical for successful host defense against infection. The mechanisms underlying leukocyte recruitment have relevance to a broad range of important health-related issues, including not only host defense, but a wide variety of inflammatory pathologies, stem cell therapy, and yet-to-be-developed systems for targeted drug delivery. During the inflammatory response, leukocytes are recruited to the vascular endothelium through a sequence of regulated adhesive events that involves multiple adhesion molecules with specific functional roles. The principal molecules involved in the process are the family of selectins (E-, P-, and L-selectin) and their counterreceptors, and several different integrins and their counterreceptors, principally ICAM-1 and VCAM-1. Selectins mediate the initial capture of the cell from the free stream and subsequent rolling interactions between the cell and the endothelium (Springer, 1994). Integrins may exist in at least two different conformations with different levels of affinity for their counterreceptors. Upregulation of integrin affinity (Dransfield et al., 1992), possibly accompanied by changes in mobility and distribution of these molecules (Kucik et al., 1996), is caused *in vivo* by chemokines (e.g., interleukin-8 (IL-8) or platelet-activating factor (PAF)) that are generated by inflamed endothelium and bind to receptors on the neutrophil surface, setting off signaling cascades within the cell (van Kooyk and Figdor, 2000). The resulting increase in integrin binding accounts for the firm attachment of the cell to the endothelial wall and its subsequent migration into the tissue. This multistep pattern of molecular recognition and intracellular signaling broadly defines the transition from selectin-dependent cell capture to shear-resistant firm adhesion mediated by activated integrins (Springer, 1994).

A role for mechanical force in leukocyte recruitment was indicated in some of the earliest studies attempting to characterize the affinity of leukocytes for different substrates. Lawrence and Springer (1991) demonstrated that rolling interactions mediated by selectins are, in most cases, a prerequisite for subsequent firm attachment mediated by integrins. The mechanism that accounts for this enhancement of integrin binding has been the subject of much conjecture. It was postulated that the slowing of the cell provides longer duration of contact with the endothelium and, therefore, could allow the slower integrin bonds to form and (*in vivo*) prolong interaction with chemotactic signals, causing upregulation of integrin affinity and subsequent attachment. Recent evidence suggests that selectin ligation itself may lead to activation of integrin binding (Green et al., 2003). Another possibility is that the rolling interaction generates compressive loading at the leading edge of the cell that facilitates integrin bond formation. All these mechanisms are likely to have some role in the transition between cell rolling and firm attachment, but it is this last mechanism that is the primary focus of this chapter. We begin with a brief review of the principal molecular players and their binding characteristics.

II. LEUKOCYTE ADHESION MOLECULES

A. Selectins

Remarkably, the family of selectin molecules includes only three known members. They are identified by letter according to the cell on which they were first identified: L-selectin (leukocyte), P-selectin (platelet), and E-selectin (endothelium). L-selectin is in fact found on leukocytes, but P-selectin, in addition to being found on platelets, is expressed on endothelium, along with E-selectin. During the inflammatory response, P-selectin is expressed first, because it is present in submembranous stores that are released onto the surface as a result of an inflammatory stimulus (Bonfanti et al., 1989). E-selectin appears more slowly, it is thought, because it is not stored, but must be synthesized prior to expression on the surface (Bevilacqua et al., 1989). L-selectin is present on "passive" neutrophils, but may be cleaved and shed on neutrophil activation (Borregaard et al., 1994). Identification of the counterreceptors for selectins has proven to be a significant challenge. The first molecule identified as a selectin ligand was the sialylated sugar, sialyl-Lewis[x] (sLex). This molecule binds to selectins by itself and forms a part of the binding site on naturally occurring ligands (Somers et al., 2000). The best characterized selectin receptor is PSGL-1, which is constitutively expressed on neutrophils. It is the principal counterreceptor for P-selectin and also binds to E-selectin (Asa et al., 1995). Naturally occurring receptors for L-selectin are GlyCAM-1 and CD34 (Watson et al., 1990), but for all of the selectins there are likely to be multiple glycosylated proteins that can serve as binding sites. For example, L-selectin can bind to E-selectin, thus indicating a potentially dual role for E-selectin in the adhesion process (Zollner et al., 1997).

The ability of selectins to mediate cell rolling is a direct consequence of the kinetics of the interactions between the selectins and their ligands. This was conclusively demonstrated in a series of studies by Hammer and colleagues, who recreated rolling behavior in a cell-free system (Brunk and Hammer, 1997). Thus, the relatively rapid on- and off-rates for selectin binding appear to have been tuned to mediate rolling interactions. Indeed, the exceptions to the general rule that selectins mediate rolling and integrins mediate firm attachment are invariably cases in which integrins either mediate or modulate rolling behavior (Alon et al., 1995).

B. Integrins

Most integrins found on neutrophils are members of the β_2-integrin family (CD18). The principal members of this group are LFA-1 and Mac-1. These bind to the endothelial ligand ICAM-1, a member of the immunoglobulin superfamily that is widely expressed on the vascular endothelium in response to stimulatory agents (Dustin et al., 1986). An important step in integrin-mediated adhesion is the change in conformation of integrins that increases the affinity of binding to ICAM-1 (Dransfield et al., 1992). *In vivo*, these "activating" conformational changes are brought about by intracellular signaling, initiated by chemokines and other stimulatory mechanisms (van Kooyk and Figdor, 2000). In addition to changing the affinity of the integrin, these intracellular signaling events can lead to cytoskeletal reorganization, changes in the association between integrins and actin-binding proteins, and possible alterations in the distribution of integrins on the cell surface (Leitinger et al., 2000). Similar conformational changes (without the additional effects on cytoskeletal organization) can be induced using divalent cations. Calcium stabilizes a low-affinity form of LFA-1, inhibiting ICAM-1 binding, while manganese or magnesium in the presence of the calcium chelator EGTA promotes the LFA-1–ICAM-1 interaction (Dransfield et al., 1992). It is important to note that the effects of divalent cations on the affinity state of Mac-1 are different from their effects on LFA-1. Whereas calcium and/or magnesium are essential for integrin affinity changes induced by inflammatory mediators, manganese, but not magnesium, induces the active form of Mac-1 in the absence of other cell activators (Diamond and Springer, 1993). Although unlikely to be important physiologically, these divalent ion effects provide a useful method for examining the effects of integrin affinity changes on cell adhesion without resorting to full-scale activation of the cell by inflammatory mediators. For example, we have demonstrated that Mg^{2+} increases the likelihood of integrin-mediated homotypic adhesion between neutrophils (Spillmann et al., 2002) and enables adhesion between neutrophils and immobilized ICAM-1 (Lomakina and Waugh, 2004).

III. ADHESION KINETICS

The bond formation reaction in which an integrin or selectin associates with its counterreceptor can, in a basic sense, be thought of as a simple first-order kinetic reaction,

$$A + B \underset{k_r}{\overset{k_f}{\rightleftharpoons}} A \otimes B, \tag{5.1}$$

where k_f is the forward rate constant, k_r is the reverse rate constant, and the symbol $\otimes$ indicates a bond formed. The process of two molecules interacting to form a bond can be thought of as involving two steps. First, the molecules must diffuse or be convected to a distance that allows them to interact, then a reaction takes place to form the bond. Consideration of these two facets of bond formation is particularly important for understanding the complexities of adhesion between cells. This was first delineated in a landmark paper by Bell (1978). To emphasize the distinct parts of the process, Bell wrote the reaction as

$$A + B \underset{d_-}{\overset{d_+}{\rightleftharpoons}} AB \underset{r_-}{\overset{r_+}{\rightleftharpoons}} A \otimes B, \tag{5.2}$$

where d_+ and d_- are the rates of formation and breakage of the encounter complex, and r_+ and r_- are the forward and reverse rates for bond formation once the encounter complex is formed, sometimes called the intrinsic rates of bond formation and breakage. Bell argues that for many cases, the concentration of the encounter complex is small compared with concentrations of other moieties, and its concentration may be considered constant. In this case the net forward and reverse rate constants are given as

$$k_f = \frac{d_+ r_+}{d_- + r_+} \tag{5.3}$$

and

$$k_r = \frac{d_- r_-}{d_- + r_+}. \tag{5.4}$$

In typical reaction schemes in three dimensions, diffusive transport is rapid compared with intrinsic rates of reaction. In this case, $d_- >> r_+$, implying $k_r \approx r_-$, and $k_f \approx r_+ d_+ / d_-$. Bell showed that in the case where receptors are localized to a membrane surface, even an ideal membrane, the opposite is true. Unless intrinsic reaction rates are extremely slow, the measured rate of reaction is dominated by the lateral diffusion of molecules into sufficiently close proximity to form a reactive complex. Formally this is the case when $r_+ >> d_-$, and we can approximate $k_f \approx d_+$ and $k_r \approx d_- \, r_-/r_+$. Thus, when both reactants are confined to a membrane, the slower rate of diffusion of molecules in membranes generally results in slower forward and reverse rates of reaction than when one or both of the same species are free to diffuse in solution.

An important additional consideration is the possibility that the formation of encounter complexes may be facilitated by relative motion between the contacting surfaces. This convective contribution to the overall forward rate of reaction was identified and analyzed by Chang and Hammer (1999). The relative importance of convective versus diffusive transport in the formation of a reactive complex is characterized in terms of the dimensionless Peclet number: $Pe = |\bar{v}| \, a/D$, where $|\bar{v}|$ is the magnitude of the relative velocity between the surfaces, a is the encounter distance, and D is the diffusion coefficient for molecules in the surface. For diffusion coefficients on the order of 3×10^{-10} cm^2/s (typical of mobile receptors in membranes) and an encounter distance of 10 nm, the relative velocity between surfaces would need to be approximately 3 μm/s for diffusion and convection to make comparable contributions to the formation of the encounter complex. This is well within the range of velocities found between cells interacting at physiological shear rates, and indicates a potentially important role for convective transport, particularly in the case of selectin binding, where initial relative velocities may be quite high. For a complete discussion of the relative importance of convection and diffusion, including quantitative predictions for the dependence of the reaction rate on Peclet number, see Chang and Hammer (1999).

A. Bond Formation and Adhesion Probability

Experimentally, what is observed in the laboratory is not bond formation per se, but rather the likelihood that an adhesive contact is formed when two surfaces interact. The probability of adhesion is, of course, related to the formation of bonds in the contact interface. If the formation of bonds is not correlated with prior bond formation (i.e., Poisson statistics apply), the probability of adhesion P_{adh} is related to the expected number of bonds $<n>$ formed in the interface by (Chesla et al., 1998)

$$<n> = -\ln(1-P_{adh}). \qquad (5.5)$$

The expected bond number is in turn related to the reaction kinetics. If it is assumed to be first-order, the reaction is (Chesla et al., 1998)

$$<n> = A_c \rho_1 \rho_2 \, K_a \left(1 - e^{-k_r t}\right) \qquad (5.6)$$

where ρ_1 and ρ_2 are the surface concentrations of molecules, $K_a = k_f/k_r$ is the equilibrium association constant, and t is the duration of contact. Although the applicability of these formulations to specific molecular interactions must be carefully evaluated, the preceding equations provide a rational framework within which we can begin to evaluate the effects of mechanical force on adhesion. Mechanical force can affect the adhesion process in multiple ways. In the next section, we consider its effects on rates of reaction.

IV. EFFECTS OF FORCE ON ADHESION KINETICS

The most obvious potential role for mechanical force in adhesion is the facilitation of bond breakage. A model of the dependence of bond breakage on loading was proposed by Bell (1978):

$$k_r = k_{r,o} \, e^{\gamma_o f / k_b T} \tag{5.7}$$

where $k_{r,o}$ is the unstressed rate for bond breakage (off-rate): f is the force on the bond, k_b is Boltzmann's constant, T is temperature, and γ_o is the reactive compliance, a parameter that characterizes how strongly the off-rate is affected by force. To satisfy thermodynamic self-consistency, the formulation of the off-rate according to Eq. (5.7) requires the following form for the bond formation (on) rate (Bhatia et al., 2003):

$$k_f = k_{f,o} \exp\left[\sigma |x_b - \lambda| \left(\gamma_o - \frac{1}{2} |x_b - \lambda| \right) / k_b T \right]. \tag{5.8}$$

Here, σ is the molecular spring constant, and $|x_b - \lambda|$ is the absolute value of the deviation in bond length relative to its preferred length λ. Note that force does not appear explicitly in the expression for the forward rate. This is consistent with the obvious point that a bond cannot support a load before it is actually formed. Rather, the forward rate is affected by the separation distance of the molecules. This depends on the distance the molecule extends from the surface and the separation distance between the opposing membranes. Although molecules extending out from a surface will have a preferred length, they may exist at other lengths with a likelihood given by a Boltzmann distribution that depends on the energy cost of achieving those other lengths. Other formulations for the dependence of bond formation and breakage have been proposed (e.g., see Dembo et al., 1988), but experimental evidence indicates that Eq. (5.7) agrees well with the behavior of several molecular pairs that have been investigated (Evans et al., 2001).

Experimental efforts to explore the effects of force on bond breakage have employed two principal approaches, atomic force microscopy (AFM) (Zhang et al., 2002) and the biomembrane force probe (BFP) (Evans et al., 2001). (BFP is a technique in which a red blood cell held in a micropipet is used as a kind of spring to measure forces applied to bonds. A ligand-coated bead is first attached to the red cell and then brought into contact with the receptor located on a cell or substrate. A piezoelectric device is used to retract the cell at controlled rates, and the force at which bonds break can be determined from the cell deformation.) These approaches take advantage of the dependence of bond lifetime on loading rate, and have been used to obtain measures of the reactive compliance, which characterizes the force dependence of bond breakage (Evans, 2001). The relationship between the most probable force at which a bond will break (f_{crit}) and the rate of loading (R_L) is (Evans et al., 2001)

$$f_{crit} = \frac{k_b T}{\gamma_o} \ln \frac{R_L \gamma_o}{k_{ro} \, k_b T}. \tag{5.9}$$

Experimentally, the force at which a bond breaks is measured repeatedly to obtain the most probable force of rupture (f_{crit}) from the distribution of rupture forces for a given loading rate. This is repeated for several different loading rates, and f_{crit} is plotted as a function of the logarithm of the loading rate. A linear relationship is expected, the slope of which includes the reactive compliance, γ_0. Evans and colleagues have tested a variety of receptor–ligand pairs with the BFP method. For example, as the loading rate on a bond between biotin and streptavidin increased from 10^{-1} to 10^5 pN/s, the critical force at which the bonds broke increased from 5 to 170 pN. The same approach was used subsequently to characterize bonds between PSGL-1 or sLex and L-selectin, lipid anchoring in membranes (Evans et al., 2001), P-selectin–PSGL-1 bonds (Heinrich et al., 2005), and E-cadherin self-association (Perret et al., 2004). Other groups have used AFM to control loading rates and observe bond dissociation. Molecular pairs investigated using this approach include cadherins (Baumgartner et al., 2000), P-selectin–PSGL-1, (Marshall et al., 2005), and both LFA-1–ICAM-1 (Zhang et al., 2002) and VLA-4–VCAM-1 (Zhang et al., 2004) interactions. In many of these cases, multiple pathways to bond breakage have been observed, as reflected in piecewise linear regimes in the force-versus-loading rate curves. For example, Tees et al. (2001) used a microcantilever technique to apply force to bonds between E-selectin and sLe^x. E-selectin was adsorbed onto a glass fiber and sLex was coupled to latex microspheres. At low loading rates (200–1000 pN/s), the reactive compliance was estimated to be 0.048 nm and the unstressed reverse reaction rate was 0.72 s^{-1}, whereas at higher rates (1000–5000 pN/s), the corresponding values were 0.016 nm and 2.2 s^{-1}. Zhang et al. (2002) reported that LFA-1/ICAM-1 bonds also exhibit different kinetic behavior at slow ($<10^4$ pN/s) and fast ($>10^4$ pN/s) loading regimes.

In contrast to the attention that has been paid to the role of force in bond breakage, there have been few if any studies that examine the role of force on bond formation. This is almost certainly due to the fact that force effects on bond formation are indirect. Recall that force does not appear explicitly in the forward rate expression Eq. (5.8). What does contribute directly is the distance between molecules on opposing surfaces. An analysis of the effects of repulsive forces on bond formation at equilibrium was delineated in a series of papers by Dembo and Bell (Bell, 1978; Bell et al., 1984; Dembo et al., 1988). At equilibrium, the number of bonds formed is a function of the dependence of the repulsive force on separation distance, the spring constants of the bonds, and the bond length. In dynamic situations, particularly at the initiation of adhesion, repulsive forces restrict the ability of bonds to form and must be overcome by other mechanisms. As described in detail in Chapter 7, the repulsive interactions may arise from stationary negative charges on the cell surface or from steric effects due to the glycocalyx. When the distance of approach between membranes is limited, forces of impingement can have a significant effect on bond formation by overcoming repulsive interactions between surfaces and changing the energy landscape that molecules must negotiate to form bonds. Some example calculations reveal how significant this effect could be on bond formation. Consider the case shown in Fig. 5.1, in which two opposing membranes are held apart because of the presence of the gylcocalyx on the cell surfaces. If the natural bond length differs from the intermembrane separation distance, the effects on the forward and reverse rates can be estimated via Eqs. (5.7) and (5.8). As a case study we examine the interaction of PSGL-1 and P-selectin. The molecular spring constant for P-selectin has been estimated from AFM measurements to be $\sim$1 pN/nm. Therefore, for an example case in which the

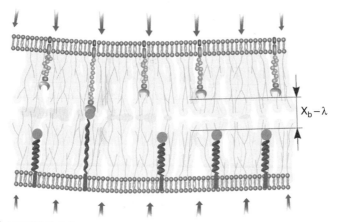

$X_b - \lambda$

FIGURE 5.1 Conceptual picture of steric limitations on bond formation. Glycosylation of membrane proteins creates a repulsive layer of polysaccharide on the cell surface. If adhesion molecules have a shorter extended length than the thickness of the layer, the layer must be compressed for opposing molecular pairs to come into close contact. The instantaneous length of a molecule extending from the surface is a stochastic quantity that depends on the length and flexibility of the molecule. Thus, the probability of bond formation depends on the length and flexibility of the adhesion molecules, the membrane separation distance, and mechanical forces acting to overcome the surface repulsion. (See Color Plate 5)

membrane interaction distance differs from the natural bond length by just 5 nm, the force on the bond should be approximately 5 pN. This force turns out to have a relatively small effect on the off-rate. The value of the reactive compliance (γ_o) is 0.039 nm (Smith et al., 1999), yielding a value for the ratio of the stressed and unstressed off-rates (see Eq. 5.7) of ~1.05. In contrast, there is a significant effect of the added separation distance on the forward rate. Taking $|x_b - \lambda|$ to be 5 nm, and σ to be 1.0 pN/nm, the ratio of the forward rates is ~0.05, giving us a decrease in the apparent association constant for bond formation of about 20-fold. If the difference between separation distance and bond length is increased to 10 nm, the ratio of the equilibrium constants decreases to 4×10^{-6}! Clearly, limitations on the close approach of molecular pairs in the interface can have a dominant effect on bond formation, and mechanical force can have a significant role in overcoming these limitations.

Several studies in which changing molecular length or the thickness of a repulsive layer has had a significant effect on cell adhesion are presented in Chapter 7. Another example in which this mechanism is likely to have played a role is the study in which an increasing strength of cell adhesion was observed when two adherent cells were peeled apart (Evans et al., 1991). The physical explanation hypothesized for this is that the bending of the membrane at the edge of the contact zone forced adjacent regions farther into the contact region into closer contact, facilitating additional bond formation. This mechanism was supported by another experiment in which colloidal adhesive forces induced by the presence of dextran in solution also strengthened adhesive interactions between contacting cells (Evans et al., 1991). Thus, there is ample evidence that force can increase bond formation by altering the distances between surface molecules over which bonds can form. How large the forces must be is more difficult to say. Theoretical estimates of the compressibility of the glycocalyx are presented in Fig. 7.8. Based on these calculations, it appears that compressive stresses in the range 1.0 to 50 N/m² could produce substantial compression of this layer. To determine whether such stresses might exist at the interface between a rolling leukocyte and the endothelium, we proceed to our next topic, macroscopic effects of force on the adhesive interface.

V. FORCE, CONTACT AREA, AND SURFACE TOPOGRAPHY

In addition to affecting bond formation rates at a molecular level, mechanical force can have a significant effect on meso- and macroscopic parameters that affect adhesion. An obvious effect of increasing force is to increase the size of the contact area between the cell and the substrate. From Eq. (5.6) it is evident that this will have a direct effect on the rate of bond formation. In addition, the actual area of membrane in intimate contact with substrate is affected by the microtopography of the cell surface within the macroscopic contact region (Fig. 5.2), and this in turn can have significant effects on cell adhesion (Williams et al., 2000). The relationships presented earlier apply directly to smooth membrane surfaces in intimate contact, but the surfaces of leukocytes are far from smooth. Undulations in the surface microtopography (the microvilli) contain enough surface area to double the area of the membrane bilayer compared with the surface area of a sphere with a diameter equal to that of the cell (Ting-Beall et al., 1993). The ability of both the microvilli and the cell as a whole to undergo significant deformation has significant influence on neutrophil adhesion to substrate. Compression of the microvilli on the neutrophil surface may increase cell adhesion by at least two mechanisms. First, it should increase the percentage of cell surface area in the contact zone that comes into intimate contact with the substrate. Second, if the molecules are distributed nonuniformly on the cell surface, it could affect the concentration of the molecules capable of binding to the substrate. Ultrastructural evidence indicates that the distribution of the major adhesion molecules on the neutrophil surface is, in fact, nonuniform. Selectins, known to mediate cell rolling, are clustered primarily on the tips of microvilli

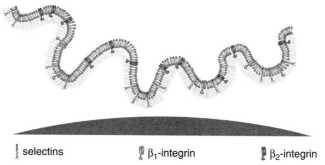

‡ selectins   ⌇ β₁-integrin   ⌇ β₂-integrin

FIGURE 5.2 Surface topography may also act to limit close approach of molecules for bond formation. As little as 5 to 10% of a macroscopic contact region may actually be in physical contact with an opposing surface. The microvilli that cover the cell surface are deformable and are expected to change shape under mechanical loading. Thus, mechanical forces may facilitate bond formation by increasing the area of close contact between the cell membrane and an adhesive substrate. (See Color Plate 6)

(Erlandsen et al., 1993; Bruehl et al., 1996). In contrast, integrins on the neutrophil surface (LFA-1 and Mac-1) appear to be randomly distributed on the nonvillus cell body (Erlandsen et al., 1993; Fernandez-Segura et al., 1996). Thus, changes in surface topography brought about by cell deformation under constant loads are likely to have substantial effects on the formation of adhesive bonds.

A. Cell Mechanical Behavior

Understanding the influence of force on adhesion at these length scales requires an appreciation of mechanical behavior of the cells. The rheological properties of neutrophils and other leukocytes have received considerable attention because of their role as circulating corpuscles, and a number of mechanical models that approximate the cellular response to applied forces have been proposed. Observations of large cellular deformations as cells were aspirated into micropipets revealed that the behavior of leukocytes is fundamentally fluid-like. Cells deform continuously into micropipets in response to a constant aspiration pressure (Evans and Yeung, 1989). In addition, a small threshold pressure required to initiate cell entry was found to be approximately constant no matter how far into the pipet the cell is drawn. These observations led to the model of the cell as a highly viscous fluid drop with a constant cortical tension, the so-called "liquid drop model" (Evans and Yeung, 1989). In the simplest form of this model, the neutrophil interior is modeled as a Newtonian fluid. Closer scrutiny of cell behavior, particularly at different rates of deformation, revealed additional complexities of the cellular response to applied forces, and this has led to several proposed refinements in the basic liquid drop model. One of these is a proposal that the interior of the cell is nonuniform in its properties, and that there is proportionally greater resistance to flow in the cell periphery than in the interior (Evans and Yeung, 1989; Yeung and Evans, 1989; Drury and Dembo, 2001). It has also been demonstrated that cellular viscosity decreases with increasing shear rate, leading to a model of the cell interior as a shear-thinning power-law fluid (Tsai et al., 1993; Drury and Dembo, 2001). Rapid aspiration and expulsion of the cell into a pipet leaves the cell in a wedge-shaped rather than cylindrical profile, suggesting that there is a very short-lived elastic component to the cellular response. This supports models of the cell as a Maxwell fluid (Dong and Skalak, 1992). Most recently, consideration has been given to the possibility that the cell interior may exhibit properties of two-phase flow, with a less viscous fluid percolating through a more viscous polymer phase (Herant et al., 2003). Although there is experimental evidence supporting all of these refinements, the essential behavior of the cell remains that of a fluid drop, with equilibrium shapes determined principally by a constant contractile force resultant in the cell cortex. Generally, the pressure difference across the membrane, ΔP, and the membrane curvature are related by the law of Laplace for liquid interfaces,

$$\Delta P = T_{cort}\left(\frac{1}{R_1} + \frac{1}{R_2}\right), \tag{5.10}$$

where T_{cort} is the cortical tension of the cell and R_1 and R_2 are the principal radii of curvature for the surface.

B. Experimental Approaches to Altering Contact Area and Topography

Flow channel studies have provided much of what is currently known about the dynamic interactions of leukocytes with endothelial monolayers or immobilized endothelial ligands. However, the complexity of the interactions of the cell with a surface under shear flow makes it difficult to draw definitive conclusions about molecular kinetics. Hammer and colleagues have developed "adhesive dynamics" simulations of a cell interacting with an adherent surface under flow (Chang et al., 2000). These models have provided considerable insight into how the kinetic properties of the interacting molecules can modify the behavior of a cell interacting with surface ligands. Notably, it has been shown that the kinetic behavior of selectin bonds directly accounts for their ability to mediate rolling interactions, as opposed to firm attachments. Indeed, the predictions of the model have been confirmed experimentally in a reconstituted, cell-free system (Brunk and Hammer, 1997). Extracting information about molecular kinetics from flow channel studies involving cells is substantially more difficult. As of this writing, there are no model calculations that fully account for the coupling between bond formation at the interface, cell deformation, and force-dependent effects on bond formation. Estimates of kinetic coefficients have been made from flow channel studies (Alon et al., 1997), but these have required simplifying assumptions that affect the accuracy of the estimates.

Alternative approaches for evaluating force effects on bond formation involve the use of laser tweezers (Dai and Sheetz, 1995) or micropipets to control the interaction of cells with adherent beads coated with endothelial ligands (Shao and Hochmuth, 1996; Spillmann et al., 2004). These approaches provide the advantage of a simple geometry of interaction and the ability to control both the force and the duration of contact. Controlling the chemistry of coating the beads used in the experiments allows both the type of ligand and the ligand density also to be controlled. In addition, the mean density of specific ligands on the neutrophil surface (e.g., LFA-1 and Mac-1) can be measured using immuno-labeling and flow cytometry. Thus, the principal variables in the kinetic equation can be controlled.

C. Relationship between Force, Contact Area, and Contact Stress

Predictions for the dependence of contact area on applied force can be made based on our understanding of cell properties and a simple equilibrium analysis (Fig. 5.3A)

The force balance for a cell interacting with a rigid spherical surface of radius R_b requires

$$f = \left[(P_c - P_o) + \frac{2T_{cort}}{R_b} \right] \pi R_{con}^2, \tag{5.11}$$

where f is the force, P_c and P_o are the pressures inside and outside of the cell, and R_{con} is the radius of the contact region between the cell and the surface. Note that the force is balanced by the pressure within the cell (relative to the surrounding pressure) plus a contribution due to the cortical tension acting over the total curvature of the bead. The pressure inside the cell is, in turn,

A

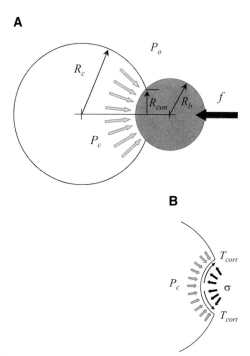

B

FIGURE 5.3 (A) Schematic illustration of the external forces acting on the contact zone. The pressure in the cell, P_c, is taken relative to the pressure in the suspending phase, P_o. The force f acting on a bead of radius R_b causes it to indent the cell surface to create a region of contact with radius R_{con}. The radius of the cell is R_c. (B) A more detailed picture of the force balance in the contact zone. The contact stress σ must balance the pressure in the cell, P_c, as well as a resultant from the membrane cortical tension, T_{cort}.

related to the cortical tension via Eq. (5.10). Approximating the shape of the cell by a sphere of radius R_c, we obtain (Lomakina *et al.*, 2004)

$$f \approx 2T_{cort}\left[\frac{1}{R_c} + \frac{1}{R_b}\right]\pi R_{con}^2. \tag{5.12}$$

The contact area A_{mac} is related to the radius of the contact zone R_{con} by simple geometry:

$$A_{mac} = 2\pi R_b \left(R_b - \sqrt{R_b^2 - R_{con}^2}\right). \tag{5.13}$$

As indicated earlier, mechanical forces should also affect adhesion via alterations of the microtopography of the cell membrane in the contact zone. To what degree force may affect microtopography depends on the load supported by individual microvilli within the contact zone. Therefore, it is not the total force that is of interest, but rather the force per unit area (contact stress) within the macroscopic contact region. To determine how the contact stress changes under different loading conditions, we apply an equilibrium analysis based on the model of the cell as a fluid droplet (Fig. 5.3B) This analysis leads to the relationship (Lomakina et al., 2004)

$$\sigma = 2T_{cort}\left(\frac{1}{R_b} + \frac{1}{R_c}\right), \tag{5.14}$$

where R_b is the radius of the substrate and R_c is the radius of the cell. Note that the contact stress depends only on the cortical tension of the cell and the curvatures of the cell and bead, and is independent of the total impingement force. Experimentally, the contact stress can be altered by using beads with different surface curvatures.

An important limiting case occurs when the substrate is flat, as is the case in flow channel studies and (approximately) *in vivo*. For cells attached to endothelium under flow, there will be some effect on cell curvature of the fluid shear, but an estimation based on the resting cell curvature provides a reasonable approximation. Measured values of cortical tension fall in the range 16–25 pN/μm (Needham and Hochmuth, 1992; Lomakina et al., 2004). Using these values and estimating a cell radius of 4.5 μm yield a contact stress in the range 7–11 pN/μm². These values can be significantly larger in dynamic situations when the cell is not in equilibrium or residual elastic stresses may persist. An analysis of the dynamics of impingement suggests that peak stresses during the first 200 ms of contact under a constant applied pressure could produce stresses that approach twice the equilibrium value (Lomakina et al., 2004). A more significant effect may result from the microvillus structure of the cell surface. Depending on the precise shape and distribution of microvilli on the surface, it is estimated that just 5 to 10% of the macroscopic area may be in physical contact with the substrate. Thus, in these localized regions of contact, stresses may be 10- to 20-fold higher than the macroscopic estimate.

Based on estimates of the resistance of the glycocalyx to compression presented in Fig. 7.8, we estimate that such stresses could reduce the thickness of the glycocalyx by 50 to 75%. This is a substantial distance compared with the extended lengths of adhesion molecules, which range from 20 to 50 nm, and illustrates the importance not only of the compressive loads that occur at the interface between rolling cells and substrate, but also of the load concentrations that are expected to occur as a result of irregular surface topography.

D. Experimental Findings

The expected effects of increasing both contact area and contact stress have been verified experimentally for interactions between neutrophils and ICAM-1 immobilized on bead surfaces (Spillmann et al., 2004) (Fig. 5.4). Contact area was increased by increasing impingement force, while contact stress was held constant by maintaining constant bead diameter for all measurements. Although there was considerable variability in adhesion probability from cell to cell, all cells showed a more or less linear increase in expected bond number $<n>$ with increasing contact area, as expected. In the same study, a significant effect of contact stress on adhesion was also observed. In this case, the force was adjusted to maintain constant contact area, and the contact stress was altered by performing experiments with beads of different diameter. Interestingly, the dependence of expected bond number on contact stress was also linear over the experimental range of contact stresses, 10–40 Pa (1.0 Pa = 1.0 N/m² = 1.0 pN/μm²) (Spillmann et al., 2004). (See Fig. 5.5.)

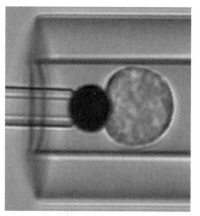

FIGURE 5.4 Experimental approach originally developed by Shao and Hochmuth (1996) for testing the force dependence of bond formation. The image is a video micrograph of a human neutrophil, inside the lumen of a large micropipet, being pushed into contact with a smaller bead held with a second, smaller micropipet. The force of contact is regulated by adjusting the upstream pressure in the large pipet. The pressure can be reversed to pull the cell away from the bead, at which point the existence of an adhesive bond can be detected. The diameter of the cell is approximately 9.0 μm, and the diameter of the bead is approximately 4.5 μm.

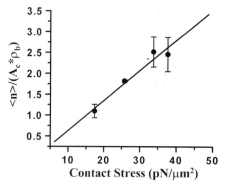

FIGURE 5.5 Dependence of bond formation rate on contact stress. Neutrophils were suspended in physiological buffer containing Mg^{2+} plus EGTA to put LFA-1 into a high affinity conformation. Beads of different diameter were coated with soluble recombinant ICAM-1, and cells and beads were brought into repeated contact to determine the probability that an adhesive contact was formed. The bond formation rate was calculated according to Eqs. (5.5) and (5.6), and then normalized by the contact area and the density of ligand on the beads as determined by flow cytometry. Contact stress was calculated according to Eq. (5.14) using a characteristic value of the cortical tension for neutrophils (20 pN/μm). There appears to be a linear increase in bond formation with increasing contact stress. Data from Spillmann et al. (2004) are replotted.

VI. SUMMARY AND CONCLUSIONS

Forces acting to compress the contact region between cells adhering to a substrate may affect bond formation in multiple ways. At the molecular level, repulsive forces due to electrostatic charge or steric hindrance may limit the close approach of the adhesion molecules. Mechanical forces can have a significant effect on adhesion by overcoming these repulsive forces and enabling bonds to form. The presence of microvilli on the surface may significantly enhance this effect by concentrating loads (and increasing contact stress) in the local regions where microvilli contact the substrate. At the macroscopic level, the deformability of the cell results in an increase in the area of contact between the cell and the substrate, thus increasing the area over which bonds can form. In addition, the stress in the contact zone may result in deformation of microvilli on the cell surface, increasing the percentage of area within the contact zone that is in molecular close contact such that bonds can form. All of these factors may contribute to the underlying mechanisms by which cell rolling facilitates integrin bond formation and firm attachment of leukocytes to endothelium.

ACKNOWLEDGMENTS

The authors thank Professor Philip Knauf for his critical reading of the manuscript, and Christopher Spillmann for his excellent early work on the role of mechanical forces in bond formation, which served as a foundation for our activities in this area of research. The authors are supported by the NIH under Grant PO1 HL18208.

SUGGESTED READING

Aplin, A. E., Howe, A., Alahari, S. K., and Juliano, R. L. (1998) Signal transduction and signal modulation by cell adhesion receptors: The role of integrins, cadherins, immunoglobulin-cell adhesion molecules, and selectins. *Pharm. Rev.* **50**: 197–263.
   Signaling events associated with cell adhesion, including integrins and selectins, are thoroughly reviewed.

Evans, E. (2001) Probing the relation between force-lifetime and chemistry in single molecular bonds. *Annu. Rev. Biophys. Biomol. Struct.* **30**: 105–128.
   The role of mechanical forces and bond breakage, particularly as assessed by dynamic force spectroscopy and the biomembrane force probe, is detailed.

Hogg, N., Laschinger, M., Giles, K., and McDowell, A. (2003) T-cell integrins: more than just sticking points. *J. Cell Sci.* **116**: 4695–4705.
   Further information on the function of integrins in T lymphocytes can be found in this article.

Springer, T. A. (1994) Traffic signals for lymphocyte recirculation and leukocyte emigration: The multistep paradigm. *Cell* **76**: 301–314.
   The development of the basic concept of leukocyte recruitment through selectin-mediated rolling, followed by integrin-mediated firm attachment, is described.

Springer, T. A., and Wang, J.-H. (2004) The three dimensional structure of integrins and their ligands, and conformational regulation of cell adhesion. *Adv. Prot. Chem.,* **68**: 29–63.
   This article concerns recent discoveries of the structure and function of integrins.

Vestweber, D., and Blank, J. E. (1999) Mechanisms that regulate the function of the selectins and their ligands. *Physiol. Rev.* **79**: 181–213.

Details of the structure and regulation of selectin molecules and their ligands are given.

Von Andrian, U. H., and Mackay, C. R. (2000) T-cell function and migration. *N. Engl. J. Med.* **343:** 1020–1034.

This more recent summary of adhesion molecules is written from the perspective of T-lymphocyte function.

REFERENCES

Alon, R., Chen, S., Puri, K. D., Finger, E. B., and T. A. Springer, E. B. (1997) The kinetics of L-selectin tethers and the mechanics of selectin-mediated rolling. *J.Cell Biol.* **138:** 1169–1180.

Alon, R., Hammer, D. A., and Springer, T. A. (1995) Lifetime of the P-selectin–carbohydrate bond and its response to tensile force in hydrodynamic flow.[erratum appears in *Nature* 1995; **376:** 86]. *Nature* **374:** 539–542.

Asa, D., Raycroft, L., Ma, L., et al. (1995) The P-selectin glycoprotein ligand functions as a common human leukocyte ligand for P- and E-selectins. *J.Biol. Chem.* **270:** 11662–11670.

Baumgartner, W., Hinterdorfer, P., Ness, W., et al. (2000) Cadherin interaction probed by atomic force microscopy. *Proc.Natl. Acad. Sci. USA* **97:** 4005–4010.

Bell, G. I. (1978) Models for the specific adhesion of cells to cells. *Science* **200:** 618–627.

Bell, G. I., Dembo, M., and Bongrand, P. (1984) Cell adhesion: Competition between nonspecific repulsion and specific bonding. *Biophys. J.* **45:** 1051–1064.

Bevilacqua, M. P., Stengelin, S., Gimbrone, M. A., Jr. and B. Seed. (1989) Endothelial leukocyte adhesion molecule 1: An inducible receptor for neutrophils related to complement regulatory proteins and lectins. *Science* **243:** 1160–1165.

Bhatia, S., King, M., and Hammer, D. (2003) The state diagram for cell adhesion mediated by two receptors. *Biophys. J.* **84:** 2671–2690.

Bonfanti, R., Furie, B. C., Furie, B., and Wagner, D. D. (1989) PADGEM (GMP140) is a component of Weibel–Palade bodies of human endothelial cells. *Blood* **73:** 1109–1112.

Borregaard, N., Kjeldsen, L., Sengelov, H., et al. (1994) Changes in subcellular localization and surface expression of L-selectin, alkaline phosphatase, and Mac-1 in human neutrophils during stimulation with inflammatory mediators. *J. Leukoc. Biol.* **56:** 80–87.

Bruehl, R. E., Springer, T. A., and Bainton, D. F. (1996) Quantitation of L-selectin distribution on human leukocyte microvilli by immunogold labeling and electron microscopy. *J. Histochem. Cytochem.* **44:** 835–844.

Brunk, D. K. and Hammer, D. A. (1997) Quantifying rolling adhesion with a cell-free assay: E-selectin and its carbohydrate ligands. *Biophys. J.* **72:** 2820–2833.

Chang, K. C. and Hammer, D. A. (1999) The forward rate of binding of surface-tethered reactants: effect of relative motion between two surfaces. *Biophys. J.* **76:** 1280–1292.

Chang, K. C., Tees, D. F. J., and Hammer, D. A. (2000) The state diagram for cell adhesion under flow: leukocyte rolling and firm adhesion. *Proc. Natl. Acad. Sci. USA* **97:** 11262–11267.

Chesla, S. E., Selvaraj, P., and Zhu, C. (1998) Measuring two-dimentional receptor–ligand binding kinetics by micropipette. *Biophys. J.* **75:** 1553–1572.

Dai, J. and Sheetz, M. P. (1995) Mechanical properties of neuronal growth cone membranes studied by tether formation with laser optical tweezers. *Biophys. J.* **68:** 988–996.

Dembo, M., Torney, D. C., Saxman, K., and Hammer, D. (1988) The reaction-limited kinetics of membrane-to-surface adhesion and detachment. *Proc. R. Soc.London Ser. B* **234:** 55–83.

Diamond, M. S. and Springer, T. A. (1993) A subpopulation of Mac-1 (CD11b/CD18) molecules mediates neutrophil adhesion to ICAM-1 and fibrinogen. *J. Cell Biol.* **120:** 545–556.

Dong, C. and Skalak, R. (1992) Leukocyte deformability: finite element modeling of large viscoelastic deformation. *J. Theor. Biol.* **21:** 173–193.

Dransfield, I., Cabanas, C., Craig, A., and Hogg, N. (1992) Divalent cation regulation of the function of the leukocyte integrin LFA-1. *J.Cell Biol.* **116:** 219–226.

Drury, J. L. and Dembo, M. (2001) Aspiration of human neutrophils: Effects of shear thinning and cortical dissipation. *Biophys. J.* **81:** 3166–3177.

Dustin, M. L., Rothlein, R., Bhan, A. K., Dinarello, C. A., and Springer, T. A. (1986) Induction by IL 1 and interferon-gamma: Tissue distribution, biochemistry, and function of a natural adherence molecule (ICAM-1) *J. Immunol.* **137:** 245–254.

Erlandsen, S. L., Hasslen, S. R. and Nelson, R. D. (1993) Detection and spatial distribution of the beta 2 integrin (Mac-1) and L-selectin (LECAM-1) adherence receptors on human neutrophils by high-resolution field emission SEM. *J. Histochem. Cytochem.* **41**: 327–333.

Evans, E. (2001) Probing the relation between force—lifetime—and chemistry in single molecular bonds. *Annu. Rev. Biophys. Biomol. Struct.* **30**: 105–128.

Evans, E., Berk, D., Leung, A., and Mohandas, N. (1991) Detachment of agglutinin-bonded red blood cells: II. Mechanical energies to separate large contact areas. *Biophys. J.* **59**: 849–860.

Evans, E., Leung, A., Hammer, D., and Simon, S. (2001) Chemically distinct transition states govern rapid dissociation of single L-selectin bonds under force. *Proc Natl Acad Sci.* **98**: 3784–3789.

Evans, E. and Yeung, A. (1989) Apparent viscosity and cortical tension of blood granulocytes determined by micropipet aspiration. *Biophys. J.* **56**: 151–160.

Fernandez-Segura, E., Garcia, J. M., and Campos, A. (1996) Topographic distribution of CD18 integrin on human neutrophils as related to shape changes and movement induced by chemotactic peptide and phorbol esters. *Cell Immunol.* **171**: 120–125.

Green, C. E., Pearson, D. N., Christensen, N. B., and Simon, S. I. (2003) Topographic requirements and dynamics of signaling via L-selectin on neutrophils. *Am J. Physiol. Cell Physiol.* **284**: C705-C717.

Heinrich, V., Leung, A., and Evans, E. (2005) Nano- to microscale dynamics of P-selectin detachment from leukocyte interfaces: II. Tether flow terminated by P-selectin dissociation from PSGL-1. *Biophys. J.* **88**: 2299–2308.

Herant, M., Marganski, W. A. and Dembo, M. (2003) The mechanics of neutrophils: Synthetic modeling of three experiments. *Biophys. J.* **84**: 3389–3413.

Kucik, D. F., Dustin, M. L., Miller, J. M., and Brown, E. J. (1996) Adhesion-activating phorbol ester increases the mobility of leukocyte integrin LFA-1 in cultured lymphocytes. *J.Clin Invest.* **97**: 2139–2144.

Lawrence, M. B., and Springer, T. A. (1991) Leukocytes roll on a selectin at physiologic flow rates: Distinction from and prerequisite for adhesion through integrins. *Cell* **65**: 859–873.

Leitinger, B., McDowall, A., Stanley, P., and Hogg, N. (2000) The regulation of integrin function by Ca^{2+}. *Biochim. Biophys. Acta.* **1498**: 91–98.

Lomakina, E. B., Spillmann, C. M., King, M. R., and Waugh, R. E. (2004) Rheological analysis and measurement of neutrophil indentation. *Biophys. J.* **87**: 4246–4258.

Lomakina, E. B. and Waugh, R. E. (2004) Micromechanical tests of adhesion dynamics between neutrophils and immobilized ICAM-1. *Biophys. J.* **86**: 1223–1233.

Marshall, B. T., Sarangapani, K. K., Lou, J., McEver, R. P., and Zhu, C. (2005) Force history dependence of receptor–ligand dissociation. *Biophys. J.* **88**: 1458–1466.

Needham, D. and Hochmuth, R. M. (1992) A sensitive measure of surface stress in the resting neutrophil. *Biophys. J.* **61**: 1664–1670.

Perret, E., Leung, A., Feracci, H. and Evans, E. (2004) Trans-bonded pairs of E-cadherin exhibit a remarkable hierarchy of mechanical strengths. *Proc. Natl. Acad. Sci. USA* **101**: 16472–16477.

Shao, J. Y. and Hochmuth, R. M. (1996) Micropipette suction for measuring piconewton forces of adhesion and tether formation from neutrophil membranes. *Biophys. J.* **71**: 2892–2901.

Smith, M. J., Berg, E. L., and Lawrence, M. B. (1999) A direct comparison of selectin-mediated transient, adhesive events using high temporal resolution. *Biophys. J.* **77**: 3371–3383.

Somers, W. S., Tang, J. Shaw, G. D., and Camphausen, R. T. (2000) Insights into the molecular basis of leukocyte tethering and rolling revealed by structures of P- and E-selectin bound to SLe(X) and PSGL-1.[erratum appears in *Cell* 2001 Jun 29;105:971]. *Cell* **103**: 467–479.

Spillmann, C., Osorio, D., and Waugh, R. E. (2002) Integrin activation by divalent ions affects neutrophil homotypic adhesion. *Ann. Biomed. Eng.* **30**: 1002–1011.

Spillmann, C. M., Lomakina, E., and Waugh, R. E. (2004) Neutrophil adhesive contact dependence on impingement force. *Biophys. J.* **87**: 4237–4245.

Springer, T. A. (1994) Traffic signals for lymphocyte recirculation and leukocyte emigration: the multistep paradigm. *Cell* **76**: 301–314.

Tees, D. F., Waugh, R. E., and Hammer, D. A. (2001) A microcantilever device to assess the effect of force on the lifetime of selectin–carbohydrate bonds. *Biophys. J.* **80**: 668–682.

Ting-Beall, H. P., Needham, D., and Hochmuth, R. M. (1993) Volume and osmotic properties of human neutrophils. *Blood* **81**: 2774–2780.

Tsai, M. A., Frank, R. S., and Waugh, R. E. (1993) Passive mechanical behavior of human neutrophils: Power-law fluid. *Biophys. J.* **65**: 2078–2088.

van Kooyk, Y. and Figdor, C. G. (2000) Avidity regulation of integrins: the driving force in leukocyte adhesion. *Curr Opin. Cell Biol.* **12**: 542–547.

Watson, S. R., Imai, Y., Fennie, C., et al. (1990) A homing receptor-IgG chimera as a probe for adhesive ligands of lymph node high endothelial venules. *J. Cell Biol.* **110**: 2221–2229.

Williams, T. E., Nagarajan, S., Selvaraj, P., and Zhu, C., (2000) Concurrent and independent binding of Fcgamma receptors IIa and IIIb to surface-bound IgG. *Biophys. J.* **79:** 1867–1875.

Yeung, A. and Evans, E. (1989) Cortical shell-liquid core model for passive flow of liquid-like spherical cells into micropipets. *Biophys. J.* **56:** 139–149.

Zhang, X., Wojcikiewicz, E., and Moy, V. T. (2002) Force spectroscopy of the leukocyte function-associated antigen-1/intercellular adhesion molecule-1 interaction. *Biophys. J.* **83:** 2270–2279.

Zhang, X., Craig, S. E., Kirby, H., Humphries, M. J., and Moy, V. T. (2004) Molecular basis for the dynamic strength of the integrin {alpha}4ß1/VCAM-1 interaction. *Biophys J.* **87:** 3470–3478.

Zollner, O., Lenter, M. C., Blanks, J. E., et al. (1997) L-selectin from human, but not from mouse neutrophils binds directly to E-selectin. *J. Cell Biol.* **136:** 707–716.

6

DYNAMICS OF THE NEUTROPHIL SURFACE DURING EMIGRATION FROM BLOOD

THOMAS R. GABORSKI and **JAMES L. MCGRATH**

Department of Biomedical Engineering, University of Rochester, Rochester, New York, USA

Neutrophils move from flowing blood to extravascular tissue by dynamically regulating the presentation of endothelial receptors on their surface. Receptor number and accesibility are both controlled in this process, but the relative importance of these parameters and the timing of surface changes has been difficult to discern. In this chapter we review what is known about the dynamics of the neutrophil membrane during it emigration from blood, and we explore the techniques of modern light microscopy that should allow us to answer remaining questions through direct measurements on living neutrophils.

I. INTRODUCTION

As one of the earliest events in the inflammation response, neutrophils leave the blood so they can police tissues and consume bacterial invaders (Fig. 6.1).

To emigrate from blood, neutrophils execute a sophisticated program of changes to their surface that causes them to roll first, halt, and then migrate on and through the endothelial barrier that separates blood and tissue. The full progression of events that transform neutrophils from quiet passengers in the blood to aggressive stalkers in tissue is most commonly called the *recruitment cascade*.

This chapter is concerned with how changes in the neutrophil membrane facilitate neutrophil recruitment, and has two main objectives. The first objective is to review what is known about biophysical changes of the neutrophil surface during emigration so as to identify the key questions that remain. This material is covered in Sections III through V. When possible, results from human neutrophils in the literature are highlighted. As these terminally differentiating cells do not lend themselves to genetic manipulations, data on neutrophil-like cell lines and related white blood cells are also included. We see that although biochemical, genetic, and biophysical manipulations of neutrophils and their

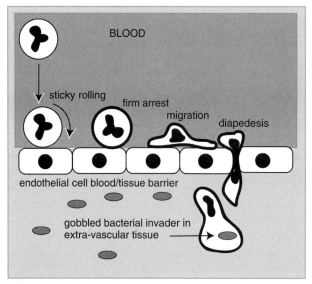

FIGURE 6.1 Neutrophil recruitment cascade. In response to inflammatory signals or infection, neutrophils leave the blood to patrol the extravascular space. The neutrophils proceed through discrete phases in which they exhibit rolling attachment to endothelium, firmly arrest atop the endothelium, and then migrate over and through the endothelial cells and into the extravascular tissue.

fellow leukocytes have taught much about the emigration program, the definitive lessons have most often come from looking directly at the living neutrophil surface. Thus, the second objective of this chapter is to explore the potential of modern and future techniques in light microscopy to answer the key questions, which is the topic of Section VI. Before taking a detailed look at the changing neutrophil surface during emigration, we begin with an overview of the cascade and an introduction to the adhesion molecules that make it work.

II. OVERVIEW OF THE NEUTROPHIL RECRUITMENT CASCADE

The movement of neutrophils from the blood to extravascular tissue is mediated by sequential phases of adhesion (Fig. 6.2).

Initially, an inactive neutrophil sticks loosely to the apical surface of the endothelium, and shear forces imparted by flowing blood cause the cell to roll. During this phase, adhesion between the endothelium and the neutrophil is mediated by glycoproteins on both surfaces. On the neutrophil surface, the key glycoproteins are PSGL-1 and L-selectin. Despite their similar functions, PSGL-1 and L-selectin are structurally very different. PSGL-1 is an integral membrane proteoglycan with many sugar-rich branches radiating from a random coil protein core. L-selectin is a member of the selectin family of integral membrane proteins. The extracellular portions of selectins contain a chain of repeat domains capped by a carbohydrate-binding lectin domain. PSGL-1 binds to the lectin domains of endothelial P-selectin and E-selectin, and the lectin domain of L-selectin binds to transmembrane proteoglycans on endothelial cells. Thus, PSGL-1 and L-selectin are actually members of complementary classes of glycoproteins that exist on both endothelial and neutrophil membranes to create velcro-like adhesion during rolling.

A

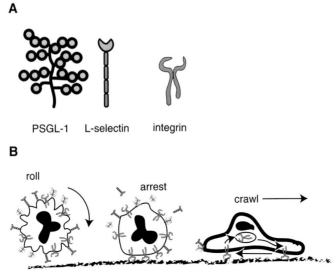

PSGL-1   L-selectin   integrin

B

roll

arrest

crawl ⟶

FIGURE 6.2 Key adhesion molecules for neutrophil recruitment. (A) Schematic structures of the glycoproteins P-selectin glycoprotein ligand (PSGL-1) and L-selectin and the integrins. (B) The glycoproteins facilitate rolling by localizing to the tips of neutrophil membrane folds. On activation, integrins first cluster at the site of adhesion to arrest rolling. The integrins then become part of a treadmilling adhesion machine, where new adhesions formed near the leading edge remain attached to the substrate as the cell glides over them. Eventually the integrins must be recycled to avoid accumulating at the back of the cell. In contrast, the glycoproteins are found in the uropod at the trailing edge of crawling cells.

The intimacy of the neutrophil and endothelial surfaces during rolling ultimately leads to neutrophil activation and arrest. Chemokines are specific carbohydrate sequences found in endothelial cell glycoproteins and are often the triggers for neutrophil activation. Many soluble factors secreted by previously activated endothelial and blood cells can also activate neutrophils. With time the activated neutrophil adheres via β_2-integrins (LFA-1 and Mac-1) to the endothelial IgG-family intercellular adhesion molecule-1 (ICAM-1). The strength of integrin-based adhesions grows until the neutrophil can adhere firmly and begin to actively spread and migrate over the endothelial surface. Integrins are essential anchors to the substrate during migration and will stay affixed to the substrate as the neutrophil advances. These dynamics naturally lead to the accumulation of integrins at the rear of the cell and suggest the need for a mechanism of integrin recycling to the leading edge during steady-state crawling (Fig. 6.2B). Although β_2-integrins are important for crawling over endothelium, it is the β_1- and β_3-integrins that bind extracellular matrix and allow the cell to crawl in the extravascular space (Table 6.1).

III. RECEPTORS AND MEMBRANES BEFORE ACTIVATION

A. Topology and Evidence of Receptor Segregation

The plasma membrane of the resting neutrophil has numerous folds that give the cell's surface the topology of a mountainous landscape. So folded is the membrane that roughly half of it must be sucked into a pipette before the remaining membrane is taut (Ting-Beall et al., 1993). The topological

TABLE I   Properties of Neutrophil Adhesion Molecules

Adhesion molecule (alternative name)	Ligand	Estimated Concentration (rest/activated)	Localization (rest/activated)
L-selectin (CD62L)	Sialyl-Lewis X	~45,000 (Rebuck and Finn, 1994); 20–100% loss with shedding depending on treatment and cell type (Ivetic et al., 2004; Phong et al., 2003)	Tips of membrane folds in unactivated cells (Erlandsen et al., 1993); uropod in crawling cells (Junge et al., 1999)
PSGL-1 (CD162L)	P-selectin, also L-selectin and E-selectin	25,000 binding sites for P-selectin on neutrophils (Ushiyama et al., 1993)	Tips of membrane folds in unactivated cells (Moore et al., 1995); uropod in crawling cells (Alonso-Lebrero et al., 2000)
$\alpha_L \beta_2$ (LFA-1; CD11a/CD18)	ICAM-1	~25,000 (Bikoue et al., 1997)/~ no change with activation (Tandon et al., 2000)	Between membrane folds in unactivated cells (Tohya and Kimura, 1998); at sites of adhesion in activated cells
$\alpha_M \beta_2$ (MAC-1; CD11b/CD18)	ICAM-1, Fibrinogen	~45,000 (Bikoue et al., 1997)/~150+% increase with activation (Tandon et al., 2000)	Between membrane folds in unactivated cells (Erlandsen et al., 1993); at sites of adhesion in activated cells
$\alpha_4 \beta_1$ (VLA-4; CD49d/CD29)	VCAM-1, fibronectin	Very little expression on neutrophils (Johnston and Kubes, 1999)	Tips of membrane folds in unactivated cells (Berlin et al., 1995); at sites of adhesion in crawling cells

variation of the neutrophil's surface is key to the rolling phase of the adhesion cascade. Numerous studies have revealed that L-selectin (Bruehl et al., 1996; Erlandsen et al., 1993; Ivetic et al., 2004; Pavalko et al., 1995; Tohya and Kimura, 1998) and PSGL-1 molecules (Moore et al., 1995) reside almost exclusively at the tips of membrane folds. There is also evidence – although considerably less extensive – that β_2 integrins reside in the valleys between folds (Erlandsen et al., 1993; Tohya and Kimura, 1998). In contrast, the α_4 subfamily of integrins can be found at the tips of lymphoid cells (Berlin et al., 1995). This topological separation of adhesion molecules should facilitate selectin and α_4-integrin initial adhesion while it sequesters β_2-integrins away from the endothelial surface (Fig. 6.2) Later in the cascade when the neutrophil activates and spreads on the endothelium, the surface likely flattens so that these integrins gain access to their counterreceptors.

The evidence for a topological partitioning of glycoproteins (L-selectin and PSGL-1) and integrins in resting neutrophils comes from immunogold labeling and electron microscopy. In this technique, neutrophils are labeled with antibody-coupled gold beads to create an electron-dense marker of adhesion molecules. In studies of this type, images invariably show L-selectin and PSGL-1 markers clustered on the tips of membrane protrusions (Erlandsen et al., 1993; Moore et al., 1995). One concern is that antibody-coupled beads might bind the tips of ruffles for the same reason that endothelial cells do — because they are the most topologically accessible portions of the neutrophil membrane. However, the evidence that integrins lie in the complementary membrane domains on the body is derived from the same methods, and so it appears that immunogold labels can access the full membrane in these studies. Because of the finger-like appearance of membrane protrusions in thin cell sections prepared for transmission electron microscopy (TEM), the membrane folds of neutrophils are frequently called microvilli in the literature. However, scanning electron microscopy (SEM), which gives a complete picture of the neutrophil surface, reveals that virtually all the folds on a neutrophil membrane are ridgelike ruffles rather than microvilli (Majstoravich et al., 2004).

B. Potential Mechanisms of Adhesion Molecule Positioning

What factors position L-selectin and PSGL-1 at the tips of membrane folds? In lymphocytes the answer appears to be members of the ezrin–radixin–moesin (ERM) family of proteins. These widely expressed submembraneous proteins have been most studied in epithelial cells, where they directly link membrane receptors to the parallel actin filaments within microvilli (Bretscher et al., 2002). Unlike neutrophils, the surface of lymphocytes is covered with finger-like protrusions that appear to be true microvilli. Recent studies have found that the ERM proteins ezrin and moesin bind directly to L-selectin (Ivetic et al., 2002) and are required for positioning L-selectin at the tips of lymphocyte microvilli (Ivetic et al., 2004). Leukocyte studies investigating the redistribution of PSGL-1 in response to chemoattractant found interactions with moesin and ezrin as well (Alonso-Lebrero et al., 2000). Members of the ERM family, however, have not yet been localized to neutrophil membrane folds. In fact, ruffles and microvilli are structures controlled differently by members of the Rho family of GTPases; active RhoA is required for microvillus formation (Oshiro et al., 1998), whereas active Rac1 is needed for ruffles and is even reported to collapse the microvilli of lymphocytes (Nijhara et al., 2004). Thus, although L-selectin is positioned at the tips of membrane folds in both lymphocytes and neutrophils, the mechanism of anchoring may be different. Another possible protein anchor for L-selectin in neutrophils is α-actinin, an actin-crosslinking protein that has been shown to associate directly with the cytoplasmic domain of L-selectin. Data on lymphocytes have shown that α-actinin is not required for microvillus positioning of L-selectin (Pavalko et al., 1995), possibly because of redundant function with ERM proteins. The α-actinin anchor may have a more essential role for positioning on the different types of protrusions in neutrophils. Although there are limited data on PSGL-1–cytoskeleton interactions related to topological positioning, the similarities with L-selectin suggest that related mechanisms are possible.

The functional importance of L-selectin positioning was well demonstrated by von Andrian and colleagues (1995). They found that although L-selectin was found primarily on microvillus tips of a leukocyte cell line, another adhesion receptor, CD44, was concentrated on the body. A chimera consisting of the L-selectin extracellular domain and the CD44 transmembrane and cytoplasmic domains localized to the cell body like CD44, and a second chimera with a CD44 extracellular domain and L-selectin transmembrane and cytoplasmic domains positioned to the microvilli. While the "misplaced" L-selectin/CD44 chimera was capable of adherence under static conditions, it could not support initial attachment for cell rolling as well as normal L-selectin under shear conditions. This work demonstrated both the importance of microvillus positioning for L-selectin-based rolling and the key role of the L-selectin cytoplasmic domain.

What sequesters β_2-integrins to regions between membrane folds? Again, cytoskeletal connectors may provide part of the answer. The actin-binding proteins filamin, α-actinin, and talin are found in neutrophils and can each simultaneously bind to an actin filament and to the cytoplasmic span of a β_2 subunit (Sampath et al., 1998; Watts and Howard, 1994). Investigating only two of these candidates, Pavalko and colleagues found that talin, but not α-actinin, was constitutively associated with β_2 subunits in resting neutrophils (Sampath et al., 1998). Given different cytoskeletal connectors for L-selectin and β_2-integrins, it is possible that the neutrophil cytoarchitecture segregates L-selectin from β_2-integrins; however, we have yet to consider the important ability of the plasma membrane to control the location of its constituent proteins.

The revolutionary description of the plasma membrane as a two-dimensional liquid has been modified by the understanding that not all of the membrane's regions are created equal. Specifically, sphingolipids and cholesterol associate into cohesive domains called *rafts* that resist extraction by cold detergent and emerge as floating fractions in sucrose gradients (Simons and Vaz, 2004) (see Chapter 9). These operational definitions have generated considerable controversy, but there is strong evidence that these detergent-resistant domains are specialized platforms that organize signaling proteins and other molecules (Simons and Ikonen, 1997). Studies in leukocytes indicate that both L-selectin (Phong et al., 2003) and LFA-1 experience structural changes that promote their inclusion within rafts after cell activation (Leitinger and Hogg, 2002), but not before. Although there is little evidence for raft organization of these receptors in resting neutrophils, immunogold L-selectin clusters are found even after microvilli are dissolved by neutrophil treatment with cytochalasin B (Finger et al., 1996). This result suggests some organization of the receptors is independent of the cytoskeleton.

C. L-Selectin Shedding

Soon after activation, the extracellular domains of L-selectin molecules are cleaved by contact with a member of the disintegrin and metalloprotease (ADAM17) family of surface proteases (Borland et al., 1999; Peschon et al., 1998). Cleavage can occur only if L-selectins and proteases are within molecular reach of each other. One study in lymphocytes reports the movement of L-selectin to protease-rich lipid rafts after activation (Phong et al., 2003). This suggests a mechanism wherein L-selectin and ADAM17 are segregated in the resting neutrophil, and domain mixing stimulates shedding on activation. In contrast, a genetic study in lymphocytes shows that when the ERM binding domains of L-selectins are deleted, L-selectins are not only displaced from microvillus tips, but also shed less (Ivetic et al., 2002). This suggests that ADAM17 is already co-localized with L-selectin in microvilli and that smaller-scale movements trigger cleavage. Too little is known about the regulation and localization of ADAM17 activity to reconcile the apparent contradiction

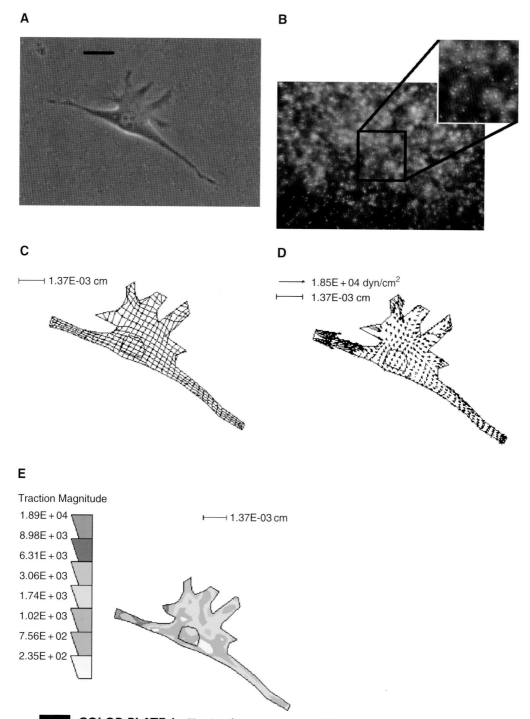

COLOR PLATE I Traction force microscopy calculates forces exerted by a cell on its substrate based on the displacements of marker beads within the surface of the substrate. (A) An image of a single cell is taken (bar = 20 μm) followed by (B) images of the bead field in its stressed state, due to the adherent cell (red), and its unstressed state, after the cell is removed (green). Yellow beads have not displaced from one image to the next. Inset: Magnified view of the bead field with the background subtracted for visual clarity (image processing courtesy of M. Mancini). (C) The cell is divided into a mesh, and (D) traction forces are calculated at each node of the mesh. (E) Color contour plot of the traction field. Note: Substrate depicted is polyacrylamide. (See Figure 1.2)

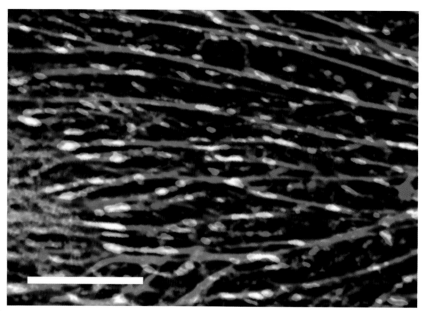

COLOR PLATE 2 Adhesion sites connect the intracellular architecture to the ECM to serve as mechanochemical links between the inside and outside of the cell. In a subregion of an endothelial cell, discrete adhesion sites containing vinculin (green) orchestrate a continuous connection from the actin cytoskeleton (red) to the fibronectin matrix (blue). Bar = 10 μm. (See Figure 2.1)

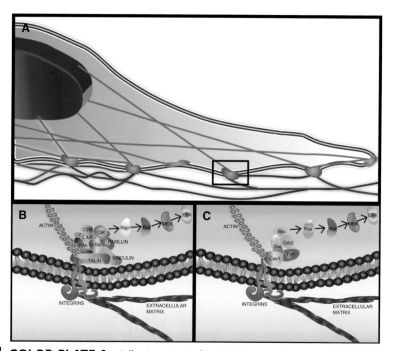

COLOR PLATE 3 Adhesion sites induce MAPK signaling networks. (A) Transmission of tension from the actin cytoskeleton (red) to the ECM (purple) through focal adhesion sites (green) represents one mechanism of mechanotransduction. Complex interactions among the molecular components of a single adhesion site (black box) lead to ERK activation through at least two distinct signaling networks. (B) FAK activation serves as the critical initial step for assembly of myriad adapter and signaling proteins. (C) Assembly of the Shc–Grb2–Sos complex serves to activate Ras and Raf. (See Figure 2.5)

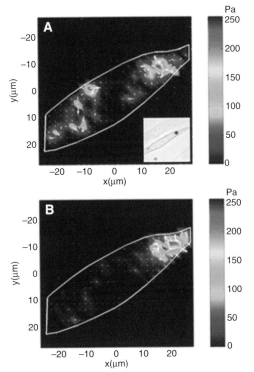

COLOR PLATE 4 Evidence of action at a distance affecting living cells. Stress maps in living human airway smooth muscle cells were obtained using the intracellular stress tomography technique (Hu et al., 2003). Stress was applied to the cell by twisting a ferromagnetic bead bound to integrin receptors on the apical cell surface. The bead position is shown on the phase-contrast image of the cell (A, inset); the black dot on the image is the bead. The white arrows indicate the direction of the stress field; the color map represents the magnitude of stress. Stress does not decay inversely with increasing distance from the bead center. Appreciable islands of stress concentration could be seen >20 μm from the bead (A), consistent with the action-at-a-distance effect. After disruption of the actin network by cytochalasin D (1 μg/ml), the stress concentration could be detected only in the local vicinity of the bead (B), demonstrating that connectivity of the actin CSK lattice plays an important role in intracellular stress distribution. Reprinted, with permission, from Hu et al. (2003). (See Figure 4.3)

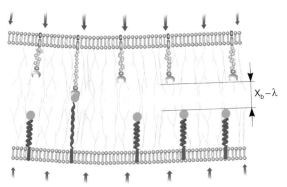

COLOR PLATE 5 Conceptual picture of steric limitations on bond formation. Glycosylation of membrane proteins creates a repulsive layer of polysaccharide on the cell surface. If adhesion molecules have a shorter extended length than the thickness of the layer, the layer must be compressed for opposing molecular pairs to come into close contact. The instantaneous length of a molecule extending from the surface is a stochastic quantity that depends on the length and flexibility of the molecule. Thus, the probability of bond formation depends on the length and flexibility of the adhesion molecules, the membrane separation distance, and mechanical forces acting to overcome the surface repulsion. (See Figure 5.1)

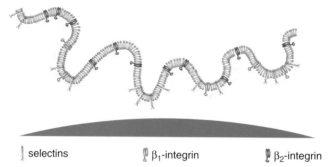

| selectins | β_1-integrin | β_2-integrin |

COLOR PLATE 6   Surface topography may also act to limit close approach of molecules for bond formation. As little as 5 to 10% of a macroscopic contact region may actually be in physical contact with an opposing surface. The microvilli that cover the cell surface are deformable and are expected to change shape under mechanical loading. Thus, mechanical forces may facilitate bond formation by increasing the area of close contact between the cell membrane and an adhesive substrate. (See Figure 5.2)

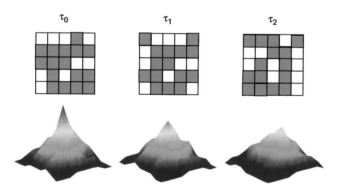

COLOR PLATE 7   Image cross-correlation spectroscopy (ICCS). ICCS quantitatively determines the similarity of two images through cross-correlation. The green checkerboards are an example of a punctate distribution of molecules, which varies over time due to diffusion. Each subsequent frame varies more from the first as the domains move from their original locations (τ_0–τ_1 vs τ_0–τ_2) The mathematical similarity between the images is related to the amplitude of the cross-correlation function. Analysis of the amplitude decay can determine the diffusion rate. ICCS can also be used to compare degrees of co-localization. If τ_1 and τ_2 were different receptors expressed on the cell surface, ICCS would show that τ_1 is more co-localized with τ_0, than is τ_2. (See Figure 6.4)

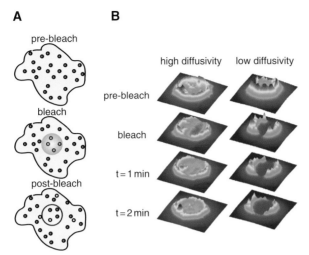

COLOR PLATE 8 Fluorescence recovery after photobleaching (FRAP). (A) Schematic illustrating a cell with a uniform distribution of surface molecules that are fluorescently labeled. A high-intensity light source such as a focused laser is used to photobleach a specific region of fluorophores in the image, typically a circlular (A) or rectangular (B) band. The rate of recovery of fluorescence intensity can be used to mathematically determine the lateral diffusion coefficient of the labeled species. (B) A resting neutrophil is labeled with fluorescent antibodies for two adhesion molecules and imaged near the center plane. The color relief represents the fluorescence intensity as recorded by a cooled CCD camera. This neutrophil is photobleached with a rectangular band of laser excitation so that diffusion can be modeled as one-dimensional. By the end of the time series, it is clear that the first species (left) is highly mobile compared with the second (right). (See Figure 6.6)

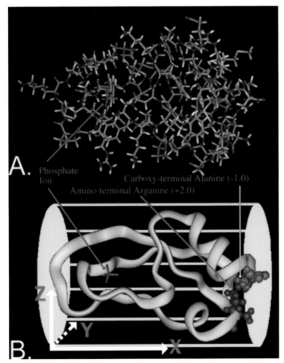

COLOR PLATE 9 Standard wireframe (A) and ribbon (B) representations of the bovine pancreatic trypsin inhibitor (BPTI) protein from the "noninteracting system" at $t = 0$. In (B), the phosphate ion has been left in wireframe format, whereas the amino-terminal arginine (red) and carboxy-terminal alanine (purple) are shown in CPK format and have valence charges of $+2.0$ and -1.0, respectively. The generalized cylindrical shape is shown in (B), with the long axis considered as the X direction, whereas the short axis is taken to be in the Y direction. (See Figure 8.1)

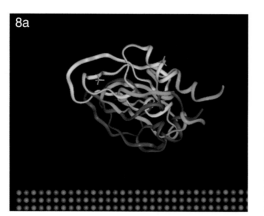

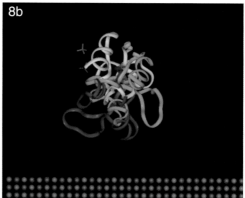

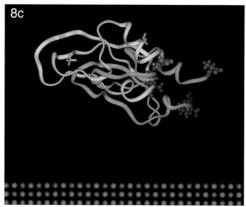

COLOR PLATE 10 (a) Ribbon overlay, in the XZ plane of interacting system 2 at 1040 ps (blue) and the "noninteracting" system at 0 ps (yellow ribbon) versus 1040 ps (green ribbon). The "noninteracting" system and interacting system 2 are seen to merge toward a similar configuration with respect to the MgO surface. (b) Ribbon overlay, in the YZ plane, identical to (a) except that the BPTI systems have been rotated by 90°, system 2 is at 1040 ps (blue), and the "noninteracting" system is at 0 ps (yellow) versus 1040 ps (green). It is apparent that the "noninteracting" system has rotated by about 90° in the direction of interacting system 2. (c) Rotation of the "noninteracting" system around the Y axis results in movement of the protein's charged termini toward the MgO surface. The yellow ribbon represents $t = 0$; the green ribbon represents $t = 1040$ ps. The amino-terminal arginine (red CPK) and carboxy-terminal alanine (purple CPK) have valence charges of +2.0 and −1.0, respectively. As discussed in the text, extending the simulation to 1040 ps has caused us to reevaluate this system. It now appears that interaction occurs even though the protein was initially placed 20 Å from the MgO surface to be a "noninteracting" control. In both the "noninteracting" system and system 2, the protein molecule rotates about its short axis so as to orient its charged amino and carboxy termini toward the MgO surface. (See Figure 8.8)

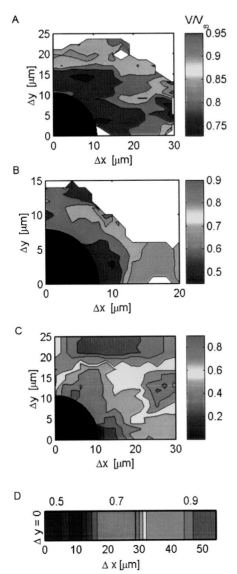

COLOR PLATE 11 (A) Rolling velocity of pairs of sLe^x-coated microspheres (diameter = 10.9 μm) flowing over a P-selectin–presenting surface in a parallel-plate flow chamber. (B) Rolling velocity of pairs of leukocytes in a mouse venule (diameter = 37 μm). (C) MAD simulations for the same conditions as in (A). (D) Rolling velocity of pairs of leukocytes in a hamster venule (diameter = 22 μm). V/V_∞ represents the dimensionless rolling velocity calculated from the rolling velocity of an isolated bead (or leukocyte), V_∞. (See Figure 12.6)

between these recent studies. PSGL-1 is also shed from the neutrophil surface on activation. Interestingly, inhibitors of L-selectin shedding do not inhibit PSGL-1 downregulation (Davenpeck et al., 2000). So, although PSGL-1 is positioned similarly to L-selectin in resting cells and also shed on activation, the shedding mechanisms may differ.

IV. CHANGES IN INTEGRIN-BASED ADHESION

The purpose of early neutrophil activation is to arrest rolling by creating strong adhesion between the neutrophil β_2-integrins and their endothelial ICAM-1 counterreceptors. In principle, adhesion can be enhanced either by increasing the number of integrins available for binding without changing the state of individual integrins or by improving the ability of individual integrins to bind without changing their numbers. Although neutrophils and the other leukocytes clearly use these modes in combination to promote arrest, the individual mechanisms have been called the *valency* and *affinity* modes of regulation, respectively (Carman and Springer, 2003). As *valency* is more traditionally used to describe the number of binding sites on individual molecules, we substitute the term *availability* for this mode of adhesion modulation. The total adhesion potential of a surface is the product of *availability* and *affinity*, commonly referred to as *avidity*.

In this section we examine the events of early neutrophil activation. We use the categories of *availability* and *affinity* to organize the major mechanisms thought to shift from selectin-based to β_2-integrin-based adhesion. Our primary focus is on the changes in receptor distributions, mobility, and membrane topology that lead to increased availability of integrins for binding (van Kooyk and Figdor, 2000). In contrast to these biophysical changes, the modulation of integrin affinity occurs primarily through signaling pathways that change the conformational state of integrins. Given the biophysical focus of this review, our discussions of affinity modulation are brief. For a more complete discussion of the signaling mechanisms regulating integrin affinity, the reader is referred to one of the many excellent reviews (see Suggested Reading).

A. Mechanisms That Increase Integrin Availability

1. Changes in Surface Morphology and Expression

Given the topological assignment of integrins to domains between membrane folds, one obvious means for enhancing integrin access to the endothelium is to collapse the folds. Even with a uniform distribution of receptors, two smooth surfaces will contact over a significantly greater area than a pair of rough surfaces. Thus, membrane flattening at neutrophil adhesion zones should both present previously sequestered integrins *and* increase the surface area available for adhesion.

A recent study by Zhu and colleagues nicely illustrates the enhanced adhesion potential of smooth contacting surfaces (Williams et al., 2001). In their study, a particular F_c receptor was inserted into the smooth membranes of red blood cells (RBCs), and into the rough membranes of two established cell lines. Using IgG-coated red cells as the target surface, the group found that the apparent affinity for RBC–RBC binding was 50-fold higher than for binding between the RBCs and either cell line, providing direct evidence for a greater adhesion potential between flattened surfaces.

Another important mechanism of availability regulation is the direct increase in the total number of β_2-integrins expressed on the neutrophil surface. In one set of estimates, the number of β_2-integrins on the surface of a neutrophil increases twofold from the resting state, with virtually all of the new integrins being Mac-1 (Tandon et al., 2000). New integrins are inserted into the plasma membrane by storage granule secretions that occur within minutes of activation (Gamble et al., 1985).

2. Clustering

In addition to changing the total number of integrins available on the neutrophil surface, there is widespread evidence that activated neutrophils concentrate integrins in local patches or "clusters." The potential for strong adhesion follows if clusters occur in zones of adhesive contact (Ward et al., 1994). In practice, clusters of two different sizes can be defined: *macroclusters,* which are detected as a micron-scale local increase in fluorescence markers on the neutrophil surface in light microscopy; and *microclusters,* which occur below the resolution of light microscopy (<200 nm) but can be detected by suboptical techniques such as fluorescence resonance energy transfer (FRET) (see Section VI). Importantly, microclusters have been detected by FRET under conditions where macroclusters are not seen by microscopy (Kim et al., 2004). One could also expect macroclusters to exist independent of microclusters because a patch of fluorescence in a light micrograph does not require that the constituent fluorophores be spaced within the 1–10 nm required for a FRET signal.

By what mechanisms might integrins organize into clusters? One possibility is that the receptors segregate into membrane rafts differently before and after activation. Unfortunately, existing data on lymphocytes appear to disagree over whether LFA-1 is associated with rafts before activation. In their study, Leitinger and Hogg (2002) conclude that LFA-1 co-localizes with raft markers after activation, but in resting cells the molecules are excluded from rafts by cytoskeletal associations. This leads to the idea that cytoskeletal reorganization on neutrophil activation could liberate LFA-1 for inclusion in rafts (Anderson et al., 2000). In contrast to the findings of Leitinger and Hogg, Krauss and Altevogt (1999) reported that LFA-1 is in the raft fractions of *resting* T lymphocytes. By chelating cholesterol, both groups find that raft integrity is required for LFA-1-based adhesion; however, this finding is contradicted by a third study (Shamri et al., 2002). These differing results may arise from differences in cell types and experimental conditions, but they may also arise from strong differences in the methods used to detect the association of LFA-1 with rafts. The fact that lipid rafts are composite structures and procedurally defined implies that their identification is highly sensitive to the details of methodology. The lack of standardized methods may contribute to the confusion over the relationship between β_2-integrins and lipid rafts.

Because integrins bind through linker molecules to actin filaments, a second possible clustering mechanism involves actin remodeling during activation to create concentrated sites for integrin binding. This idea is consistent with the finding in many cell systems that integrin ligation can trigger signals that promote actin polymerization (Shattil et al., 1998). Thus, a small zone of adhesion could grow through positive feedback, where integrin ligation drives actin assembly, and actin assembly captures diffusing integrins into the contact zone.

Unlike raft-based clusters, the cytoskeletal clustering mechanism requires integrin ligation. In fact, the idea that clusters may form in the absence of adhesion is controversial (Carman and Springer, 2003). The evidence for adhesion-independent clustering comes from fluorescence images of individual cells isolated from suspensions in which they were labeled, activated, and fixed. Because leukocytes that express LFA-1 also typically express ICAM, its ligand, many experiments in which leukocytes are activated in concentrated suspensions should naturally lead to homotypic adhesion between cells (Kim et al., 2004). If the cells are fixed in this state but then mechanically processed so as to break adhesions, macroclusters will appear on isolated cells at the former contact zones. In another class of experiments, clusters are induced through antibody-based crosslinking. Artificial clusters will be naturally created whenever multivalent and polyclonal antibodies bind mobile receptors. These clusters may mimic the crowding of integrins at zones of contact, but they cannot be taken as evidence of natural adhesion-independent cluster mechanisms.

3. Diffusion

The lateral mobility of adhesion molecules is paramount for cell–cell or cell–substrate adhesion (Bell, 1978). We have already alluded to the need to recruit integrins to clusters via membrane diffusion. When a neutrophil makes contact, there are a finite number of adhesion molecules that populate the contact zone. Before the cell rolls or translates away, there is a period during which highly mobile receptors can diffuse into the contact zone. With enough ligand present, the substrate will act like a sink, drawing in more integrin receptors. Clearly, the faster that integrins move laterally within the neutrophil membrane, the more likely it is that adhesion will strengthen and cause arrest.

Despite the importance of β_2-integrin mobility for adhesion, it is not immediately obvious that the integrins should diffuse. Data indicate that the cytoplasmic tails of integrins can physically link to the cytoskeleton in both the resting and activated states (Lub et al., 1997; Sampath et al., 1998; van Kooyk et al., 1999), and such anchors would seem to render the receptors immobile. However, direct measures of LFA-1 mobility in lymphocytes (Kucik et al., 1996) reveal two things: (1) only a fraction of the integrins can be described as "immobile" in the resting state, and (2) on activation, the fraction that is clearly mobile not only increases, but also diffuses an order of magnitude faster than in the resting state. Thus, integrins appear to release from their cytoskeletal tethers to facilitate adhesion. One idea is that cell-activated pathways break cytoskeletal anchors to promote integrin diffusion. Sampath et al. (1998) provide compelling biochemical support for this hypothesis. Their study in neutrophils demonstrated that β_2-integrins associate with the actin-crosslinking protein talin prior to activation, but with α-actinin afterward. They further showed that talin is proteolytically cleaved by the calcium-dependent protease calpain. Although the group did not perform direct measurements of integrin mobility, they did demonstrate that calpain activity was required for integrin clusters to form.

B. Regulation of Integrin Affinity

As mentioned, the balance of the literature on integrin adhesion regulation is dedicated to understanding affinity modulation (Harris et al., 2000). The core findings are that (1) individual integrins clearly demonstrate both low and high affinity for ligands, (2) intracellular signals are responsible for the change from low to high affinity (Constantin et al., 2000; van Kooyk and Figdor, 2000), and (3) affinity increases arise because conformational changes are mechanically transduced from the cytoplasmic to the extracellular domains of the integrins. Most relevant for this review is that β_2-integrins on resting leukocytes are locked in the low-affinity conformation and that activation compels a switch to high-affinity binding of ICAM-1. Indeed, the modulation of receptor availability by clustering, diffusion, or surface modulation appears to be of little use if individual integrins are experimentally locked out of high-affinity conformations (Carman and Springer, 2003). Interestingly, in studies of β_1- and β_3-integrins, talin is essential for integrin activation. Current models suggest that cleavage by calpain or binding by the lipid PtdIns(4,5)P$_2$ causes integrins to switch to their active conformations (Calderwood, 2004). Similarly, L-plastin, a leukocyte-specific actin-binding protein of the α-actinin family, is phosphorylated in response to inflammatory stimulation leading to Mac-1 activation (Jones et al., 1998). Thus, although affinity modulation may be essential for strong adhesion, in normal circumstances the mechanisms for modulating receptor availability and affinity may be inextricably linked.

V. CRAWLING NEUTROPHILS

On activation, the neutrophil membrane becomes highly dynamic. The numerous but small membrane folds of the resting surface give way to a few large protrusions as the cell spreads to firmly adhere to, and then migrate on the endothelium (Zigmond and Sullivan, 1979). Driving these changes in cell shape are dynamic events in the actin cytoskeleton. Within 10 min of activation, the amount of actin polymerized in the neutrophil first doubles and then returns to basal levels around the time a cell actively crawls (Howard and Oresajo, 1985; Wallace et al., 1984). Thus, periods of actin assembly and disassembly are both needed to transform from the resting to the crawling cell morphology.

Not all L-selectin and PSGL-1 molecules are shed from the neutrophil surface with activation, and the surviving receptors typically pool into a single, subhemispherical "cap" on the activated cell (Alonso-Lebrero et al., 2000; Green et al., 2004; Junge et al., 1999). In migrating cells, L-selectin and PSGL-1 molecules localize exclusively to the trailing membrane or *uropod*. Uropod "cap" structures are a well-known hallmark of the polarized morphology of crawling leukocytes, and recent work localizes numerous raft protein and lipid markers to these structures (Pierini et al., 2003). Furthermore, the major raft constituent cholesterol is essential for uropod cap formation and cell migration (Pierini et al., 2003). It seems likely that L-selectin and PSGL-1 are just additional constituents of the raftlike cap that is the uropod. New studies examining the co-localization of L-selectin and PSGL-1 with raft constituents and the cholesterol requirements for L-selectin caps would confirm this.

Once the neutrophils transmigrate, β_1- and β_3-integrins are essential for adhesion and migration in the extracellular matrix of the basement membrane. These integrins bind to fibronectin, vitronectin, and laminin (Sixt et al., 2001). Active maintenance of the integrins at the leading edge of a migratory neutrophil is necessary. A recent study has shown oriented endocytic recycling of $\alpha_5\beta_1$ from the rear to the leading edge in neutrophils crawling on fibronectin-coated coverslips (Pierini et al., 2000). Recycling also plays an important role in β_2-integrin mediated migration. Mutation of the β subunit of LFA-1 impairs detachment at the rear of crawling cells and results in highly stretched morphologies. An antibody that recognizes activated LFA-1 shows a high concentration in the rear of the mutant-expressing cells, whereas wild type–expressing cells showed a high concentration at the front and body of the motile cells, supporting the notion of active recycling to the leading edge (Tohyama et al., 2003).

VI. QUESTIONS FOR LIGHT MICROSCOPY

Despite the impressive progress made in understanding the mechanisms of neutrophil recruitment, there are many questions that biochemical fractionation and traditional light and electron microscopy cannot answer. Is L-selectin really immobilized at the tips of membrane folds, or is it moving dynamically while concentrated there? Are β_2-integrins truly confined to valleys, or are the limited data on integrin position an artifact of the fixation and other processing required for electron microscopy? What molecules reside in clusters and how can we compare degrees of clustering? What is the relative impact of affinity versus the availability regulation of integrins? To answer these and other questions, the techniques that have brought the field this far must be augmented with methods that allow investigators to "see" the surface of *living* neutrophils with increasing spatial and temporal resolution. To represent a true advance, the new techniques must minimize the likelihood that the cell has been changed from its natural state. In this section we review quantitative techniques from light microscopy that fit these criteria and explore how they can be applied to achieve the next level of understanding in neutrophil recruitment.

A. Distribution and Co-localization

An ideal technique for determining which receptors reside on the ridges of live neutrophils is total internal reflection fluorescence microscopy (TIRFM) (Fig. 6.3).

In this imaging modality, only fluorophores within approximately 200 nm of the coverslip surface are excited. The laser light source is aligned such that the beam strikes the coverslip–water interface at an angle of incidence greater than the critical angle for total internal reflection. An induced evanescent wave at this interface, which decays exponentially, excites only the fluorophores within molecular reach of the coverslip, leaving all other fluorescent molecules in the cell unexcited (see Suggested Reading).

In an inverted microscope, gravity naturally holds cells against the coverslip floor of observation chambers. Thus, if integrins and glycoproteins are labeled with fluorescent antibodies and TIRFM imaging is used, only those molecules at the tips of the ruffled neutrophil membrane should fluoresce. TIRFM should be useful in determining the times after activation when β_2-integrins are capable of reacting with their ligands and in discovering if fluorescent raft components and glycoproteins are co-localized to ruffle tips in resting neutrophils. Using a general membrane label and standard fluorescence imaging in combination with

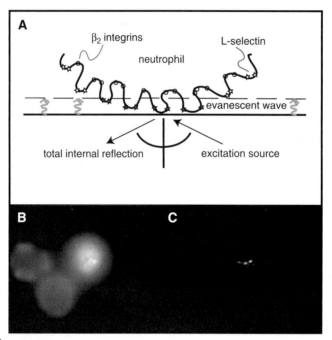

FIGURE 6.3 Total internal reflection microscopy (TIRFM). (A) Schematic illustrating how TIRFM can be used to image the ridges of a resting neutrophil. The excitation light source has a high incident angle and is totally internally reflected at the coverglass–water interface. An evanescent wave propagates in the direction of the cell with exponential decay. In a typical TIRFM setup, only fluorophores within ~200 nm of the glass surface are excited. (B) Epifluorescence image of a whole-blood dilution labeled with a reactive fluorophore. The labeled proteins on the membrane surface give near-uniform intensity. The image was taken while focusing close to the coverslip at the bottom of the resting neutrophil. Two red blood cells can be seen to the left, floating just above the focal plane. The light source was then switched to a laser aligned for TIRFM (C) Distinct punctate fluorescent image demonstrating excitation of only a few areas, probably corresponding to neutrophil ridges. Fluorescence signal from the cell body is eliminated (also note the absence of signal from the out-of-focus red cells).

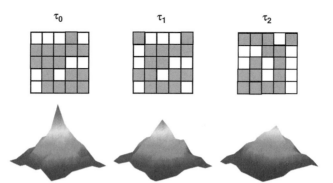

FIGURE 6.4 Image cross-correlation spectroscopy (ICCS). ICCS quantitatively determines the similarity of two images through cross-correlation. The green checkerboards are an example of a punctate distribution of molecules, which varies over time due to diffusion. Each subsequent frame varies more from the first as the domains move from their original locations (τ_0–τ_1 vs τ_0–τ_2) The mathematical similarity between the images is related to the amplitude of the cross-correlation function. Analysis of the amplitude decay can determine the diffusion rate. ICCS can also be used to compare degrees of co-localization. If τ_1 and τ_2 were different receptors expressed on the cell surface, ICCS would show that τ_1 is more co-localized with τ_0, than is τ_2. (See Color Plate 7)

TIRFM, one could infer the locations of both ridges and valleys. Furthermore, with live cells, TIRFM can capture changes in topology and adhesion molecule localization throughout neutrophil activation.

Another intriguing possibility is to combine TIRFM with image analysis to correlate receptor co-localization during neutrophil activation with membrane topology. Image cross-correlation spectroscopy (ICCS) mathematically compares two or more images for similarity (Fig. 6.4).

The cross-correlation function's shape and maximum are directly related to the pixel-by-pixel similarity of the images (Petersen et al., 1993). ICCS can quantitatively compare co-localization of multiple receptors. With this technique, changes in co-localization or macroclustering during neutrophil activation can be precisely defined. Quantitative co-localization data would be valuable in comparing rates of redistribution with various cell activation stimuli. Dose dependence curves for co-localization and redistribution could be created for the first time, helping to reveal downstream events of activation-induced signal transduction pathways. Additionally, such a rich set of images may help to elucidate how L-selectin and PSGL-1 meet their fate during shedding. Are they redistributed to domains containing sheddases as β_2-integrins take over the primary role of adhesion?

B. Clustering

Studies of integrin adhesion regulation are often plagued by vague descriptions of clustering prior to and after activation (Carman and Springer, 2003). ICCS is well suited for determining the correlation between images with

A

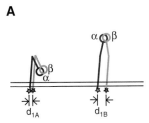

B

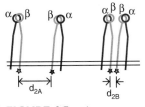

FIGURE 6.5 Fluorescence resonance energy transfer (FRET). FRET is commonly used to determine the distance between fluorescent probes tagged to biological molecules. (A) Illustration of the use of FRET to determine the conformation state of a β_2-integrin molecule. In the low-affinity state, the cytoplasmic α and β subunits are in close proximity (d_{1A}). On activation, the subunits move apart (d_{1B}) to create the new conformation, which will result in significantly lower energy transfer between the two fluorescent probes (Kim et al., 2003). (B) The β subunit of the integrin is labeled with either a green or red fluorescent probe. After any variance from equal distribution is controlled for, energy transfer can be used to examine the incidence of microclustering. In a random distribution on the surface, the molecules are spaced far enough apart that there is minimal energy transfer (d_{2A}) On clustering (d_{2B}), the amount of energy transfer between integrins increases dramatically.

punctate domains, as well as for quantifying changes in degree of clustering following stimulation. This technique should be used to quantify images that have been previously analyzed only qualitatively. However, the limit of ICCS occurs at the optical resolution of a light microscope (~200 nm). Clusters spaced less than this distance may not be resolvable as individual domains. To explore the potential involvement of lipid partitions in integrin and selectin organization, several laboratories have turned to a technique with superior spatial resolution.

In fluorescence resonance energy transfer (FRET) (Fig. 6.5), there are commonly two different fluorophore populations, referred to as the donor and acceptor. When the donor fluorophore is excited and in close proximity to an acceptor, it can nonradiatively transfer its energy to the acceptor. This process, known as Förster energy transfer, is inversely proportional to the sixth power of the separation distance. By providing spatial information with suboptical resolution, FRET is ideal for probing microclustering, dimer formation, and conformational changes of β_2-integrins on the neutrophil surface. Adhesion molecules tagged with donor and acceptor fluorophores should exhibit energy transfer when they approach one another in spatially confined compartments. Alternatively, some investigators probe the proximity of receptors with raft constituents. Combining FRET with ICCS would be a novel way of capturing dynamic movements of molecules on both cellular and molecular length scales.

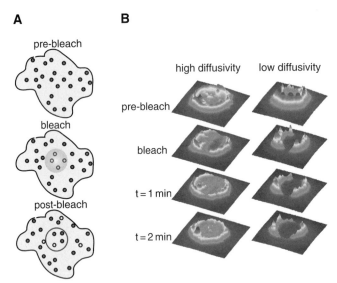

FIGURE 6.6 Fluorescence recovery after photobleaching (FRAP). (A) Schematic illustrating a cell with a uniform distribution of surface molecules that are fluorescently labeled. A high-intensity light source such as a focused laser is used to photobleach a specific region of fluorophores in the image, typically a circlular (A) or rectangular (B) band. The rate of recovery of fluorescence intensity can be used to mathematically determine the lateral diffusion coefficient of the labeled species. (B) A resting neutrophil is labeled with fluorescent antibodies for two adhesion molecules and imaged near the center plane. The color relief represents the fluorescence intensity as recorded by a cooled CCD camera. This neutrophil is photobleached with a rectangular band of laser excitation so that diffusion can be modeled as one-dimensional. By the end of the time series, it is clear that the first species (left) is highly mobile compared with the second (right). (See Color Plate 8)

C. Diffusion

Neutrophil firm adhesion is dependent on the availability of β_2-integrins to bind ligand on the endothelial surface, and availability is governed in part by the rate of diffusion into a contact region. A widely applied method for quantitatively measuring lateral mobility is fluorescence recovery after photobleaching (FRAP) (Fig. 6.6).

In this technique, a short, high-energy laser pulse is used to irreversibly bleach fluorescently labeled molecules within a defined region (often a circular or rectangular band). By observation of the recovery of fluorescence signal with a low-energy light source in the region of interest, the diffusion coefficient of the mobile species can be deduced (Axelrod et al., 1976). FRAP has become essential for measuring and comparing lateral mobility rates of specific receptors on the surface of a cell. Models have been developed that account for mobile and immobile populations, as well as flux rates between these populations. Fluorescent F_{ab} fragments or monoclonal antibodies are often used to tag proteins of interest. Because these labels bind one or two receptors, respectively, they minimize artificial clustering that can restrict lateral mobility. Lateral mobility data on β_2-integrins will reveal the potential for receptor clustering and adhesion strengthening during neutrophil–substrate interactions. As outlined earlier, TIRFM experiments can provide information on which receptors reside on exposed neutrophil membrane folds. Combining spatial information with quantitative mobility data could result in a comprehensive understanding of receptor dynamics on the most exposed regions of the neutrophil surface. TIRFM–FRAP accomplishes this by extending the lateral mobility measurements of FRAP with the specific imaging capability of TIRFM (Sund and Axelrod, 2000). This approach can answer questions pertaining to domain continuity and receptor mobility between neutrophil membrane ridges, as well as whether selectins and integrins intermix between the folds and body of the cell. Just as promising is the idea of using correlation spectroscopy to track receptor domains over time. In ICCS, time-lapse images are mathematically compared with an original image. As the clusters diffuse from their original location, cross-correlation with the original image decreases (see Suggested Reading). FRAP is ideal for measuring diffusion of individual molecules on a surface, but a relatively small bleach band can result in skewed results for domain diffusion. With domain-like distributions of L-selectin and β_2-integrins, it is important to consider not only molecular diffusion, but also cluster movements in the membrane.

FRAP can also address the hypothesis that cytoskeletal release and retethering are involved in integrin conformational changes. Cytoskeletal association may help stabilize the high-affinity conformation or be responsible for anchoring the integrin on binding its counterreceptor during firm adhesion. In either case, direct evidence of such a cytoskeletal restraint release has not yet been presented. Because of advances in low light–cooled CCD technology, it is now possible to capture extremely rapid events by fluorescence microscopy not only to detect the release, but also to quantify its duration and effect on clustering. By use of a laser light source and a high-speed camera, it would be possible to use FRAP to detect changes in lateral mobility due to release of cytoskeletal tethers. Timing the photobleach to coincide with various stages of activation would be important. High-speed image acquisition following the bleach should show two distinct rates of recovery if there is a temporary period of "free" integrin before retethering by the cytoskeleton. Previous work in this area has been performed with single-particle tracking (SPT). With this tool, individual particles, commonly small fluorescent or gold beads, are tracked across the surface of the cell to gather information about lateral mobility. The particles are often functionalized with antibodies specific to a target molecule. Tracking displacements of the marker between time points will elucidate when a mobility restraint is released as well as when tethering recommences. Use of a multivalent bead, however, will result in the potential binding of several receptors. Changes in lateral mobility will be apparent only if all of the receptors bound to the bead undergo cytoskeletal rearrangement at the same time. Recent work has focused on tracking single F_{ab} fragments that bind only one target molecule (Georgiou et al., 2002). Increases in camera sensitivity will make this approach more feasible and a valuable quantitative tool for lateral mobility measurements.

D. Affinity State

The integrin class of adhesion molecules is widely known to adopt multiple conformational arrangements. These conformational states are indicative of the affinity for ligand. It is electron microscopy and crystal structures that have led investigators to the discovery of these multiple conformational states. These techniques, however, require the integrin to be outside of the native cellular environment for characterization. FRET has become the standard for detecting integrin conformational changes on live cells. The proteins are modified such that a fluorescent tag is placed in a specific location so that when the integrin changes shape, its distance from a reference fluorophore can be inferred. This is commonly done in two ways: In the first case, the donor and acceptor fluorophore are on the same protein (see Fig. 6.5), or the reference fluorophore is present on a nearby protein or molecule that is in a close and known association with the protein of interest. In the second case, the reference molecule is often a fluorescence-labeled fatty acid, residing in the plasma membrane. Further development of acceptor–donor pairs will increase the sensitivity of this assay. Placement of the fluorophores in multiple locations on the receptor will also lead to additional structural information and a better understanding of the different conformational states. Combining these FRET techniques with quantitative imaging methods will help further the understanding of the neutrophil's dependence on affinity and availability during recruitment.

VII. CONCLUSIONS

Neutrophils play an essential role in mediating the inflammatory response and ridding the body of foreign invaders. Through the regulation of surface adhesion molecules, these dynamic cells leave the circulation via a regulated process involving rolling, firm adhesion, and migration. The transition through these stages involves complex dynamics of the neutrophil surface that determine the type and extent of neutrophil adhesion. Receptors move, cluster, and shed, and membranes expand and spread. More than two decades of investigation by biochemistry and light and electron microscopy have revealed the main adhesion molecules and suggested a variety of mechanisms that regulate adhesion. Definitively establishing these mechanisms now requires that we observe the dynamics of molecules and membranes over a range of temporal and length scales using the ever-increasing power of quantitative light microscopy.

SUGGESTED READING

Integrin Valency and Affinity

Carman, C. V., and Springer, T. A. (2003) Integrin avidity regulation: Are changes in affinity and conformation underemphasized? *Curr. Opin. Cell Biol.* **15**: 547–556.

This review is a recent addition to the long debate on the relative importance of studying integrin affinity versus availability. The authors note that the literature on availability has been primarily qualitative or ambiguous and emphasize the role of integrin affinity in neutrophil recruitment.

Integrin Signaling

Harris, E. S., McIntyre, T. M., Prescott, S. M., and Zimmerman, G. A. (2000) The leukocyte integrins. *J. Biol. Chem.* **275**: 23409–23412.

This minireview provides basic background on integrin structure, ligand recognition, and signaling. Inside-out and outside-in signaling with respect to leukocyte integrins is discussed.

Optical Microscopy Reviews and Online Tutorials

The Olympus website is an invaluable resource for introductory through moderately advanced understanding of optical microscopy. There are reviews and interactive tutorials for numerous imaging techniques.

Modern Microscopy and Fluorescence

4 April 2003, *Science* **300**

While not entirely devoted to quantitative light microscopy, this special issue of *Science* highlights the advances in fluorescence microscopy in the study of dynamic biological events.

Total Internal Reflection Fluorescence Microscopy (TIRFM)

Axelrod, D. (2001) Total internal reflection fluorescence microscopy in cell biology. *Traffic* **2**: 764–774.

The pioneer of TIRFM, Daniel Axelrod, explains the applications of TIRFM and importantly how to implement this modality in a standard fluorescence microscope.

Image Cross-Correlation Spectroscopy (ICCS)

Srivastava, M., and Petersen, N. O. (1998) Diffusion of transferrin receptor clusters. *Biophys Chem* **75**: 201–211.

This study applies the ICCS method to study diffusion of receptor clusters in a cell membrane. It discusses how the diffusion coefficient measurement differs between FRAP and ICCS.

Fluorescence Recovery after Photobleaching (FRAP)

Reits, E.A., and Neefjes, J.J. (2001) From fixed to FRAP: Measuring protein mobility and activity in living cells. *Nat. Cell Biol.* **3**: E145–E147.

This review gives a short historical account of the introduction of FRAP into cell biology and a basic primer to the mathematical analysis of mobility data.

Fluorescence Resonance Energy Transfer (FRET)

Jares-Erijman, E.A., and Jovin, T.M. (2003) FRET imaging. *Nat Biotechnol* **21**: 1387–95.

The authors of this review article provide a comprehensive overview of numerous potential methods of studying resonance energy transfer in biological systems including the mathematical basis.

REFERENCES

Alonso-Lebrero, J. L., Serrador, J. M., Dominguez-Jimenez, C., Barreiro, O., Luque, A., del Pozo, M. A., Snapp, K., Kansas, G., Schwartz-Albiez, R., Furthmayr, H., et al. (2000) Polarization and interaction of adhesion molecules P-selectin glycoprotein ligand 1 and intercellular adhesion molecule 3 with moesin and ezrin in myeloid cells. *Blood* **95**: 2413–2419.

Anderson, S. I., Hotchin, N. A., and Nash, G. B. (2000) Role of the cytoskeleton in rapid activation of CD11b/CD18 function and its subsequent downregulation in neutrophils. *J. Cell Sci.* **113** (Pt. 15): 2737–45.

Axelrod, D., Koppel, D. E., Schlessinger, J., Elson, E., and Webb, W. W. (1976) Mobility measurement by analysis of fluorescence photobleaching recovery kinetics. *Biophys. J.* **16**: 1055–1069.

Bell, G. I. (1978) Models for the specific adhesion of cells to cells. *Science* **200**: 618–627.

Berlin, C., Bargatze, R. F., Campbell, J. J., von Andrian, U. H., Szabo, M. C., Hasslen, S. R., Nelson, R. D., Berg, E. L., Erlandsen, S. L., and Butcher, E. C. (1995) Alpha 4 integrins mediate lymphocyte attachment and rolling under physiologic flow. *Cell* **80**: 413–422.

Bikoue, A., D'Ercole, C., George, F., Dameche, L., Mutin, M. and Sampol, J. (1997) Quantitative analysis of leukocyte membrane antigen expression on human fetal and cord blood: normal values and changes during development. *Clin. Immunol. Immunopathol.* **84**: 56–64.

Borland, G., Murphy, G., and Ager, A. (1999) Tissue inhibitor of metalloproteinases-3 inhibits shedding of L-selectin from leukocytes. *J. Biol. Chem.* **274**: 2810–2815.

Bretscher, A., Edwards, K., and Fehon, R. G. (2002) ERM proteins and merlin: Integrators at the cell cortex. *Nat. Rev. Mol. Cell Biol.* **3**: 586–599.

Bruehl, R. E., Springer, T. A., and Bainton, D. F. (1996) Quantitation of L-selectin distribution on human leukocyte microvilli by immunogold labeling and electron microscopy. *J. Histochem. Cytochem.* **44**: 835–844.

Calderwood, D. A. (2004) Integrin activation. *J. Cell Sci.* **117**: 657–666.

Carman, C. V., and Springer, T. A. (2003) Integrin avidity regulation: Are changes in affinity and conformation underemphasized? *Curr. Opin. Cell Biol.* **15**: 547–556.

Constantin, G., Majeed, M., Giagulli, C., Piccio, L., Kim, J. Y., Butcher, E. C., and Laudanna, C. (2000) Chemokines trigger immediate beta2 integrin affinity and mobility changes: Differential regulation and roles in lymphocyte arrest under flow. *Immunity* **13**: 759–769.

Davenpeck, K. L., Brummet, M. E., Hudson, S. A., Mayer, R. J., and Bochner, B. S. (2000) Activation of human leukocytes reduces surface P-selectin glycoprotein ligand-1 (PSGL-1, CD162) and adhesion to P-selectin *in vitro*. *J. Immunol.* **165**: 2764–2772.

Erlandsen, S. L., Hasslen, S. R., and Nelson, R. D. (1993) Detection and spatial distribution of the beta 2 integrin (Mac-1) and L-selectin (LECAM-1) adherence receptors on human neutrophils by high-resolution field emission SEM. *J. Histochem. Cytochem.* **41**: 327–333.

Finger, E. B., Bruehl, R. E., Bainton, D. F., and Springer, T. A. (1996) A differential role for cell shape in neutrophil tethering and rolling on endothelial selectins under flow. *J. Immunol.* **157**: 5085–5096.

Finn, A. and Rebuck, N. (1994) Measurement of adhesion molecule expression on neutrophils and fixation. *J. Immunol. Methods* **171**: 267–270.

Gamble, J. R., Harlan, J. M., Klebanoff, S. J., and Vadas, M. A. (1985) Stimulation of the adherence of neutrophils to umbilical vein endothelium by human recombinant tumor necrosis factor. *Proc. Natl. Acad. Sci. USA* **82**: 8667–8671.

Georgiou, G., Bahra, S. S., Mackie, A. R., Wolfe, C. A., O'Shea, P., Ladha, S., Fernandez, N., and Cherry, R. J. (2002) Measurement of the lateral diffusion of human MHC class I molecules on HeLa cells by fluorescence recovery after photobleaching using a phycoerythrin probe. *Biophys. J.* **82**: 1828–1834.

Green, C. E., Pearson, D. N., Camphausen, R. T., Staunton, D. E., and Simon, S. I. (2004) Shear-dependent capping of L-selectin and P-selectin glycoprotein ligand 1 by E-selectin signals activation of high-avidity beta2-integrin on neutrophils. *J. Immunol.* **172**: 7780–7790.

Harris, E. S., McIntyre, T. M., Prescott, S. M., and Zimmerman, G. A. (2000) The leukocyte integrins. *J. Biol. Chem.* **275**: 23409–23412.

Howard, T. H., and Oresajo, C. O. (1985) The kinetics of chemotactic peptide-induced change in F-actin content, F-actin distribution, and the shape of neutrophils. *J. Cell Biol.* **101**: 1078–1085.

Ivetic, A., Deka, J., Ridley, A., and Ager, A. (2002) The cytoplasmic tail of L-selectin interacts with members of the ezrin–radixin–moesin (ERM) family of proteins: Cell activation-dependent binding of moesin but not ezrin. *J. Biol. Chem.* **277**: 2321–2329.

Ivetic, A., Florey, O., Deka, J., Haskard, D. O., Ager, A., and Ridley, A. J. (2004) Mutagenesis of the ezrin–radixin–moesin binding domain of L-selectin tail affects shedding, microvillar positioning, and leukocyte tethering. *J. Biol. Chem.* **279**: 33263–33272.

Johnston, B. and Kubes, P. (1999) The alpha4-integrin: an alternative pathway for neutrophil recruitment? *Immunol. Today* **20**: 545–550.

Jones, S. L., Wang, J., Turck, C. W., and Brown, E. J. (1998) A role for the actin-bundling protein L-plastin in the regulation of leukocyte integrin function. *Proc. Natl. Acad. Sci. USA* **95**: 9331–9336.

Junge, S., Brenner, B., Lepple-Wienhues, A., Nilius, B., Lang, F., Linderkamp, O., and Gulbins, E. (1999) Intracellular mechanisms of L-selectin induced capping. *Cell Signal.* **11**: 301–308.

Kim, M., Carman, C. V., and Springer, T. A. (2003) Bidirectional transmembrane signaling by cytoplasmic domain separation in integrins. *Science* **301**: 1720–1725.

Kim, M., Carman, C. V., Yang, W., Salas, A., and Springer, T. A. (2004) The primacy of affinity over clustering in regulation of adhesiveness of the integrin αLβ2. *J. Cell Biol.* **167**: 1241–1253.

Krauss, K., and Altevogt, P. (1999) Integrin leukocyte function-associated antigen-1-mediated cell binding can be activated by clustering of membrane rafts. *J. Biol. Chem.* **274**: 36921–36927.

Kucik, D. F., Dustin, M. L., Miller, J. M., and Brown, E. J. (1996) Adhesion-activating phorbol ester increases the mobility of leukocyte integrin LFA-1 in cultured lymphocytes. *J. Clin. Invest.* **97**: 2139–2144.

Leitinger, B., and Hogg, N. (2002) The involvement of lipid rafts in the regulation of integrin function. *J. Cell Sci.* **115**: 963–972.

Lub, M., van Kooyk, Y., van Vliet, S. J. and Figdor, C. G. (1997) Dual role of the actin cytoskeleton in regulating cell adhesion mediated by the integrin lymphocyte function-associated molecule-1. *Mol. Biol. Cell* **8**: 341–351.

Majstoravich, S., Zhang, J., Nicholson-Dykstra, S., Linder, S., Friedrich, W., Siminovitch, K. A., and Higgs, H. N. (2004) Lymphocyte microvilli are dynamic, actin-dependent structures that do not require Wiskott–Aldrich syndrome protein (WASp) for their morphology. *Blood* **104**: 1396–1403.

Moore, K. L., Patel, K. D., Bruehl, R. E., Li, F., Johnson, D. A., Lichenstein, H. S., Cummings, R. D., Bainton, D. F. and McEver, R. P. (1995) P-selectin glycoprotein ligand-1 mediates rolling of human neutrophils on P-selectin. *J. Cell Biol.* **128**: 661–671.

Nijhara, R., van Hennik, P. B., Gignac, M. L., Kruhlak, M. J., Hordijk, P. L., Delon, J., and Shaw, S. (2004) Rac1 mediates collapse of microvilli on chemokine-activated T lymphocytes. *J. Immunol.* **173**: 4985–4993.

Oshiro, N., Fukata, Y., and Kaibuchi, K. (1998) Phosphorylation of moesin by rho-associated kinase (Rho-kinase) plays a crucial role in the formation of microvilli-like structures. *J. Biol. Chem.* **273**: 34663–34666.

Pavalko, F. M., Walker, D. M., Graham, L., Goheen, M., Doerschuk, C. M., and Kansas, G. S. (1995) The cytoplasmic domain of L-selectin interacts with cytoskeletal proteins via alpha-actinin: Receptor positioning in microvilli does not require interaction with alpha-actinin. *J. Cell Biol.* **129**: 1155–1164.

Peschon, J. J., Slack, J. L., Reddy, P., Stocking, K. L., Sunnarborg, S. W., Lee, D. C., Russell, W. E., Castner, B. J., Johnson, R. S., Fitzner, J. N., et al. (1998) An essential role for ectodomain shedding in mammalian development. *Science* **282**: 1281–1284.

Petersen, N. O., Hoddelius, P. L., Wiseman, P. W., Seger, O., and Magnusson, K. E. (1993) Quantitation of membrane receptor distributions by image correlation spectroscopy: concept and application. *Biophys. J.* **65**: 1135–1146.

Phong, M. C., Gutwein, P., Kadel, S., Hexel, K., Altevogt, P., Linderkamp, O., and Brenner, B. (2003) Molecular mechanisms of L-selectin-induced co-localization in rafts and shedding [corrected]. *Biochem. Biophys. Res. Commun.* **300**: 563–569.

Pierini, L. M., Eddy, R. J., Fuortes, M., Seveau, S., Casulo, C., and Maxfield, F. R. (2003) Membrane lipid organization is critical for human neutrophil polarization. *J. Biol. Chem.* **278**: 10831–10841.

Pierini, L. M., Lawson, M. A., Eddy, R. J., Hendey, B. and Maxfield, F. R. (2000) Oriented endocytic recycling of alpha5beta1 in motile neutrophils. *Blood* **95**: 2471–2480.

Sampath, R., Gallagher, P. J., and Pavalko, F. M. (1998) Cytoskeletal interactions with the leukocyte integrin beta2 cytoplasmic tail: Activation-dependent regulation of associations with talin and alpha-actinin. *J. Biol. Chem.* **273**: 33588–33594.

Shamri, R., Grabovsky, V., Feigelson, S. W., Dwir, O., Van Kooyk, Y., and Alon, R. (2002) Chemokine stimulation of lymphocyte alpha 4 integrin avidity but not of leukocyte function-associated antigen-1 avidity to endothelial ligands under shear flow requires cholesterol membrane rafts. *J. Biol. Chem.* **277**: 40027–40035.

Shattil, S. J., Kashiwagi, H., and Pampori, N. (1998) Integrin signaling: The platelet paradigm. *Blood* **91**: 2645–2657.

Simons, K., and Ikonen, E. (1997) Functional rafts in cell membranes. *Nature* **387**: 569–572.

Simons, K., and Vaz, W. L. (2004) Model systems, lipid rafts, and cell membranes. *Annu. Rev. Biophys. Biomol. Struct.* **33**: 269–295.

Sixt, M., Hallmann, R., Wendler, O., Scharffetter-Kochanek, K., and Sorokin, L. M. (2001) Cell adhesion and migration properties of beta 2-integrin negative polymorphonuclear granulocytes on defined extracellular matrix molecules: Relevance for leukocyte extravasation. *J. Biol. Chem.* **276**: 18878–18887.

Sund, S. E., and Axelrod, D. (2000) Actin dynamics at the living cell submembrane imaged by total internal reflection fluorescence photobleaching. *Biophys. J.* **79**: 1655–1669.

Tandon, R., Sha'afi, R. I., and Thrall, R. S. (2000) Neutrophil beta2-integrin upregulation is blocked by a p38 MAP kinase inhibitor. *Biochem. Biophys. Res. Commun.* **270**: 858–862.

Ting-Beall, H. P., Needham, D., and Hochmuth, R. M. (1993) Volume and osmotic properties of human neutrophils. *Blood* **81**: 2774–2780.

Tohya, K., and Kimura, M. (1998) Ultrastructural evidence of distinctive behavior of L-selectin and LFA-1 (alphaLbeta2 integrin) on lymphocytes adhering to the endothelial surface of high endothelial venules in peripheral lymph nodes. *Histochem. Cell Biol.* **110**: 407–416.

Tohyama, Y., Katagiri, K., Pardi, R., Lu, C., Springer, T. A., and Kinashi, T. (2003) The critical cytoplasmic regions of the alphaL/beta2 integrin in Rap1-induced adhesion and migration. *Mol. Biol. Cell* **14**: 2570–2582.

Ushiyama, S., Laue, T. M., Moore, K. L., Erickson, H. P. and McEver, R. P. (1993) Structural and functional characterization of monomeric soluble P-selectin and comparison with membrane P-selectin. *J. Biol. Chem.* **268**: 15229–15237.

Van Kooyk, Y., and Figdor, C. G. (2000) Avidity regulation of integrins: The driving force in leukocyte adhesion. *Curr. Opin. Cell Biol.* **12**: 542–547.

Van Kooyk, Y., van Vliet, S. J., and Figdor, C. G. (1999) The actin cytoskeleton regulates LFA-1 ligand binding through avidity rather than affinity changes. *J. Biol. Chem.* **274**: 26869–26877.

Von Andrian, U. H., Hasslen, S. R., Nelson, R. D., Erlandsen, S. L., and Butcher, E. C. (1995) A central role for microvillous receptor presentation in leukocyte adhesion under flow. *Cell* **82**: 989–999.

Wallace, P. J., Wersto, R. P., Packman, C. H., and Lichtman, M. A. (1984) Chemotactic peptide-induced changes in neutrophil actin conformation. *J. Cell Biol.* **99**: 1060–1065.

Ward, M. D., Dembo, M., and Hammer, D. A. (1994) Kinetics of cell detachment: Peeling of discrete receptor clusters. *Biophys. J.* **67**: 2522–2534.

Watts, R. G., and Howard, T. H. (1994) Role of tropomyosin, alpha-actinin, and actin binding protein 280 in stabilizing Triton insoluble F-actin in basal and chemotactic factor activated neutrophils. *Cell Motil. Cytoskel.* **28**: 155–164.

Williams, T. E., Nagarajan, S., Selvaraj, P., and Zhu, C. (2001) Quantifying the impact of membrane microtopology on effective two-dimensional affinity. *J. Biol. Chem.* **276**: 13283–13288.

Zigmond, S. H., and Sullivan, S. J. (1979) Sensory adaptation of leukocytes to chemotactic peptides. *J. Cell Biol.* **82**: 517–527.

7
GYCOCALYX REGULATION OF CELL ADHESION

PHILIPPE ROBERT, LAURENT LIMOZIN, ANNE-MARIE BENOLIEL, ANNE PIERRES, and PIERRE BONGRAND

Laboratoire d'Immunologie INSERM U600 – CNRS FRE2059, Hôpital de Sainte-Marguerite, Marseille, France

Cell adhesion is a process of prominent importance that influences essentially all steps of cell life. Also, all living cells are found to be surrounded by a polysaccharide-rich layer called the *glycocalyx*. The first purpose of this chapter is to review experimental evidence supporting the concepts that (1) glycocalyx components have the potential to impair cell–cell or cell–surface approach, and this repulsion is dependent on contact kinetics, and (2) synthesis, removal, or displacement of glycocalyx elements is commonly used by living cells to initiate, terminate, or prevent adhesion. The second purpose of the chapter is to provide the reader with numerical estimates of cell surface parameters as a starting point for further studies of the mechanical forces exerted by glycocalyx elements on cells encountering each other. As a conclusion, we suggest potential biological functions for the glycocalyx and promising avenues of research to answer questions raised in the review.

I. INTRODUCTION

A. Cell Adhesion Is a Paramount Process That Requires Tight Regulation

Most phases of cell development and function are highly dependent on adhesive interactions (see Pierres et al., 2000, for a review). Thus, cell survival and proliferative capacity usually require the availability of a suitable adherent surface, demonstrating anchorage dependence. Cell differentiation is dependent on both contact interaction and soluble stimuli. Cell migration is essential to early development as well as function of mature organisms; a well-known example is the continuous circulation of immune cells throughout higher organisms. Also, the triggering of active cell functions such as phagocytosis is dependent on both qualitative and quantitative features of phagocyte-to-target adhesion.

In all the aforementioned situations, cell adhesion must be both specific and accurately regulated. Thus, phagocytic cells must discriminate between foreign and/or dangerous particles and normal components. Cell migration on a surface requires efficient attachment of a lamellipodium sent forward, and detachment of posterior structures. Indeed, both excessive adhesion and lack of attachment may hamper cell progression (Palecek et al., 1997).

Thus, understanding adhesion is a major challenge for biologists. In addition to merely theoretical interest, such an understanding might help predict and even manipulate important pathological processes. For example, an early step in atherosclerosis, a major cause of mortality in developed countries, may well be the attachment of blood monocytes to endothelial cells. Metastasis is dependent on detachment of cancer cells from a primitive tumor and their subsequent attachment to a target organ. Infection is usually triggered by the attachment of bacteria or viruses to target tissues or cells. Finally, successful use of biomaterials requires tight control of adhesion, to achieve suitable integration into tissues and/or prevent attachment and activation of inflammatory cells as well as bacterial binding.

B. The Glycocalyx Is the Outermost Cell Region

As depicted in Fig. 7.1, the lipid bilayer is often considered a natural boundary between the cell interior and exterior. However, many cell-bound molecules may extend several tens or even hundreds of nanometers outward. These include:

- The extracellular domains of intrinsic membrane proteins are typically 5–20 nm in size. These proteins are usually studded with oligosaccharide chains of typically 5–20 residues often terminating with a negatively charged sialic acid moiety. These carbohydrate chains represent from a few percent to more than half the protein weight.
- Extrinsic membrane molecules may be released with relatively mild treatment, for example, with surface-active agents or divalent cation chelators.
- Proteoglycans are made of a core protein bearing a few polysaccharide chains that may be several hundreds of nanometers long. Specific examples are given later (e.g., episialin). In some cases, the core protein may bear a transmembrane and intracytoplasmic moiety. Otherwise, proteoglycans may behave as extrinsic membrane proteins.

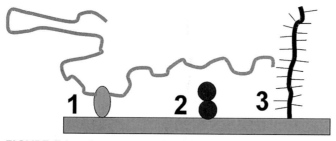

FIGURE 7.1   Schematic view of the glycocalyx. The bilayer (about 4.5 nm thick) is covered with different molecular species, including (from left to right) (1) proteoglycans, with a proteic core bearing very long polysaccharide chains, usually flexible repeats of typical disaccharide units longer than several hundreds nanometers; (2) glycoproteins typically 10–30 nm in size and a few oligosaccharide chains; and (3) mucin-like molecules such as CD43, with an extended peptide backbone bearing tens of oligosaccharide chains, often terminating with negatively charged sialic acid residues.

Thus, the outermost cell region is a fairly carbohydrate-rich zone that may be called the *glycocalyx* (Bennett, 1963), *cell coat, fuzzy coat* or *pericellular matrix*. Two remarks may be drawn from the preceding description.

1. Due to molecular heterogeneity, there is no clear-cut glycocalyx boundary, because molecular density is probably continously decreasing as distance to the bilayer increases.
2. The difference between cellular and extracellular structures may not be obvious. Thus, fibronectin may be considered an extracellular matrix protein as well as a pericellular protein that may be bound by a variety of cell-specific receptors, including integrins and nonintegrin molecules.

C. The Glycocalyx May in Principle Modulate Cell Adhesion through Different Nonexclusive Mechanisms Described in This Review

Clearly, the glycocalyx may modulate adhesion through a variety of nonexclusive mechanisms, including (1) generation of repulsive forces impeding cell–cell or cell–surface approach, (2) masking of cell adhesive sites, or, alternatively, (3) expression of molecular sites specifically bound by membrane receptors of neighboring cells. Also, these mechanisms may be permanent or triggered by behavioral stimuli. Admittedly, it may be difficult to discriminate between different possibilities. Thus, only detailed molecular knowledge of a given interaction may help determine whether adhesion is prevented by nonspecific repulsion or specific masking of a particular binding site.

In the present chapter, we first remind the reader of alternate mechanisms used by cells to regulate adhesion. Then we sequentially review experimental evidence supporting a role for the cell glycocalyx in adhesion, and selected theoretical or experimental data allowing some mechanistic insight into these processes. Finally, we briefly mention other possible functions of the glycocalyx, and present problems.

II. CELL ADHESION IS DETERMINED BY A VARIETY OF COMPLEMENTARY MECHANISMS THAT MUST BE WELL UNDERSTOOD TO ASSESS THE SPECIFIC ROLE OF THE GLYCOCALYX

A. Cell Adhesion Is Mediated Mainly by Specific Ligand–Receptor Interactions

Much evidence has shown that a limited number of specific ligand–receptor interactions may be both necessary and sufficient to mediate cell–cell adhesion. Thus, it is likely that a few hundred specific antibodies bound to membrane Fc receptors may impart to phagocytic cells the capacity to bind particles recognized by these antibodies. Also, it is often possible to prevent cell–cell adhesion by blocking or removing a restricted population of membrane receptors. However, these experiments do not provide a formal proof that a few molecules account for all adhesion, as it is difficult to exclude the possibility that adhesion might result from the triggering of nonspecific processes by a few specific interactions.

Thus, consideration of experiments done with better controlled model systems is warranted; a few or even a single antibody molecule may agglutinate osmotically sphered erythrocytes (Tha et al., 1986). A few selectin molecules are able to mediate leukocyte attachment to artificial surfaces under hydrody-namic flow (Alon et al., 1995).

Because specific ligand–receptor interactions usually involve a molecular contact area smaller than a few square nanometers, it is concluded that inter-cellular adhesion may involve much less than 1/1000th of the total cell area (which is on the order of 1000 μm^2). Thus, a major question concerns the potential influence of the glycocalyx on association between cell membrane molecules rather than bulk surface parameters.

B. Receptor-Mediated Cell Adhesion May Be Regulated by a Variety of Mechanisms Unrelated to the Glycocalyx

It is essential to be aware of the multiple parameters with the potential to influence adhesion, to assess the specific influence of the glycocalyx. A brief description is given later and summarized in Fig. 7.2.

The major mechanism used by living cells to control adhesion is probably the modulation of *membrane receptor expression*. Thus, when endothelial cells are activated by a suitable inflammatory stimulus, P-selectin molecules may be transferred from intracellular vesicles to the plasma membrane within less than a few minutes. Simultaneously, protein synthesis is triggered, leading to the membrane expression of E-selectin 4–6 h later and VCAM-1 within a day. Conversely, membrane receptor expression may be downregulated by a variety of processes: receptors may be released in soluble form through proteolytic cleavage, or they may be endocytosed and either destroyed or recycled toward the membrane after a passage in storage vesicles. Leukocyte L-selectin provides a good illustration of this downward regulation: cell activation may result in rapid release of these molecules through proteolytic cleavage of a specific site.

Another important regulatory mechanism involves *receptor conformational change*. This possibility is especially well illustrated by integrins, a major family of cell surface receptors that mediate, in particular, adhesion to extracellular matrix proteins (fibronectin, laminin, collagen) or immunoglobulin superfamily molecules (including endothelial VCAM-1 and the ubiquitous ICAM-1). It has long been known that integrins can undergo major structural changes triggered by extracellular parameters, such as divalent cations and monoclonal antibodies. Alternatively, integrin activity may be triggered by intracellular signals (inside-out signaling). It has long been reported that integrins can display multiple affin-ity states. The structural basis of these changes has recently been elucidated as a consequence of remarkable crystallographic studies (Xiong et al., 2003).

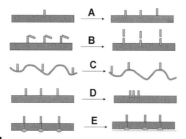

FIGURE 7.2 Mechanisms used by living cells to regulate adhesive behavior. (A) Change of receptor density. (B) Change of receptor conformation and affinity. (C) Receptor concentration on the tips of microvilli. (D) Receptor clustering. (E) Modulating receptor interaction with cytoskeletal elements and, therefore, lateral mobility and anchoring strength.

In addition to the aforementioned major regulatory mechanisms, many less well-known processes may dramatically regulate adhesion. Thus, *receptor localization* may be used as a rapid way of modulating cell adhesiveness. First, as shown in Fig. 7.2C, cell encounter with a foreign surface often involves limited regions corresponding to the tips of microvilli and/or most convex parts of the cell membrane. It appears that many adhesion molecules may be concentrated on the tips of microvilli. Further, this localization may be required for functional activity (von Andrian et al., 1995), and it is liable to modulation by cell-activating stimuli (Erlandsen et al., 1993). A second mechanism that has drawn much attention during the last 10 years is receptor *clustering*. This was advocated as the main way of activating integrins following the initial observation that complement CRIII receptor activation on neutrophils is correlated to a limited gathering of receptors in small clusters of two to six molecules (Detmers et al., 1987). This clustering may result in easier formation of multimolecular attachments, which might be required to mediate adhesion when a single ligand–molecular interaction involves the initial formation of a transient complex that is too short-lived to allow membrane immobilization and additional bond formation (Vitte et al., 2004).

Another important parameter of receptor adhesion efficiency is *lateral mobility on the cell membrane*. This point was demonstrated in very elegant experiments by Kucik et al. (1996), who used single-particle tracking to assess the mobility of lymphocyte integrins. They showed that in some cases, cytoskeletal impairment might increase integrin activity through an increase in mobility.

The influence of *receptor–cytoskeleton interactions* on receptor activity is, however, less straightforward than suggested by Kucik and colleagues' experiments. Indeed, lack of receptor connection with the cell interior may be detrimental to adhesion efficiency for at least two reasons. First, receptor uprooting may prevent strong adhesion, and cytoskeletal attachment may be required to resist high disruptive forces. Second, strong adhesion usually requires the formation of a fairly extensive contact area, which may necessitate active cell spreading on neighboring surfaces. This spreading often requires specific receptor-mediated signaling that is somewhat related to receptor connection with intracellular structures.

In conclusion, experimental data obtained on many cellular systems are often difficult to interpret in view of the multiplicity of mechanisms potentially involved in the regulation of adhesion. It is therefore difficult to rule out the possibility that a given cell treatment aimed at modulating glycocalyx structure might alter adhesion through unexpected glycocalyx-independent processes.

III. CELL GLYCOCALYX AND ADHESION: SELECTED EXPERIMENTAL DATA

A. Morphological Evidence: Cell–Cell or Cell–Particle Approach Is Opposed by a Negatively Charged Carbohydrate-Rich Layer That May Reduce Contact Areas

The simplest evidence that cell contact with foreign particles is prevented by an external structure was provided by electron microscopy. Clarris and Fraser (1968) incubated fibroblasts or synovial cells with particles such as erythrocytes, then monitored samples with electron microscopy. They found that cells were separated from particles with an electron-light layer that might be several micrometers thick. This layer resisted extensive washing and treatment with cation chelators such as EDTA or with enzymes such as DNase and RNase. It was decreased after trypsinization and essentially removed with hyaluronidase. Also, electron micrographs frequently revealed that lipid bilayers remained separated by an electron-light gap 10–20 nm wide (Martinez-Palomo, 1970), suggesting the presence of intervening material.

The presence of carbohydrates was demonstrated with specific procedures such as periodic acid–Schiff (PAS) staining. Indeed, the presence on the cell surface of a PAS-positive region led Bennett (1963) to coin the name *glycocalyx* for this structure (he also suggested *sweet husk*, which met with less success).

Another common way of staining the glycocalyx for electron microscopy was through the use of cationic dyes such as colloidal iron, ruthenium red, and alcian blue. This emphasizes the presence of negative charges that were often decreased or removed with sialidases (or neuraminidases).

Some quantitative data were obtained by analyzing electron micrographs (Foa et al., 1996). Rat macrophages were incubated with sheep erythrocytes that had been treated with glutaraldehyde, which caused them to be bound (and subsequently ingested) by phagocytes. Samples were studied with electron microscopy. The mean width of the apparent gap between macrophages and red cells was 21.5 ± 0.5 nm (SE). When erythrocytes were treated with neuraminidase, which removed most surface negative charges, the gap width was decreased to 18.1 ± 0.65 nm. Experiments were repeated after exposing cells to cationic dyes. With this treatment, the cells appeared to have a dark boundary, the thickness of which strongly depended on the staining procedure. Thus, the boundary of macrophages (i.e., lipid bilayer + cell coat) was shown to be 18.5 ± 1.4 and 39.2 ± 5 nm thick after ruthenium red and alcian blue staining, respectively. Further, when adhesion zones were observed, they appeared entirely dark, suggesting that they are filled with carbohydrate material. Interestingly, the cell coat appeared reduced in contact areas, as the thickness of the stained line separating macrophages from erythrocytes was comparable to the thickness of free regions, suggesting that adhesion might result in 50% reduction of glycocalyx thickness. It is, however, useful to emphasize three points of caution before interpreting the aforementioned quantitative data:

- First, sample processing for electron microscopy may alter the size of structures such as the glycocalyx, and it would be highly desirable to compare distance estimates obtained with different techniques. In this respect, it may be worth mentioning that electron microscopy and interference reflection microscopy yielded similar estimates for the distance between spreading cells and glass surfaces (Heath, 1982).

- Second, in line with the previous remark, we must be aware that the plane of section may not be perpendicular to cell membranes, yielding an artificially increased intercellular gap. As suggested by Foa et al. (1996), assuming that the orientation of the section plane is random, the average apparent gap should be twice the actual gap.

- Third, although the boundaries of stained glycocalices often seem quite clear-cut, the choice of the boundary is somewhat arbitrary and most thresholding procedures are only qualitative. This may provide an explanation for the fairly large difference in thickness estimates obtained with ruthenium red and alcian blue in the aforementioned exeriments.

In conclusion, morphological evidence suggests that cells are coated with a negatively charged polysaccharidic layer opposing particle approach. However, these results do not allow unambiguous identification of repulsive structures. Thus, it might be argued that cell–cell or cell–particle repulsion is mediated by a few stiff protein structures, and that sugars may only fill available space. It is therefore interesting to quote limited evidence suggesting that cell adhesion results in substantial egress of glycocalyx elements from contact areas.

Energy filtering transmission electron microscopy (EFTEM) was used to quantitatively study contacts formed between macrophages and altered erythrocytes (Soler et al., 1998). The basic principle consisted of labeling cells with terbium, a cation that was expected to label cell surface mucopolysaccharides, some phospholipids, and some protein cation binding sites. In contrast with lanthanum, terbium seems to be excluded from cell cytoplasm, resulting in highly contrasted glycocalyx images. Elemental analysis suggested that total label displayed an at least twofold decrease in contact areas, in accordance with the aforementioned qualitative observations.

Other experiments were planned to study a representative glycocalyx constituent rather than heterogeneous carbohydrate populations. Leukosialin (CD43) is a major constituent of leukocyte membranes. The extracellular part resembles a rigid rod about 45 nm long, with 239 amino acid residues and about 80 O-linked oligosaccharides. Therefore, this molecule might be expected to participate in glycocalyx-mediated repulsion. Thus, Soler et al. (1997) studied the distribution of CD43 molecules on monocyte-like human THP-1 cells. Using immunofluorescence, they found that CD43 was homogenously distributed on isolated THP-1 cells. However, crosslinking CD43 with antibodies resulted in rapid surface aggregation (i.e., patching) and concentration in a pole of the cell (i.e., capping). Further, when THP-1 cells were made to bind altered erythrocytes, CD43 was often redistributed out of contact areas. This redistribution was specific, because β_2-integrins were not excluded from contact areas.

Electron microscopy was also used together with colloidal gold labeling: when isolated THP-1 cells were studied, CD43 was fairly concentrated on the tips of microvilli. However, when cells were made to bind erythrocytes, CD43 was excluded from the tips of microvilli in contact areas.

B. Functional Evidence: Adhesion Efficiency Is Inversely Related to Glycocalyx Thickness

Significant data support the concept that removal of glycocalyx elements increases cell adhesiveness, and increasing cell coat thickness impedes adhesion. Representative examples are now discussed.

1. Adhesion May Be Enhanced by Removing Glycocalyx Elements

Removal of cell surface negative charges is often considered to facilitate adhesion. We describe a representative example (Capo et al., 1981). As indicated before, phagocytes do not adhere to fresh erythrocytes, but have been found to bind and subsequently ingest glutaraldehyde-treated red cells. When fresh human erythrocytes were depleted of sialic acid residues with neuraminidase, or neutralized with positive polylysine, only minimal uptake was observed after incubation with rat macrophages. However, when glutaraldehyde-treated red cells were exposed to neuraminidase or polylysine, adhesion to macrophages exhibited sevenfold and sixfold increases, respectively. The simplest interpretation of these experiments is that the binding of (hydrophobic) glutaraldehyde-treated red cells by macrophages might be hampered by repulsive forces generated by negative charges of the cell coat.

Another well-studied example is the repulsive potential of polysialic acid (Rutishauser et al., 1988). The neural cell adhesion molecule NCAM may bear varying amounts of huge polysialic residues. This expression is developmentally regulated and may provide a convenient way for neural cells to control adhesion. This view is strongly supported by the finding that polysialic acid removal greatly enhances neural cell agglutination with wheat germ agglutinin (WGA). However, this treatment did not increase cell capacity to bind radioactive WGA, suggesting that the antiadhesive potential of polysialic acid was not due to mere masking of binding sites. This control is important, because glycocalyx alteration may enhance adhesion by generating new adhesion sites. Thus, neuraminidase treatment of many cells will uncover glucose or galactose groups that may be recognized by phagocyte glucan receptors. Another example illustrating the intricacy of data interpretation is a report from Razi and Varki (1998), whose study focused on CD22, a so-called siglec, a sialic acid binding receptor found on B lymphocytes. This may be blocked by *cis* interaction with cell sialic acid moieties that must be released to allow *trans* interaction with other cells.

An elegant way of overcoming the unmasking hypothesis is to prevent the membrane expression of repulsive structures with genetic techniques. Thus, lymphocytes from CD43 knockout mice displayed increased interaction with endothelial walls (Stockton et al., 1998). Also, when expression of the large episialin/MUC1 molecule was inhibited with siRNA, cells displayed increased cadherin-mediated aggregation. These examples both support the concept that glycocalyx elements act as an antiadhesive barrier, and suggest that a few molecular species may dominate repulsion.

2. Adhesion May Be Prevented by Increasing the Expression of Glycocalyx Constituents

Conversely, inducing the expression of large carbohydrate molecules by target cells could decrease adhesive properties. Wesseling et al. (1995) transfected melanoma cells with MUC1/episialin gene, which resulted in inhibition of integrin-mediated adhesion to extracellular matrix proteins. Interestingly, inhibition was reversed when transfected cells were treated with Ts2/16 monoclonal antibody, which is known to increase the affinity of β_2-integrins. Thus, the antiadhesive effect was quantitative rather than qualitative. Additional experiments were performed to test the hypothesis that this effect is related to molecule size. Episialin is endowed with an extracellular domain made of 30–90 repeats bearing O-linked carbohydrates. Wesseling et al. (1996) made murine L929 cells express mutated episialin with a variable number of extracellular repeats. They concluded that molecular length is the main determinant of inhibition of cell aggregation that could be mediated by the homophilic E-cadherin adhesion molecule. Indeed, aggregation was efficiently prevented by a full-length molecule bearing 36 repeats (estimated length: 175–200 nm), whereas an 8-repeat molecule was much less efficient. Further, when sialic acid was removed with neuraminidase, the full-length molecule retained its capacity to prevent adhesion, whereas the antiadhesive effect of the 8-domain molecule was markedly decreased. Similar results were obtained with other molecules: B lymphocytes from transgenic mice expressing increased amounts of leukosialin (CD43) displayed decreased homotypic aggregation and interaction with T lymphocytes. Ostberg et al. (1996) concluded that these phenomena resulted from a general antiadhesive effect of membrane mucins. Similarly, increased polysialylation of NCAM in transfected PC12 cells was associated with decreased integrin-mediated adhesion to collagen (Horstkorte et al., 1999).

3. Glycocalyx-Mediated Inhibition of Adhesion May Decrease When Cell–Cell Contact Is Prolonged

The results we have described provide convincing evidence that in some cases the presence on cell surfaces of huge hydrophilic molecules such as proteoglycans is per se sufficient to hamper adhesive interactions. However, reliable experimental data also show that cell adhesion can be mediated by membrane receptors much thinner than the cell coat, demonstrating that the repulsive barrier is by no means impenetrable. This apparent paradox was resolved by consistent reports suggesting that glycocalyx repulsion is time-dependent and may vanish after a sufficient contact time. This point is important, as most tests performed *in vitro* to assess adhesion are done under static conditions, for example, by letting cells or particles sediment over other cells and allowing contact to last several minutes. These conditions may be different from physiological situations, in which cell or fluid motion makes contact much more transient. We now describe, first, experimental data and, then, suggested mechanisms.

The initial interaction between flowing leukocytes and activated endothelium may be mediated by particularly long molecules. Indeed, a common ligand–receptor couple comprises endothelial P-selectin and leukocyte ligand PSGL-1. These molecules are each about 40 nm long, and they are often concentrated on microvilli. Interaction may thus be initiated at a fairly long distance. In a very clever set of experiments, Patel et al. (1995) explored the importance of P-selectin molecular length. The extracellular part of the molecule comprises a terminal lectin group and epidermal growth factor homology domain that seem to be involved in binding. This structure is connected to a stalk made of nine repeats. Thus, Patel et al. made CHO cells express wild-type or mutated P-selectin with a various number of repeats, and they measured the transfectant capacity to bind polymorphonuclear neutrophils either under static conditions or under dynamic conditions: Static adhesion was determined by allowing 15 min of contact between sedimented neutrophils and CHO cells plated in culture wells, followed by gentle washing and quantification of the number of bound cells. Dynamic adhesion was studied in a flow chamber mimicking blood cell–endothelium interaction. Results were as follows:

- Under static conditions, all constructs bound neutrophils with comparable efficiency.
- Under dynamic conditions (wall shear rate = 200 or 400 s^{-1}), adhesive efficiency was comparable to that of P-selectin molecules containing between six and nine repeats. Adhesion was less efficient with shorter molecules, and it was abolished when mutant P-selectin contained three or fewer repeats.
- When transfected cells had decreased glycocalyx due to an enzymatic defect, the binding efficiency of shortened molecules tested under flow was somewhat restored.

Thus, this study provides compelling evidence that glycocalyx repulsion is particularly efficient under flow. Also, it provides an order of magnitude for the dynamic cell coat thickness: estimating at 3.8 nm the length contribution of a repeat segment, as suggested by the authors, it is concluded that dynamic adhesion is hampered when P-selectin is shortened by 15.2 nm, not 11.4 nm, yielding a dynamic endothelial glycocalyx thickness on the order of 25 nm.

Other experiments reported the same year by Sabri et al. (1995) show that dynamic repulsion may be demonstrated at a wall shear rate at least tenfold lower than found in blood. These authors studied the interaction between cells from the human THP-1 phagocytic cell line and microspheres bearing monoclonal antibodies specific for different THP-1 membrane molecules. Under static conditions, adhesion was very efficient, as between 40% and more than 80% of spheres sedimenting on a THP-1 cell remained bound after 15 min of incubation. Further, this efficiency was not significantly changed when cells were treated with neuraminidase. In contrast, when sphere–cell contact occurred under moderate shear flow (22 s^{-1}), binding efficiency was very poor, as less than 5% of contacts were conducive to adhesion, and this parameter was dramatically increased (up to 10-fold or more) when phagocytes were treated with neuraminidase. Similar conclusions were obtained by Foa et al. (1996) when bound phagocytes were made to interact with red cells that had been coated with opsonizing antibodies with or without treatment with neuraminidase.

4. Biological Evidence Is Supported by Experiments Done with Artificial Glycocalices

Although the aforementioned experiments may be considered fairly convincing, it is useful to emphasize that similar conclusions are suggested by experiments performed with biomimetic systems. Indeed, data obtained with simplified systems are expected to allow straightforward interpretation. Also, they may suggest applications of practical importance.

Holland et al. (1998) engineered a so-called artificial glycocalyx by preparing surfactant polymers that were adsorbed on artificial surfaces through alkanoyl chains, and exposed solvated dextran to the surrounding medium. This resulted in the formation of a stable layer 0.7–1.2 nm thick that efficiently prevented protein adsorption.

Cell adhesion could also be inhibited using this strategy: Houseman and Mrksich (2001) succeeded in preventing integrin-mediated cell adhesion to surfaces coated with RGD integrin ligand by adding oligo(ethyleneglycol) groups of varying size. Interestingly, adhesion was inversely related to the size of grafted antiadhesive groups.

To rule out the possibility that adhesion-related cell properties might be altered on cell encounters with glycocalyx-like elements, thus hampering the mechanical interpretation of glycocalyx antiadhesive properties, some experiments were done in our laboratory with a fully controlled artificial model: Glass slides were coated with a layer of streptavidin, then a mixture of biotinylated murine immunoglobulin and potentially antiadhesive molecules such as dextran (a branched D-glucose homopolymer), polyethyleneglycol, and hyaluronic acid (a linear polymer made of alternating N-acetylglucosamine and glucuronic acid residues; this may be considered as the simplest glycosaminoglycan). Slides were exposed to flowing microspheres coated with anti-murine immunoglobulin antibodies. As shown in Fig. 7.3, hyaluronic acid, but not dextran, efficiently inhibited sphere–surface adhesion. Enzyme linked immunoassay was used to check

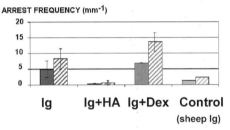

ARREST FREQUENCY (mm⁻¹)

Ig   Ig+HA Ig+Dex Control
(sheep Ig)

FIGURE 7.3 Effect of repeller molecules on adhesion under flow. Microspheres coated with anti-murine immunoglobulin were deposited on different surfaces under flow. Bead motion was monitored to determine the number and duration of arrests. The figure shows the frequency of binding events (full bars: arrest longer than 2 s, hatched bars: all arrests). Dex, dextran (MW 500,000); HA, hyaluronic acid (MW 700,000); Ig, murine immunoglobulin. Bar = 2 SE.

that binding inhibition was not due to a reduction of the number of glass-bound immunoglobulins. It was then hypothesized that the antiadhesive influence of hyaluronan was related to its capacity to prevent sphere–surface approach. This was tested by depositing glass spheres (about 10 μm in diameter) on slides and determining surface–sphere distance with quantitative treatment of sphere images obtained with interference reflection microscopy/reflection interference contrast microscopy (IRM/RICM) (Curtis et al., 1964; Kuhner et al., 1996). Results are summarized in Fig. 7.4. Clearly, the distance between spheres and hyaluronic acid–coated surfaces remained larger than the distance measured on dextran-coated surfaces by about 5–60 nm. Thus, it seemed reasonable to ascribe the antiadhesive potential of hyaluronic acid to higher size and capacity to repel adhesive spheres.

In conclusion, cell glycocalyx may hamper adhesive interactions, and this effect is more important when contacts are transient. Three questions are raised by these results:

- Is this concept of merely theoretical interest? In other words, is it useful only to let us understand why cells are coated with a glycocalyx, or is this process of physiological relevance?
- Is the influence of glycocalyx on adhesion a simple consequence of steric hindrance by bulky molecules, or are more complex cellular processes involved?
- Is the influence of glycocalyx on adhesion liable to biophysical modeling?

These questions are addressed in the next three sections of this chapter.

Thickness (nanometer)

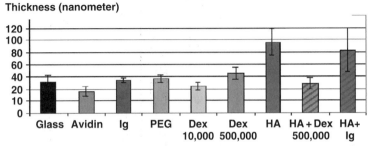

Glass Avidin   Ig   PEG   Dex   Dex   HA   HA+Dex HA+
                        10,000 500,000   500,000   Ig

FIGURE 7.4 Estimating the thickness of model surface layers. Reflection interference contrast microscopy was used to determine the distance between hovering spheres and surfaces coated with different molecular species: Dex, dextran (MW 10,000 or 500,000); HA, hyaluronic acid; Ig, immunoglobulin; PEG, polyethyleneglycol. Bar = 2x SE.

IV. GLYCOCALYX CHANGES MAY INFLUENCE IMPORTANT PHYSIOLOGICAL OR PATHOLOGICAL PROCESSES

A. Phagocytosis

Phagocytosis is a basic function of plants as well as animals, both unicellular and multicellular organisms. It is obviously essential for phagocytes of higher organisms to discriminate between autologous and foreign structures. For this purpose, they are endowed with an impressive variety of recognition structures, including pattern recognition receptors and immunoglobulin receptors, that enable them to bind and subsequently ingest antibody-coated particles. However, in view of the wide diversity of biomolecules, it is not obvious that a faultless recognition system might exist, and recent evidence suggests that in addition to molecular recognition, immune cells can detect warning messages that increase their level of sensitivity and activity (Matzinger, 2002). Indeed, it has long been known that many resting phagocytes are fairly inefficient and that they require proper activation to efficiently bind and destroy appropriate prey. How can this activation be achieved? One mechanism is to increase the density of a variety of membrane receptors through exocytosis and neosynthesis. However, recent evidence suggests that a much more economical way of increasing phagocyte activity might consist of reducing glycocalyx-mediated repulsion, thus simultaneously enhancing the activity of all membrane receptors.

This concept is supported by experimental evidence. First, phagocyte activation was reported to result in shedding of glycocalyx elements such as the aforementioned CD43/leukosialin. This involves both proteolytic cleavage (Rieu et al., 1992; Remold-O'Donnell and Parent, 1994) and sialic acid removal (Cross and Wright, 1991). These are phenomena associated with increased cell spreading or aggregation. Further, experimental evidence supports the hypothesis that both phenomena are causally related. Thus, while tumor necrosis factor α induces CD43 release and spreading in neutrophils, both responses are prevented when elastase and sialidase are blocked with human albumin (Nathan et al., 1993). More recently (Sabri et al., 2000), when THP-1 cells were activated with physiological stimuli such as γ interferon and plating of fibronectin-coated surfaces, they displayed increased capacity to bind antibody-coated spheres under flow, and this increase exhibited the same kinetics as CD43 dyssialylation, whereas no upregulation of membrane immunoglobulin receptors was observed.

Thus, much evidence supports the concept that a rapid consequence of phagocyte activation is a partial shedding of glycocalyx components, resulting in increased capacity to bind to foreign particles through various membrane receptors.

B. Adaptative Immune Response

The triggering of the adaptive immune response relies on adhesive events. Indeed, following the uptake of foreign antigens by dendritic cells or other phagocytes, T lymphocytes are stimulated by formation of a specific contact with antigen-presenting cells. Further, they may activate other T or B lymphocytes with both soluble mediators and contact interactions. Therefore, it is not surprising that immune cell activation is often reported to induce marked intercellular aggregation. The following two examples strongly suggest that this response is related to the glycocalyx and has functional relevance.

Treating B lymphocytes with bacterial lipopolysaccharide was found to enhance their capacity to adhere to T lymphocytes, which may be of obvious use in the development of antibody responses. Guthridge et al. (1994) demonstrated that adhesion is not related to increased expression of LFA-1 and ICAM-1 receptors, which are likely candidates for mediating attachment. However, a neuraminidase inhibitor was found to prevent this adhesive response.

More recently, Pappu and Shrikant (2004) reported that (1) decreasing surface sialylation of T lymphocytes with neuraminidase increases their capacity to be activated by antigen-presenting cells, and (2) activating T lymphocytes triggers both surface desialylation and proliferative response.

In conclusion, immune cell aggregation was found to favor immune response development. This aggregation is enhanced by glycocalyx degradation, and this degradation was indeed found to occur on lymphocyte activation. Thus, lowering the anti-adhesive barrier may be a general means of enhancing immune responses. In this respect, it is interesting to mention that the major histocompatibility complex, which is deeply connected to immune responses, was reported to harbor a sialidase that might be directly involved in immune regulation (Milner et al., 1997).

C. Leukocyte–Endothelium Interaction: Inflammation and Atherosclerosis

Atherosclerosis is a major cause of morbidity and mortality, and was recently recognized as an inflammatory disease (Ross, 1999). An early step of atherosclerosis is the attachment of monocytes to the blood vessel walls, with subsequent transformation into foam cells and formation of so-called fatty streaks. Monocyte–endothelium attachment is mediated by a variety of membrane receptors, many of which have been identified during the last 15 years. It is easily conceivable that the contact between these molecules may be somewhat impaired by the endothelial glycocalyx, which may be more than 100 nm thick (Squire et al., 2001). The validity of this fairly intuitive hypothesis was indeed demonstrated by experimental data. Thus, altering endothelial glycocalyx by *in vivo* heparitinase injection increased leukocyte–endothelium attachment in mouse cremaster muscle, and reciprocally, heparin sulfate or heparin decreased this interaction (Constantinescu et al., 2003).

Interestingly, oxidized lipoproteins, which are considered atherogenic, were demonstrated to decrease endothelial glycocalyx thickness by 60% in only 10 min (Vink et al., 2001), and these proteins are known to induce leukocyte–endothelium attachment (Constantinescu et al., 2003). Thus, glycocalyx maintenance or degradation may play an important role in the development of atherosclerosis.

D. Tumor Cell Metastasis

Cancer is also a major cause of morbidity and mortality. The main pathogenic process is metastatic invasion, which involves detachment of cancer cells from a primitive tumor, followed by their migration through various routes and binding to a target organ for metastasis formation. Thus, it is not surprising that many studies have revealed a strong relationship between tumor cell adhesive behavior and invasive capacity (Benoliel et al., 2003).

Many reports support the concept that (1) tumor cells often display marked glycocalyx abnormalities, and (2) cancer cell antiadhesive potential plays an important role in tumor development by influencing adhesion to target organs, and also by imparting tumor cells the capacity to resist adhesion and subsequent destruction by cells from the immune system such as cytotoxic T lymphocytes and natural killer cells. We now provide a few examples.

McFarland et al. (1995) showed that the sensitivity of tumoral T lymphocytes to destruction by cytolytic cells was increased when leukosialin/CD43 expression was prevented by gene targeting.

As emphasized by Wesseling et al. (1996), the antiadhesive mucin episialin is overexpressed on many carcinoma cells. They suggested that anti-adhesion may provide a possible explanation for the relative lack of correlation between E-cadherin expression and cancer evolution. Indeed, much evidence suggests that E-cadherin activity should be inversely correlated to tumor cell malignity, and it would be tempting to speculate that glycocalyx impairment of E-cadherin-mediated adhesion might at least partially account for the unexpectedly bad prognosis of tumors with high E-cadherin expression.

More recently, Tanaka et al. (2000) reported a negative correlation between the invasiveness of lung cancer cells and expression of polysialic acid, the antiadhesive capacity of which was described earlier.

In conclusion, it is well recognized that tumor cell malignity is related to invasive potential. However, due to the multiplicity of independent parameters involved in both adhesion and cancer development, it is very difficult to derive and validate general rules allowing us to predict the relationship between tumor cell adhesive behavior and aggressivity (André et al., 1990). The limited experimental evidence presently available may be an incentive to further explore the potential of glycocalyx manipulation for cancer management (Tanaka et al., 2000).

V. THE INFLUENCE OF GLYCOCALYX COMPONENTS ON CELL ADHESION IS NOT A MERE CONSEQUENCE OF STERIC HINDRANCE

In view of the aforementioned evidence, it would be quite tempting to speculate that the relationship between cell adhesion and glycocalyx is quite straightforward. Indeed, it would seem quite reasonable to conclude that (1) the presence of bulky surface elements between interacting membranes is sufficient to account for the inhibition of adhesion, (2) the exit of repellers from the contact region might suffice to restore attachment, and (3) living cells might occasionally take advantage of this simple mechanism to regulate adhesion when necessary.

The aim of this section is to convince the reader that this simple view may be somewhat oversimplified. We consider a biological model that recently attracted much attention to illustrate the general concept that nonspecific physical effects and subtle biological functions are often deeply enmeshed in actual molecular systems.

An essential step of the triggering of adaptive immune responses is the activation of antigen-specific T lymphocytes by antigen-presenting cells (APCs) bearing, on their membrane, complexes formed by antigen-derived oligopeptides and major histocompatability complex (MHC) molecules. The first step is nonspecific cell interaction. Recognition of specific oligopeptide–MHC complexes

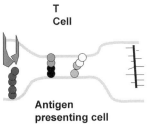

FIGURE 7.5 Immunological synapse. In a mature synapse, adhesion molecule couples are thought to be sorted according to their length (T-cell receptor + antigen or CD2 + CD58: about 16 nm; LFA-1 integrin and ICAM-1: about 40 nm; CD43: about 40 nm).

by a T lymphocyte results in contact extension with topographical sorting and arrangement of membrane molecules, leading to the formation of the so-called immunological synapse (Grakoui et al., 1999). This fosters suitable molecular redistribution, resulting in efficient lymphocyte activation.

A remarkable feature of the immunological synapse (Fig. 7.5) is the sorting of membrane molecules according to size, with a central region harboring ligand–receptor couples about 7–8 nm long and a peripheral gathering of larger molecules, including integrin–ligand couples and leukosialin with a length on the order of 40 nm (Fig. 7.5). This organization seems predictable on the basis of molecular size and the observation that alteration of molecular size through genetic engineering results in altered activation (Wild et al., 1999). It is thus appealing to consider molecular organization and adhesion as a consequence of simple physical phenomena, including steric exclusion of larger molecules from the peripheral zone. In this respect, CD43 exclusion should be required to achieve optimal adhesion and activation, and this exclusion should be a mere consequence of adhesion (Qi et al., 2001).

It is certainly useful to emphasize that this is not the whole story. Indeed, several clever experiments clearly established that (1) CD43 exclusion is driven by specific interaction with cytoskeletal elements such as moesin, rather than forces exerted on the extracellular domain (Delon et al., 2001); (2) the intracellular domain of CD43 is both necessary and sufficient to account for the effect on adhesion (Walker and Green, 1999) and, in any case, CD43 exclusion is not required for synapse formation (Savage et al., 2002). These findings are in line with numerous reports demonstrating a coupling between CD43 perturbation and generation of intracellular biochemical signals.

In conclusion, although a detailed discussion of this example does not fall into the scope of the present review, the data we describe show that the simple mechanical view of the relationship between glycocalyx properties and cell adhesion is not the whole story. Only careful analysis of well-defined models may yield an accurate understanding of the role played by glycocalyx-generated repulsion in adhesive phenomena.

VI. BIOPHYSICAL MODELING OF GLYCOCALYX REPULSION

To achieve a satisfactory understanding of the experimental data we have described, it is certainly useful to estimate the mechanical forces likely to be generated by glycocalyx elements during cell–cell contact. This problem is addressed in the present section. First, we describe quantitative properties of the cell surfaces. Second, we discuss the feasibility of relating cell surface properties to glycocalyx-generated forces.

A. Modeling the Cell Surface

The only way of predicting intercellular forces and assessing their functional importance is to rely on a quantitative model of cell surfaces. We sequentially consider static and dynamic properties. A full discussion does not fall into the scope of this chapter, and we refer the reader to previous reviews for additional details (Pierres et al., 2000).

1. A Static View of the Cell Surface

A molecular view of the cell surface was given in Fig.7. 1. There is only a need to add some quantitative estimates. The basis is the phospholipid bilayer, about 4.5 nm thick. As mentioned in the Introduction, this includes many transmembrane proteins which may occupy between about 10 and 40% of the available surface (Ryan et al., 1988). Cell surface proteins and lipids bear a number of oligosaccharide chains of typically about 10 residues. The length of a hexose group may be estimated as 0.57 nm, and the size of a protein core may vary between a few nanometers and a few tens of nanometers. Globular proteins are made of folded polypeptide chains. However, mucin-like molecules that are studded with O-linked oligosaccharide chains may be constrained to display an extended structure for structural reasons. The expected length per amino acid residue is about 0.25 nm (Jentoft, 1990).

The cell surface also includes a variety of proteoglycans that include a protein core and long polysaccharide chains composed of repeating disaccharide units. They may be several hundreds of nanometers long or longer. Thus, cells are surrounded by a polysaccharide-rich external layer, which warrants the name *glycocalyx*. As discussed earlier, the glycocalyx may be 100 to 500 nm thick, or thicker. The boundary may be difficult to determine because it may merge with the extracellular matrix. Thus, several authors have reported the external cell layer to be more than 1 μm thick.

Now, there remains the molar concentration of glycocalyx elements to estimate, which may vary widely. The glycocalyx may include a few polysaccharide residues per square nanometer, with comparable orders of magnitude for the components of oligosaccharide chains and glycosaminoglycans. Finally, some of these residues bear negative charges, including sulfate groups and sialic acid residues that often cap oligosaccharide chains. It seems acceptable to estimate surface charges at about 0.1 negative electronic charge per square nanometer.

A micrometer-scale view of the cell surface is also important. In contrast to erythrocytes, nucleated cells are often studded with protrusions of various shapes, appearing as lamellipods or cylindrical microvilli. The typical diameter of microvilli, or lamellipodium thickness, is about 0.1 μm. Their number may be estimated by considering that the bilayer area is 30–60% larger than the area of a sphere enclosing the cell volume (this estimate was obtained for blood cells). Thus, the initial intercellular contact may be modeled as initial contact between a 0.1-μm-diameter cylinder and a fairly flat surface.

2. Cell Surface Rheology and Active Movements

To estimate the potential influence of glycococalyx elements on cell–cell approach and bond formation, we need accurate information on active and passive deformations. Clearly, such information is not available at present, as there is no general mechanical model for the cell surface. A consequence of this lack of a satisfactory model is that it is very difficult to extrapolate numerical constants obtained under particular conditions to movements related to adhesion. To illustrate this point, we now quote a few interesting data, while emphasizing that the use of estimated constants may not be warranted in situations markedly different from initial observations.

a. Glycocalyx Viscosity and Mechanical Resistance

Pericellular matrix *viscosity* was estimated by Lee et al. (1993b). Using the fluorescence recovery after photobleaching (FRAP) technique, they measured the diffusion coefficient of fluorescent phosphatidylethanolamine incorporated into the plasma membrane bilayer. They also measured the diffusion coefficient of phophatidylethanolamine labeled with 30-nm-diameter colloidal gold particles. For this purpose, they used enhanced video microscopy; the diffusion coefficients were about 7.5×10^{-9} and 1.4×10^{-9} cm²/s, respectively. Ascribing the difference between these values to the viscous interaction between gold particles and the pericellular matrix, they estimated the viscosity at 0.05–0.09 Pa-s (i.e., 50- to 90-fold higher than that of water). When cells were treated with heparitinase, the apparent viscosity decreased to about 0.01 Pa-s.

Interestingly, this estimate is consistent with an earlier report from Tank et al. (1982), who showed that the diffusion coefficients of several membrane receptors in cytoskeleton-free regions was on the order of 3×10^{-9} cm²/s. Modeling membrane receptors as spheres of 2-nm radius and assuming that the extracellular part of the molecules accounted for most of the resistance to motion, due to the lack of cytosleketal action on intracellular domains, we may use Stokes law to estimate local viscosity; the value of about 1 Pa-s obtained is consistent with the aforementioned estimates within a factor of 10.

A similar estimate for the order of magnitude of the pericellular matrix may be obtained from an interesting report by Wier and Edidin (1988), who used FRAP to compare the mobility of wild-type MHC molecules and mutated molecules that have lost three glycosylation sites; the diffusion constants D were 6×10^{-10} and 17×10^{-10} cm²/s, respectively. If the decrease in the friction constant (i.e., kT/D) is ascribed to a decrease in the effective molecule radius, on the order of 1 nm, the estimated viscosity is on the order of 1 Pa-s.

Admittedly, these estimates rely on crude assumptions, and specific interactions between diffusing molecules and their environment may dramatically alter the estimated viscosity.

b. Glycocalyx Elasticity

It would be of obvious interest to know what force is required to make microvilli penetrate into cell coats to allow molecular contact between potentially adhesive molecules. Data are very scarce and difficult to interpret. A specific example is the endothelial cell coat. Much evidence (reviewed by Pries et al., 2000) suggests that the blood vessels are covered with a so-called

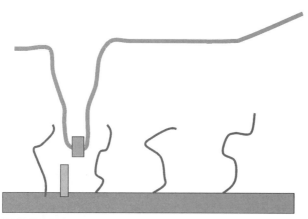

FIGURE 7.6 Endothelial glycocalyx. The endothelial cell glycocalyx may be viewed as a barrier narrowing the flowing blood column and preventing approach of the cell toward the endothelial layer.

endothelial surface layer (ESL) that prevents blood flow near cell surfaces as well as cell approach (Fig. 7.6).

The outward force exerted on erythrocytes was estimated to be about 6 N/m². Thus, the repulsion of a 0.1-μm-diameter microvillus would be on the order of 0.06 pN. However, it is very difficult to use this estimate to derive the force potentially exerted by the glycocalyx in other circumstances. Also, the mechanical properties of the ESL may be quite different from those of other glycocalices.

c. Cell Surface Passive Mechanical Properties

By use of the fairly naive model of bond formation sketched in Fig. 7.7, it is useful to try and estimate the velocity and force exerted by cell protrusions penetrating the glycocalyx. First, a convenient order of magnitude for the displacement velocity of a thin cell lamellipodium was estimated to be about 50 nm/s by Rotsch et al. (1999). These authors used atomic force microscopy to monitor the kinetic and mechanic properties of active fibroblast edges.

The force exerted by a cell to push forward a protrusion is not well known. A total force of 30 nN was required to stop a polymorphomonuclear leukocyte moving in a capillary tube (Usami et al., 1992). Unfortunately, it is difficult to relate this value to the protrusion force generated by microvilli or lamellipodia.

B. Models for Intercellular Forces Potentially Generated by the Glycocalyx

At this point in the chapter, it is highly desirable to take advantage of the aforementioned properties of cell surface structure to derive estimates for intercellular forces. Unfortunately, this is not feasible at present for two

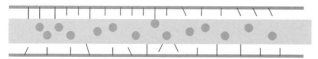

FIGURE 7.7 Model for cell–cell approach. The tip of a cell protrusion (modeled as a sphere of 0.05-μm radius) approaches a neighboring cell through repulsive glycocalyx elements.

major reasons: First, the cell surfaces properties are not known with enough accuracy to allow precise modeling. Second, even if we were able to provide a very accurate molecular model of the glycocalyx, no fully satisfactory theory would be available to predict interactions. This difficulty is illustrated in recent estimates of interaction forces between microspheres and surfaces coated with hyaluronic acid, which may be considered the simplest model of cell coat because it is an unbranched flexible polyelectrolyte chain (Albersdorfer and Sackmann, 1999).

Thus, we discuss only the expected order of magnitude of several basic interaction forces, which are considered separately. The challenge is to account simultaneously for all interactions. We consider sequentially electrodynamic, electrostatic, and steric forces *at equilibrium*. Dynamic effects are then very briefly considered. The reader is referred to more general reviews for overall information, and we essentially discuss the link between models of the cell surface and expected interactions.

1. Electrodynamic Interactions

It is well known that the interaction energy W between two half-spaces of media 1 and 2 separated by an empty gap of thickness d is

$$W = -A_{12}/12\pi d^2,$$

where A_{12} is the so-called Hamaker constant. When interaction occurs in a material medium (0), we must use an effective constant:

$$A_{102} = A_{12} - A_{10} - A_{20} + A_{00}.$$

As reviewed by Pierres et al. (2000), the Hamaker constant may be derived from material constants of interacting media, including refractive index and dielectric constant, based on Lifshitz theory. By use of structural models for the cell surface, as described earlier, the electrodynamic interaction between approaching cells was calculated by modeling these surfaces as phospholipid bilayers 4.5 nm thick covered with a 20-nm-thick glycocalyx. The estimated Hamaker constants were 3.65×10^{-21}, 0.6×10^{-21}, and 0.1×10^{-21} J, respectively, for the bilayer–bilayer, bilayer–cell coat and cell coat–cell coat interactions in water (see also Yu et al., 1998, for similar estimates and direct observation with the surface force apparatus). The interaction energy was rather low due to the small value of the Hamaker constant for glycocalyx–glycocalyx interaction energy (Pierres et al., 2000). However, a significantly higher interaction may be obtained when the inter-bilayer distance d is lower than twice the glycocalyx thickness ($d < 2L$). This may be calculated as follows:

Let n be the average number of sugar residues per unit volume in the glycocalyx, and let α/r^6 be the electrodynamic interaction energy between two sugar groups separated by distance d in water. Assuming that *this is low enough not to alter the spatial distribution of molecules,* the interaction energy between a sugar residue and surrounding chains may be obtained by straightforward integration:

$$W = \int_{2a}^{\infty} \alpha n \, dv/r^6 = \pi \alpha n/6a^3.$$

Here, a is the residue radius, and all interactions are considered additive.

Now, the relationship between α and Hamaker constant may be obtained by straightforward integration of the interaction between two half-spaces separated by a distance d, yielding $A = \pi^2 \, \alpha n^2$. When cell coats touch each other, the total interaction energy between sugars per unit area is

$$W = (1/2) \, 2Ln \, \pi\alpha n/6a^3 = AL/(3\pi a^3).$$

When the intercellular distance d is decreased, assuming that the density of cell layers is increased by a factor $2L/d$, the energy increase per unit area is (noting that A is multiplied by $(2L/d)^2$),

$$W = [AL/6\pi a^3] \, [(2L/d) - 1].$$

The attractive force F per unit area is therefore

$$F = (A/3\pi a^3)(L/d)^2.$$

Estimating at about 1 nm the minimum distance between the centers of sugar residues, which is obviously only an order of magnitude, we obtain an estimate for the electrodynamic attraction, taking $L = 20$ nm and $A = 0.1 \times 10^{-21}$ J. Numerical values are shown in Fig. 7.8.

2. Electrostatic Repulsion

According to the Debye–Hückel theory, which is approximately valid in physiological media, the interaction energy between two charges q_1 and q_2 at distance r may be written

$$W = q_1 q_2 \exp(-\kappa r)/4\pi\varepsilon r,$$

where $1/\kappa$ is the Debye–Hückel length, which is close to 0.8 nm, and ε is the dielectric permittivity, which is about 6.9×10^{-10} MKS units in water. Following a straightforward integration, the energy density in a medium containing a charge density ρ is $\rho^2/2\varepsilon\kappa^2$. This equation can be used to estimate the energy increase resulting from the interaction between two surfaces coated with a layer of charge density ρ and thickness L, and separated by a distance d smaller than $2L$:

$$W = \rho^2 L(2L/d - 1)/\varepsilon\kappa^2 \text{ (per unit area).}$$

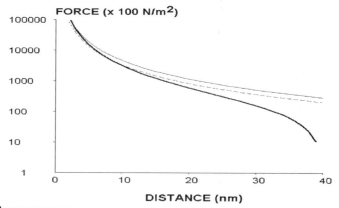

FIGURE 7.8 Order of magnitude of interactions potentially mediated by electrodynamic, electrostatic, and steric forces. Estimates were obtained with gross approximations as explained in Section VI. (B) Broken line : electrodynamic attraction. Thin continuous line: electrostatic repulsion. Thick continuous line: steric repulsion.

Further, the repulsive force per unit area is

$$F = (2\rho^2/\varepsilon\kappa^2) \, (L/d)^2.$$

Note that this formula was obtained by assuming that surface layers do not intermix and have equal thickness $d/2$ after approach. Numerical values were obtained by assuming a charge density of 1 electron charge per square nanometer. Results are shown in Fig. 7.8.

3. Steric Stabilization

It is now well recognized that surfaces bearing flexible polymeric chains repel each other as a consequence of the loss of freedom — and therefore entropy — resulting from the decrease in available space. This phenomenon may be called *steric stabilization* (Napper, 1977). Despite the major progress made because of the work of de Gennes (1979, 1987, 1988, for reviews), theories of interaction are still considered quite complex, particularly when polymers are physically adsorbed onto surfaces, and their density may vary during approach. However, the situation is much simpler when polymers are bound to the surface through a single anchoring point, and their density may be considered as constant.

It seems reasonable that in some cases the polymer chains constituting the glycocalyx are close enough to interact, which means that the average distance between anchoring points is lower than the natural radius of free chains (Lee et al., 1993a; Squire et al., 2001). In this case, chains are more extended than when they are isolated, and they may generate fairly long-range interactions. Also, it seems reasonable to consider polysaccharide hydrophilic, which means that water is a good solvent. The repulsive force F per unit area between cell surfaces coated with a glycocalyx L thick may be estimated on the basis of de Gennes' theory (de Gennes, 1987; Israelachvili, 1991) as

$$F \approx (kT/s^3) \, [(2L/d)^{9/4} - (d/2L)^{3/4}],$$

where k is Boltzmann's constant, T is absolute temperature, and $P(D)$ is the repulsive pressure. Assuming a glycocalyx thickness L of 20 nm and mean chain spacing s of 7 nm, the estimated repulsion is shown in Fig. 7.8. Remarkably, electrodynamic, electrostatic, and steric forces estimated in a fairly independent way share the same order of magnitude within a factor of 10.

As explained earlier, the actual interaction energy between approaching cell surfaces is not the sum but a complex combination of electrodynamic, electrostatic, and entropic effects. The purpose of this simplified discussion is only to look for an order-of-magnitude estimate of the expected importance of different kinds of interaction.

4. Dynamics

An additional cause of complexity is that approaching cells are out of equilibrium, as demonstrated by the aforementioned experiments on the dynamic features of glycocalyx-generated repulsion. In view of Fig. 7.7, a first approximation might consist of calculating cell–cell repulsion as the force resisting the approach of a sphere of radius $R = 0.05 \, \mu m$ (corresponding to the tip of a microvillus) approaching a surface at a distance d on the order of 40 nm (corresponding to the total glycocalyx thickness), with velocity v on the order of 50 nm/s (corresponding to the velocity fluctuation of the tip of a lamellipodium).

Although the viscoelastic properties of the glycocalyx remain poorly known, a first-order estimate might consist of assuming that the glycocalyx layer increases the medium viscosity relative to water by a factor k ranging from 5 to more than 100, depending on polymer structure (see the experiments by Kuhner and Sackmann, 1996, and Albersdorffer et al., 1999).

The expected hydrodynamic force may thus be estimated as

$$F = (6\pi\eta Rv)(R/d)(k).$$

The first term on the right-hand side represents Stokes law, the second term represents the hydrodynamic repulsion between the protrusion and the wall, and k represents the effective viscosity increase due to the polymer layer. Using numerical estimates discussed earlier, F should be lower than 0.01 pN when contact is mediated by the tip of microvilli. This estimate may be increased by 10,000-fold if contact is mediated by a smooth cell body. Cell roughness is therefore expected to play a major role in modulating glycocalyx anti-adhesive potential and adhesion efficiency.

VII. CONCLUSION

A. Multiple Roles for the Glycocalyx?

The main purpose of this review was to discuss the influence of glycocalyx on cell adhesion. However, it is in order to briefly mention other possible functions of the pericellular matrix. First, we discuss two mechanisms that are indeed part of the adhesive processes.

1. De-adhesion

Although most investigators studying cell adhesion ask how cells can manage to bind to other surfaces, it sometimes appears that it is more difficult for a cell to detach from a surface than to initiate adhesion. A well-documented example is cells crawling on a surface: many reports have convincingly demonstrated that cell motility may be impaired with excessive substrate adhesiveness. Indeed, cells may find it difficult to release their posterior part, namely, the uropod, to move forward. Detachment may be achieved by pulling (using myosin-based motors) or by proteolytic degradation of adhesion molecules. An intriguing possibility discussed by Seveau et al. (2000) consisted of concentrating anti-adhesive molecules such as CD43 in contact areas to promote separation.

2. Pro- and Anti-Adhesive Function of the Glycocalyx

Because the glyocalyx is the outermost part of the cell surface, it is well suited to prevent and to initiate adhesion. It is therefore only an apparent contradiction to state that the glycocalyx may be both pro- and anti-adhesive. Thus, mucin-like molecules such as CD43 may prevent adhesion in some circumstances. However, PSGL-1, the mucin-like ligand of P-selectin, seems to play an important role in promoting the attachment of flowing leukocytes to inflamed endothelial walls. This dual function of mucins was recently emphasized by Fukuda (2002).

3. Filter

It was emphasized that the glycocalyx acts as a sieve, with a selection of medium components reaching the cell membrane. This function, which is also ascribed to bacterial outer structures, was particularly well studied on endothelial cells (Pries et al., 2000).

4. Adhesion Sensor

As mentioned earlier, mechanical stimulation of glycocalyx elements such as CD43 triggers intracellular signaling cascades. Thus, it might be conceivable that some cell surface molecules may act as adhesion sensors that provide a general way for cells to be aware of the proximity of another surface, whatever the adhesion molecules. To our knowledge, this possibility has not been fully investigated.

5. Handle to Pull at the Membrane

Some years ago, Yonemura et al. (1993) reported the unexpected finding that CD43/leukosialin is concentrated in the cleavage furrow during cytokinesis. It is tempting to speculate that this phenomenon may be somewhat related to the physical properties of the extracellular domain. If this is the case, CD43 may help generate membrane repulsion during the separation of daughter cells. Alternatively, it may act as a handle to help cytoskeletal elements pull at the plasma membrane without tearing it. Clearly, it would be useful to address this situation on a quantitative basis to assess the potential use of bulky extracellular domains.

B. Perspective

For a physical chemist well aware of the mechanisms involved in colloid stability, it may seem fairly obvious that the glycocalyx should provide the only way of avoiding nonspecific cell adhesion in concentrated electrolyte solutions such as physiological media. In view of the aforementioned evidence, the complexity of cell biological systems is such that it seems both useful and difficult to prove the validity of this fairly naive view. In this respect, it is useful to determine an approximate estimate of the time-dependent intercellular forces generated by a membrane–membrane approach. It seems very difficult to obtain reliable theoretical estimates due to our limited knowledge of cell surface molecular structure and the complexity of polymer theories. Perhaps in the short term, the most attractive way of tackling this problem may involve the study of dynamic interactions between biomimetic systems, as illustrated by so-called colloidal probe techniques (Kühner and Sackmann, 1996; Albersdörfer and Sackmann, 1999; Picart et al., 2004). A major challenge would be building more and more realistic systems, with increasing resemblance to biological models.

SUGGESTED READING

Arnott, S., Rees, D. A., and Morris, E. R. (1984) *Molecular Biophysics of the Extracellular Matrix*. Humana Press, Clifton, NJ.

A useful complement to Hay's book, with emphasis on biophysical principles.

De Gennes, P. G. (1979) *Scaling Law in Polymer Physics.* Cornell University Press, Ithaca, NY.

Now a classic on polymer physics that lends insight into glycocalyx structure and generated forces.

De Gennes, P. G. (1988) Model polymers at interfaces. In *Physical Basis of Cell–Cell Adhesion* (P. Bongrand, Ed.), pp. 39–60. CRC Press, Boca Raton, FL.

A simplified view with emphasis on cellular systems.

Hay, E. D. (1991) *Cell Biology of Extracellular Matrix.* Plenum, New York.

A general presentation of the cell surface from a cell biologist's perspective.

Israelachvili, J. N. (1991) *Intermolecular and Surface Forces.* Academic Press, New York.

A major and very convenient reference on all aspects of intermolecular forces.

Pierres, A., Benoliel, A-M., and Bongrand, P. (2000) Cell–cell interactions. In *Physical Chemistry of Biological Interfaces* (A. Baszkin and W. Nord, Eds.), pp. 459–522. Marcel Dekker, New York.

A review including data on cell structure and adhesion molecules as well as application to cellular systems of general principles of intermolecular forces.

REFERENCES

Albersdörfer, A., and Sackmann, E. (1999) Swelling behavior and viscoelasticity of ultrathin grafted hyaluronic acid films. *Eur. Phys. J. B* **100**: 663–672.

Alon, R., Hammer, D. A., and Springer, T. A. (1995) Lifetime of P-selectin-carbohydrate bond and its response to tensile force in hydrodynamic flow. *Nature* **374**: 539–542.

André, P., Benoliel, A. M., Capo, C., Foa, C., Buferne, M., Boyer, C., Schmitt-Verhulst, A. M., and Bongrand, P. (1990) Use of conjugates made between a cytolytic T cell clone and target cells to study the redistribution of membrane molecules in contact areas. *J. Cell Sci.* **97**: 335–347.

Bennett, H. S. (1963) Morphological aspects of extracellular polysaccharides. *J. Histochem. Cytochem.* **11**: 14–23.

Benoliel, A. M., Pirro, N., Marin, V., Consentino, B., Pierres, A., Vitte, J., Bongrand, P., Sielezneff, I., and Sastre, B. (2003) Correlation between invasiveness of colorectal tumor cells and adhesive potential under flow. *Anticancer Res.* **23**: 4891–4896.

Capo, C., Garrouste, F., Benoliel, A. M., Bongrand, P., and Depieds, R. (1981) Nonspecific binding by macrophages: Evaluation of the influence of medium-range electrostatic repulsion and short-range hydrophobic interaction. *Immunol. Commun.* **10**: 35–43.

Clarris, B. J., and Fraser J. R. E. (1968) On the pericellular zone of some mammalian cells *in vitro. Exp. Cell Res.* **49**: 181–193.

Constantinescu, A. A., Vink, H., and Spaan, J. A. E. (2003) Endothelial cell glycocalyx modulates immobilization of leukocytes at the endothelial surface. *Arterioscler. Thromb.Vasc. Biol.* **23**: 1541–1547.

Cross, A. S., and Wright, D. G. (1991) Mobilization of sialidase from intracellular stores to the surface of human neutrophils and its role in stimulated adhesion responses of these cells. *J. Clin. Invest.* **88**: 2067–2076.

Curtis, A. S. G. (1964) The mechanism of adhesion of cells to glass. *J. Cell Biol.* **20**: 199–215.

De Gennes, P. G. (1979) *Scaling Law in Polymer Physics.* Cornell University Press, Ithaca, NY.

De Gennes, P. G. (1987) Polymers at an interface: A simplified view. *Adv. Colloid Interface Sci.* **27**: 189–209.

De Gennes, P. G. (1988) Model polymers at interfaces. In *Physical Basis of Cell–Cell Adhesion* (P. Bongrand, Ed.), pp. 39–60. CRC Press, Boca Raton, FL.

Delon, J., Kaibuchi, I., and Germain, R. N. (2001) Exclusion of CD43 from the immunological synapse is mediated by phosphorylation-regulated relocation of the cytoskeletal adaptor moesin. *Immunity* **15**: 691–701.

Detmers, P. A., Wright, S. D., Olsen, E., Kimball, B., and Cohn, Z. A (1987) Aggregation of complement receptors of human neutrophils in the absence of ligand. *J. Cell Biol.* **105**: 1137–1145.

Erlandsen, S. L., Hasslen, S. R, and Nelson, R. D. (1993) Detection and spatial distribution of the α2-integrin (Mac-1) and L-selectin (LECAM-1) adherence receptors on human neutrophils by high-resolution field emission SEM. *J. Histochem. Cytochem.* **41**: 327–333.

Foa, C., Soler, M., Benoliel, A. M., and Bongrand, P. (1996) Steric stabilization and cell adhesion. *J. Mater. Sci: Mater. Med.* **7**: 141–148.

Fukuda, M. (2002) Role of mucin-type O-glycans in cell adhesion. *Biochim. Biophys. Acta* **1573**: 394–405.

Grakoui, A., Bromley, S. K., Sumen, C., Davis, M. M., Shaw, A. S., Allen, P. M., and Dustin, M. L. (1999) The immunological synapse: A molecular machine controlling T cell activation. *Science* **285**: 221–227.

Guthridge, J. M., Kaplan, A. M., and Cohen, D. A. (1994) Regulation of B cell: T-cell interaction: Potential involvement of an endogenous B cell sialidase. *Immunol. Invest.* **23**: 393–411.

Heath, J. P. (1982) Adhesions to substratum and locomotory behaviour of fibroblastic and epithelial cells in culture. In *Cell Behaviour* (R. Bellairs, A. Curtis, and G. Dunn, Eds.), pp. 77–108. Cambridge University Press, Cambridge.

Holland, N.B., Qiu, Y., Ruegsegger, M., and Marchant, R. E. (1998) Biomimetic engineering of non-adhesive glycocalyx-like surfaces using oligosaccharide surfactant polymers. *Nature* **392**: 799–801.

Horstkorte, R., Lessner, N., Gerardyschahn, R., Lucka, L., Danker, K., Reutter, W. (1999) Expression of the polysialyltransferase ST8SiaIV: Polysialylation interferes with adhesion of PC12 cells in vitro. *Exp. Cell Res.* **246**: 122–128.

Houseman, B. T., and Mrksich, M. (2001) The microenvironment of immobilized Arg–Gly–Asp peptides is an important determinant of cell adhesion. *Biomaterials* **22**: 943–955.

Israelachvili, J. N. (1991) *Intermolecular and Surface Forces.* Academic Press, New York.

Jentoft, N. (1990) Why are proteins O-glycosylated? *Trends Biochem. Sci.* **15**: 291–294.

Kucik, D. F., Dustin, M. L., Miller, J. M., and Brown, E. J. (1996) Adhesion-activating phorbol ester increases the mobility of leukocyte integrin LFA-1 in cultured lymphocytes. *J. Clin. Invest.* **97**: 2139–2144.

Kühner, M., and Sackmann, E. (1996) Ultrathin hydrated dextran films grafted on glass: Preparation and characterization of structural, viscous and elastic properties by quantitative microinterferometry. *Langmuir* **12**: 4866–4876.

Lee, G. M., Johnstone, B., Jacobson, K., and Caterson, B. (1993a) The dynamic structure of the pericellular matrix on living cells. *J. Cell Biol.* **123**: 1899–1907.

Lee, G. M., Zhang, F., Ishihara, A., McNeil, C. L., and Jacobson, K. A. (1993b) Unconfined lateral diffusion and estimate of pericellular matrix viscosity revealed by measuring the mobility of gold-tagged lipids. *J. Cell Biol.* **120**: 25–35.

Martinez-Palomo, A. (1970) The surface coats of animal cells. *Int. Rev. Cytol.* **29**: 29–75.

Matzinger, P. (2002) The danger model: A renewed sense of self. *Science* **296**: 301–305.

McFarland, T. A., Ardman, B., Manjunath, N., Fabry, J. A., and Lieberman, J. (1995) CD43 diminishes susceptibility to T lymphocyte-mediated cytolysis. *J. Immunol.* **154**: 1097–1104.

Milner, C. M., Smith, S. V., Carrillo, M. B., Taylor, G. L., Hollinshead, M., and Campbell, R. D. (1997) Identification of a sialidase encoded in the human major histocompatibility complex. *J. Biol. Chem.* **272**: 4549–4558.

Napper, D. H. (1977) Steric stabilization. *J. Colloid Interface Sci.* **58**: 390–407.

Nathan, C., Xie, Q., Halbwachs-Mecarelli, L., and Jin, W. (1993) Albumin inhibits neutrophil spreading and hydrogen peroxide release by blocking the shedding of CD43 (sialophorin, leukosialin). *J. Cell Biol.* **122**: 243–256.

Ostberg, J. R., Dragone, L. L., Driskell, T., Moynihan, J. A., Phipps, R., Barth, R. K., and Frelinger, J. G. (1996) Disregulated expression of CD43 (leukosialin, sialophorin) in the B cell lineage leads to immunodeficiency. *J. Immunol.* **157**: 4876–4884.

Palecek, S. P., Loftus, J. C., Ginsberg, M. H., Lauffenburger, D. A., and Horwitz, A. F. (1997) Integrin–ligand binding properties govern cell migration speed through cell–substratum adhesiveness. *Nature* **385**: 537–539.

Patel, K. D., Nollert, M. U., and McEver, R. P. (1995) P-selectin must extend a sufficient length from the plasma membrane to mediate rolling of neutrophils. *J. Cell Biol.* **131**: 1893–1902.

Pappu, B. P., and Shrikant, P. A. (2004) Alteration of cell surface sialylation regulates antigen-induced naive CD8$^+$ T cell responses. *J. Immunol.* **173**: 275–284.

Picart, C., Sengupta, K., Schilling, J., Maurstad, G., Ladam, G., Bausch, A. R., and Sackmann, E. (2004) Microinferometric study of the structure, interfacial potential, and viscoelastic properties of polyelectrolyte multilayer films on a planar substrate. *J. Phys. Chem. B* **108**: 7196–7205.

Pierres, A., Benoliel, A-M., and Bongrand, P. (2000) Cell–cell interactions. In *Physical Chemistry of Biological Interfaces* (A. Baszkin and W. Nord, Eds.), pp. 459–522. Marcel Dekker, New York.

Pries, A. R., Secomb, T. W., and Gathtgens, P. (2000) The endothelial surface layer. *Eur. J. Physiol.* **440**: 653–666.

Qi, S. Y., Groves, J. T., and Chakraborty, A. K. (2001) Synaptic pattern formation during cellular recognition. *Proc. Natl. Acad. Sci. USA* **98**: 6548–6553.

Razi, N., and Varki, A. (1998) Masking and unmasking of the sialic acid-binding lectin activity of CD22 (Siglec-2) on B lymphocytes. *Proc. Natl Acad. Sci. USA* **95**: 7469–7474.

Remold-O'Donnell, E., and Parent, D. (1994) Two proteolytic pathways for downregulation of the barrier molecule CD43 of human neutrophils. *J. Immunol.* **152**: 3595–3605.

Rieu, P., Porteu, F., Bessou, G., Lesavre, P., and Halbwachs-Mecarelli, L. (1992) Human neutrophils release their major membrane sialoprotein, leukosialin (CD43), during cell activation. *Eur. J. Immunol.* **22**: 3021–3026.

Ross, R. (1999) Atherosclerosis: An inflammatory disease. *N. Engl. J. Med.* **340**: 115–136.

Rotsch, C., Jacobson, K., and Radmacher, M. (1999) Dimensional and mechanical dynamics of active and stable edges in motile fibroblasts investigated by using atomic force microscopy. *Proc. Natl. Acad. Sci. USA* **96**: 921–926.

Rutishauser, U., Acheson, A., Hall, A. K., Mann, D. M., and Sunshine, J. (1988) The neural cell adhesion molecule (NCAM) as a regulator of cell-cell interactions. *Science* **240**: 53–57.

Ryan, T. A., Myeres J., Holowka, D., Baird, B., and Webb, W. W. (1988) Molecular crowding on the cell surface. *Science* **239**: 61–64.

Sabri, S., Pierres, A., Benoliel, A. M., and Bongrand, P. (1995) Influence of surface charges on cell adhesion: Difference between static and dynamic conditions. *Biochem. Cell Biol.* **73**: 411–420.

Sabri, S., Soler, M., Foa, C., Pierres, A., Benoliel, A. M., and Bongrand, P. (2000) Glycocalyx modulation is a physiological means of regulating cell adhesion. *J. Cell Sci.* **113**: 1589–1600.

Savage, N. D. L., Kimzey, S. L., Bromley, S. K., Johnson, K. G., Dustin, M. L., and Green, J. M. (2002) Polar redistributuion of the sialoglycoprotein CD43: implications for T cell function. *J. Immunol.* **168**: 3740–3746.

Seveau, S., Keller, H., Maxfield, F. R., Piller, F., and Halbwachs-Mecarelli, L. (2000) Neutrophil polarity and locomotion are associated with surface redistribution of leukosialin (CD43), an anti-adhesive membrane molecule. *Blood* **95**: 2462–2470.

Soler, M., Desplat-Jego, S., Vacher, B., Ponsonnet, L., Fraterno, M., Bongrand, P., Martin, J. M., Foa, C. (1998) Adhesion-related glycocalyx study: Quantitative approach with imaging-spectrum in the energy filtering transmission electron microscope. *FEBS Lett.* **429**: 89–94.

Soler, M., Merant, C., Servant, C., Fraterno, M., Allasia, C., Lissitzky, J. C, Bongrand, P., and Foa, C. (1997) Leukosialin (CD43) behavior during adhesion of human monocytic THP-1 cells to red blood cells. *J. Leukoc. Biol.* **61**: 609–618.

Squire, J. M., Chew, M., Nneji, G., Neal, C., Barry, J., and Michel, C. (2001) Quasi-periodic substructure in the microvessel endothelial glycocalyx: A possible explanation for molecular filtering? *J. Struct. Biol.* **136**: 239–255.

Stockton, B. M., Cheng, G., Manjunath, N., Ardman, B., and Von Andrian, U. (1998) Negative regulation of T cell homing by CD43. *Immunity* **8**: 373–381.

Tanaka, F., Otake, Y., Nakagawa, T., Kawano, Y., Miyahara, R., Li, M., Yanagihara, K., Nakayama, J., Fujimoto, I., Ikenaka, K., and Wada, H. (2000) Expression of polysialic acid and STX, a human polysialyltransferase, is correlated with tumor progression in non small cell lung cancer. *Cancer Res.* **60**: 3072–3080.

Tank, D. W., Wu, E. S., and Webb, W. W. (1982) Enhanced molecular diffusivity in muscle membrane blebs: release of lateral constraints. *J. Cell Biol.* **92**: 207–212.

Tha, S. P., Shuster, J., and Goldsmith, H. L. (1986) Interaction forces between red cells agglutinated by antibody: IV Time and force dependence of break-up. *Biophys. J.* **50**: 1117–1126.

Usami, S., Wung, S. L., Skierczynski, B. A., Skalak, R., and Chien, S. (1992) Locomotion forces generated by a polymorphonuclear leucocyte. *Biophys. J.* **63**: 1663–1666.

Vink, H., Constantinescu, A. A., and Spaan, J. A. E. (2001) Oxidized lipoproteins degrade the endothelial surface layer: Implications for platelet–endothelial cell adhesion. *Circulation* **101**: 1500–1502.

Vitte, J., Benoliel, A. M., Eymeric, P., Bongrand, P., and Pierres, A. (2004) Beta 1 integrin-mediated adhesion may be initiated by multiple incomplete bonds, thus accounting for the functional importance of receptor clustering. *Biophys. J.* **86**: 4059–4074.

Von Andrian, U. H., Hasslen, S. R., Nelson, R. D., Erlandsen, S. L., and Butcher, E.C. (1995) A central role for microvillus receptor presentation in leukocyte adhesion under flow. *Cell* **82**: 989–999.

Walker, J., and Green, J. M. (1999) Structural requirements for CD43 function. *J. Immunol.* **162**: 4109–4114.

Wesseling, J., van der Valk, S. W., and Hilkens, J. (1996) A mechanism for inhibition of E-cadherin-mediated cell–cell adhesion by the membrane associated mucin episialin MUC-1. *Mol. Biol. Cell* **7**: 565–577.

Wesseling, J., Vandervalk, S. W., Vos, H. L., Sonnenberg, A., and Hilkens, J. (1995) Episialin (MUC1) overexpression inhibits integrin-mediated cell adhesion to extracellular matrix components. *J. Cell Biol.* **129**: 255–265.

Wier, M., and Edidin, M. (1988) Constraint of the translational diffusion of a membrane glycoprotein by its external domain. *Science* **242**: 412–414.

Wild, M. K., Cambiaggi, A., Brown, M. H., Davies, E. A., Ohno, H., Saito, T., and van der Merwe, P. A. (1999) Dependence of T cell antigen recognition on the dimensions of an accessory receptor–ligand complex. *J. Exp. Med.* **190**: 31–41.

Xiong, J-P., Stehle, T., Goodman, S. L., and Arnaout, M. A. (2003) New insights into the structural basis of integrin activation. *Blood* **102**: 1155–1159.

Yonemura, S., Nagafuchi, A., Sato, N., and Tsukita, S. (1993) Concentration of an integral membrane protein, CD43 (leukosialin, sialophorin), in the cleavage furrow through the interaction of its cytoplasmic domain with actin-based cytoskeletons. *J. Cell Biol.* **120**: 437–449.

Yu, Z. W., Calvert, T. L., and Leckband, D. (1998) Molecular forces between membranes displaying neutral glycosphingolipids: Evidence for carbohydrate attraction. *Biochemistry* **37**: 1540–1550.

III

ENGINEERED BIOMIMETIC SURFACES

8

ATOMISTIC MODELING OF PROTEIN ADSORPTION TO CERAMIC BIOMATERIALS IN WATER: A FIRST STEP TOWARD REALISTIC SIMULATION OF THE BIOMATERIALS SURFACE *IN VIVO*

ALAN H. GOLDSTEIN

School of Engineering, New York State College of Ceramics at Alfred University, Alfred, New York, USA

Development of modeling systems with atomic-scale resolution is universally recognized as essential to the successful fabrication of surfaces that control, modify, or measure cell function. Molecular and atomistic modeling can elucidate the mechanism of interaction between biological macromolecules and materials surfaces, which, in turn, has global applications ranging from implanted biomedical devices to DNA and protein arrays. Interfacial phenomena, especially protein adsorption, control essential biomaterials properties such as biocompatibility, and also control the efficiency of engineered structures such as biomimetic tissue scaffolds designed to promote the growth and development of specific cells. Bioengineers require unified software solutions capable of modeling realistic systems composed of biomolecules, materials, and the aqueous medium in which they interact. Significant progress was made toward creating such a system by implementing a molecular dynamics and local minimization (MD-LM) strategy to simulate a system in which a protein (bovine pancreatic trypsin inhibitor; BPTI) encounters a materials surface (MgO) in water. MD-LM simulations out to 1040 ps show reorientation of the BPTI protein toward the MgO surface, even at distances previously considered to be too far away for force field–mediated interactions. These data support the hypothesis that, at least in an aqueous environment, the solvent plays a paramount role in the initial adsorption event. This chapter describes our most recent output from the model and suggests future directions for research.

I. INTRODUCTION

Development of atomistic modeling systems is universally recognized as essential to the successful production of nanoengineered biomaterials. In this context, nanoengineering is considered to mean fabrication of molecular-scale devices with atomic resolution. What are these devices? There is no simple answer to this question because the field of bioengineering has evolved into what is perhaps the most interdisciplinary of all scientific endeavors.

On one end, bioengineering looks to molecular biology to provide tools based on a schematic of cellular structure and function. We know that the properties of living cells result from the properties of their component biological macromolecules. It is these macromolecules — nucleic acids, proteins, carbohydrates, and lipids — that form the alphabet with which the language of life is spoken. Depending on their unique shapes and other chemical properties, macromolecules interact with one another to form higher orders of structure. Biotechnology is based on our emerging understanding of the macromolecular components of the cell, their chemical structures, how they are synthesized, and how they function after they are synthesized. There are several broad categories of macromolecular function: informational, catalytic, structural, and signal sensing/transduction. These categories are, of course, not mutually exclusive. Regardless of any specific role in the living system, biological macromolecules share one crucial characteristic: they are materials. Once these biological materials have been characterized, they may be used as components of bioengineered systems.

At the other end of the spectrum, bioengineers look to materials science and engineering to provide essential components for their devices. Materials used in medicine are now known as "biomaterials," so that we have biomaterials engineers and biomaterials scientists, whose goal is to use metals, polymers, glasses, ceramics, and composites in a wide range of biomedical applications (Ratner, 1996). Biomaterials, by definition, must interact with biological materials. Our understanding of these interactions is directly proportional to our understanding of the structure and function of the interface between the living and nonliving worlds. If all materials in a specific system, both biological and nonbiological, have been characterized at the atomic and molecular levels, then it is possible to consider a mechanistic approach to total systems engineering. Integrated bioengineering of living and nonliving materials is sometimes called nanobiotechnology.

Understanding the biomolecular interface is a *de facto* goal of nanobiotechnology. We know that cellular function results from the transfer of information via a limited range of chemical phenomena; from the *de novo* polymerization of DNA to the relatively slight rearrangements in three-dimensional structures involved in some signal transduction processes (e.g., G-proteins). Biotechnology involves the manipulation of living systems via the modification of the materials from which living systems are built. Bioengineers strive to expand our ability to manipulate living systems by integrating nonliving materials (and devices made from these materials) into this scenario. As our characterization tools for living and nonliving materials converge at atomic-scale resolution, the ability to transfer chemical information between living and nonliving materials emerges. This is the basis for "biomimetic" strategies, by which nonliving materials are fabricated to mimic the structure and function of living materials. Beyond biomimetics is a realm where living and nonliving materials are both treated as components in device fabrication based on the underlying, unifying principles of physical chemistry: the realm of nanobiotechnology. Nanobiotechnology offers bioengineers the full range of material with which to build surfaces engineered to control, modify, or measure cellular function at the atomic or molecular level.

The ability to deliver on the promise of nanobiotechnology is dependent on our ability to characterize the structure and function of both living and nonliving materials. Modeling has always played a paramount role in our elucidation of the nature of the physical world. Over the past 50 years, modeling has become synonymous with computer simulation. But the emergence of nanobiotechnology has brought an important bottleneck into focus. To date, efforts to develop computer simulation strategies to predict biomolecular structures and analogous efforts to simulate the structures of nonliving materials have proceeded in an almost completely independent manner.

Proteins represent the most advanced state of computer modeling of biomolecular structures (Branden and Tooze, 1991; Cleland, 1996). The Protein Data Bank (PDB) has expanded into the Research Collaboratory for Structural Bioinformatics (RCSB), a "single worldwide repository for the processing and distribution of three-dimensional biological macromolecular structure data" (http://www.rcsb.org/pdb/) that is approaching 30,000 structures of varying degrees of resolution. Of equal importance, structure–function analyses have revealed that the virtually unlimited range of protein functions results from permutations and combinations of a relatively limited set of structural motifs (Branden and Tooze, 1991). Structural stability for a given protein is almost always imparted by some type of folding involving α helices and β sheets packed together into a stable three-dimensional conformation based on energy minimization within the protein's native environment. These folding motifs are well-characterized for a large number of protein families and, in fact, are so common that we can now recognize superfamilies with more than 200 members that use almost identical folding motifs to maintain structural stability (e.g., the globin superfamily). Atomic and molecular modeling of proteins has now reached the point where protein engineering is possible. Proteins with enhanced thermal stability (Cleland, 1996) or modified substrate specificity are now routinely produced. Catalytic antibody technology and other strategies make it possible to design protein surfaces to mimic nonbiological catalysts.

Computer simulation is playing a role of increasing importance in our attempts to understand protein folding. *De novo* prediction of the native conformation of a protein from the primary amino acid sequence remains one of the great challenges to modern biochemistry and molecular biology and will, in fact, depend entirely on computational chemistry. There is reason to be optimistic with respect to the ultimate development of an ability to predict and fabricate proteins with specific structure–function relationships (cf. Fernandez and Li, 2004; Kuhlman et al., 2003, and references therein). But even though significant barriers to the prediction of protein structure from *ab initio* calculations remain, X-ray crystal structure data (as well as data obtained by NMR and other experimental techniques) currently provide the capability to produce accurate three-dimensional pictures of a wide variety of protein structures. It is taken as a given in our work that available high-resolution structures include all spatial and electrochemical information necessary to predict how a protein will interact with a surface. What we lack are computational strategies to predict exactly how the protein's native macromolecular structure is distorted in proximity to a materials surface in a specific solution or solvent.

As our understanding of protein structure–function relationships continues to grow, protein engineering will evolve to match the more mature fields of engineering, in which fundamental rules of design permit precise molecular structures to be fabricated. A mature protein engineering capability will also allow fabrication of true composites containing both (e.g.) proteins and nonliving materials such as polymers and glasses. Such composites will often be developed as "biomaterials." However, the accepted general definition of a biomaterial, "any material, natural or artificial, that comprises whole or part of a living structure or functions in intimate contact with living tissue" (Ratner, 1996), is too broad a definition for such composites within the context of nanobiotechnology. We have proposed the term *protein–materials composite*, more generally *biomolecular–materials composite* (Cormack et al., 2004), to define this new class of materials. The gene microarray, a DNA–glass composite, dramatically demonstrates how even a relatively simple biomolecular materials composite can revolutionize an area of science and technology. The structural characterization of these biomolecular–materials composites will, in itself, require new technology. If prediction of the native conformation of a protein from primary sequence data remains one of the great challenges to modern bioinformatics, how much more challenging is the integration of biomolecular and materials modeling? Yet this is precisely the type of simulation required to design and fabricate integrated devices at the molecular scale with atomic precision; that is, design and fabricate the products of nanobiotechnology. The work summarized here indicates that, with respect to many applications, existing systems and databases will provide sufficient spatial and electrochemical information to attempt computer-aided design (CAD) of protein–materials composites with specific functions.

The ability to simulate the structure of biomolecular–materials composites via computer modeling will be crucial to the development of a mature engineering technology that, by definition, calls for functional design specifications of atomic/molecular precision. Our primary interest involves interfacial phenomena that occur between biological and nonbiological materials, so-called biosurfaces, with an emphasis on glass and ceramic substrates (cf. Clare et al. 2003; Korwin-Edson et al., 2004). Many of the most important applications of this new class of materials will be based on properties resulting from surface-mediated phenomena. Conversely, many of the most important problems faced by biomaterials scientists and engineers result from biomolecular adsorption to a materials surface (Horbett, 1996). Therefore, our focus is on adsorption processes. These should be amenable to force field–based simulations using molecular mechanics and dynamics. Although extremely high-quality "turnkey" software suites exist for materials or biopolymer modeling, equivalent systems do not exist for biomolecular–materials composite systems.

To date, little has been done to integrate and standardize software development between the fields of materials science and molecular biology. This type of software is required because it is widely recognized that bioengineering will continue to vigorously integrate information from the molecular life sciences. The result will be new hybrid systems and devices as evidenced by emerging fields such as tissue engineering and biomimetics. Many of these systems require a fusion of living and nonliving materials for the fabrication of bioactive composites. Strategies for "tissue scaffolds" and other "bioreactive" materials already employ a limited palette of biomolecular–materials composites, for example, RGD-decorated synthetic polymers or calcium phosphates containing bone morphogenic factors. With the advent of nanotechnology, this trend will only accelerate.

For this research, we use two commercially available software systems from Accelrys, Cerius and Insight, commonly used for structure–function studies of materials and biological macromolecules, respectively. These software tools are used in conjunction with Accelrys' Discover package to conduct molecular mechanics and dynamics studies. Complications arise when attempting to simulate the structure of a composite biomolecular–materials surface. Addition of water to simulate a simple aqueous environment creates a further challenge.

We have demonstrated the utility of our approach via the simulated adsorption of the protein bovine pancreatic trypsin inhibitor (BPTI) to an MgO surface using pure water as the solvent (Cormack et al., 2004). The output of the initial simulation was found to be quite reasonable, in terms of both binding energies versus protein orientation and the solvated structure of the entire system. Our results indicated that intramolecular noncovalent bonds crucial to macromolecular structure, that is, hydrogen bonding, electrostatic bonding, dipole–dipole bonds, play virtually no role in the initial adsorption of BPTI to the MgO surface. The largest share of binding energy is the result of interactions of the two materials with interfacial water molecules.

Any useful modeling system for protein adsorption must account for two universally observed phenomena: (1) Proteins bind to surfaces in virtually every nonliving "biomaterial" ever examined, *in vivo* or *in vitro* (Horbett, 1996); and (2) in aqueous solution, the physicochemical properties of both the biomaterials surface and the protein are controlled mainly by the solvent. Our initial simulations, which extended over 240 ps, were consistent with electrochemical theory insofar as initial adsorption was mediated by the solvent double-layer (Bockris and Reddy, 1970), rather than by direct protein–material contacts. The energetics of binding were in the range expected from experimental data on protein adsorption. However, as we extended the simulations into the nanosecond time scale, it became clear that larger systems, specifically greater physical dimensions and additional solvent molecules in the simulation box, are required to predict the energetics (and, therefore, the final conformation) of the complete act of protein adsorption with confidence.

What has emerged to date is a picture of adsorption in which the aqueous solvent plays a paramount role because of its ability to mediate long-range interactions that come into play long before the protein and materials surface ever "see" one another via direct interaction of their component atoms. This picture, obtained completely from simulations, is entirely consistent with bioelectrochemical theory and points to the need for workers to take the entire physical system into account when attempting to develop predictive models. This work was an important first step in the development of a "turnkey," cross-platform simulation system available to a wide range of biomaterials scientists and bioengineers for predicting the structure of the biomolecular–materials composites interface in the aqueous environment. Other workers are developing alternative strategies (cf. Agashe et al., 2005; West et al., 1997).

II. SIMULATION ALGORITHM

The computer simulation procedure to be described contains three parts: (1) derivation of interatomic potential parameters, (2) formation of simulated materials structure, and (3) evaluation of an effective simulation strategy for investigation of an adsorption event. Suitable interatomic potentials are necessary for the adequate description of the individual structures of which our system is composed: the MgO surface, the BPTI protein molecule, and water molecules. In addition, we require a reasonable description of the interactions between the three separate parts (i.e., surface–protein, surface–water, and protein–water interactions). The formal components of the simulation have been reported elsewhere (Cormack et al., 2004), but for convenience we restate the basic parameters of our simulation system here.

For the protein, we used the intermolecular potentials known as the consistent-valence force field (CVFF) (Hagler et al., 1985). We excluded bonded, cross-term interactions and intermolecular interactions between bonded neighbors and next-nearest neighbors. The potential energy of the protein molecule was modeled in accordance with standard methodology (de Leeuw, 1997; Hagler et al., 1985). Inorganic materials have also been well characterized via computer simulations. The MgO surface of interest here can be accurately modeled within the framework of the Born model of the solid, wherein the Mg and O atoms are treated as nonpolarizable point ions, with integer atomic charges of $+2.0e$ and $-2.0e$ (in units of electron charge), respectively, and with parameterized short-range potentials of the form given in the equation

$$U_{i,j}(r_{i,j}) = A_{i,j}\exp(-r_{i,j}/-\rho_{i,j}) - C_{i,j}\,r_{i,j}^{-6}, \qquad (8.1)$$

where $U_{i,j}$ is the short-range, pair-potential energy between atoms i and j at distance r, and $A_{i,j}$, $\rho_{i,j}$ and $C_{i,j}$ are parameters determined for each pair interaction given in Table 8.1.

The parameters we have used (Table 8.1) were taken from earlier work specific to nonpolarizable ions (Sanders, 1984), wherein the MgO parameters were refit from a model originally derived for a polarizable anion (Catlow et al., 1982, 1991).

In simulating protein adsorption to a surface, well-established potentials are not available to describe the interactions between the inorganic surface and the protein molecule. The simplest approach would be to consider all explicit potential interactions between the protein molecule and the magnesium atoms to be purely coulombic. In addition, the interactions between

■ **Table 8.1   Potential Parameters Used in the Present Study**

i	j	$A_{i,j}$ (kcal/mol)	$\rho_{i,j}$ (Å)	$C_{i,j}$ (kcal/mol Å^6)
O_s	O_s	5.24599×10^5	0.149000	6.42500×10^2
O_s	Mg	2.22839×10^4	0.315000	0.00000×10^0
O_s	O_w	1.78234×10^5	0.229132	1.96997×10^2
O_s	H	8.18944×10^3	0.259707	6.73483×10^1
Mg	O_w	8.66508×10^4	0.233356	4.34161×10^2
Mg	H	3.98139×10^3	0.261534	1.48432×10^2

the atoms of the protein molecule and the oxygen of the inorganic surface could be represented by a Lennard-Jones (1931) potential with a single oxygen parameter that serves the various interactions through a combinatory procedure. This tactic has been applied with some success in the case of organic molecules interacting with silica-related inorganic materials, such as zeolites (Catlow et al. 1991; Kiselev et al., 1985; Sastre et al., 2000). But, as previously discussed, attractive coulombic interactions, between the magnesium of the materials surface and the nitrogen, carboxylate oxygen, and carbonyl oxygen in the protein, produced unreasonably close binding so that the short-range potentials acting between these atoms could not be neglected (Cormack et al., 2004).

We also required a water model that functioned well with both the inorganic surface and the protein molecule. There has been substantial simulation of the interaction between water and MgO surfaces (Marmier et al., 1998; de Leeuw, 1997; Langel and Parrinello, 1995; Scamehorn et al., 1993). We used the MgO–water potential parameters of McCarthy et al. (1996), constructed for the potential given in Eq. (8.1), derived by fitting to energy surfaces obtained from a series of Hartree–Fock, quantum-mechanical calculations. The water molecule did not deviate substantially from its bulk configuration, which was an O—H bond distance of 0.9475 Å and H—O—H bond angle of 105.59°. The charges on the hydrogen and oxygen atoms of the water molecule, from Mulliken population analyses, were $+0.41e$ and $-0.82e$, respectively (Scamehorn et al., 1993). This was of great benefit because the CVFF potential offers a semirigid water model with an O—H bond distance of 0.96 Å and H—O—H bond angle of 104.5°, while the charges on the hydrogen and oxygen atoms of the water molecule are also $+0.41e$ and $-0.82e$, respectively (Dauber-Osguthorpe et al., 1988). This allowed us to use the short-range potentials derived by McCarthy et al. (1996) for the MgO surface–water interactions with the same CVFF parameters used for both the protein and water systems.

We did have to refit the MgO surface–water potentials to use Mg and O charges of ± 2.0 as opposed to ± 1.966 used by McCarthy et al. (1996). The pair-potential parameters for the MgO surface, as well as the adjusted MgO surface–water interactions employed in the present study, are given in Table 8.1, where O_s represents the surface oxygen and O_w is the oxygen within the water molecule. The parameters apply to a simple Buckingham potential model defined by Eq. (8.1).

All that remained was to establish the interaction potentials between the MgO surface and the protein molecule. A reasonable first approximation was to use the potential parameters for the MgO surface–water interactions cited in Table 8.1 for all other forms of hydrogen and oxygen within the protein. Also, interactions with nitrogen atoms can be assumed equivalent to the oxygen interactions, as has been done before (Westwood et al., 1995). Finally, all other explicit potential interactions between the surface MgO and the protein (e.g., with carbon, sulfur, and potassium atoms) have been treated as purely coulombic, whereas the single, CVFF oxygen potential represents these interactions with the surface oxygen.

The goal of this work was to simulate the interaction between a single small protein molecule and an infinite inorganic surface, within an aqueous environment. To this end, an eight-layer, 67.3380 × 67.3380-Å magnesium oxide system, with periodic boundary conditions, was chosen to serve as the surface. The bottom three layers of the system were fixed at their bulk positions. Infinite space above the surface was simulated by adjusting the Z-direction boundary to a dimension of 86.8345 Å, whereby the eventual surface–protein–water system was functionally uncoupled from its periodic images. The BPTI crystal structure designated 6PTI was downloaded in PDF format from the Protein Data Bank *(http://www.rcsb.org/pdb/)*, and the raw file was modified as necessary within InsightII (i.e., input missing residues, add hydrogen molecules, pH set to 7.0). The protein molecule was covered with a 15-Å layer of water and minimized in accordance with the procedure to be described. From a suitable inorganic surface and protein structure, the necessary simulation systems could be constructed. The general shape of the BPTI molecule itself can be taken to be roughly that of a cylinder, as shown in Fig. 8.1.

All the constructed BPTI–MgO–water systems were such that the cylinder's long (*X*) and short (*Y*) axes (Fig. 8.1) were parallel to the MgO surface. The two systems that represent the protein molecule interacting with the surface were composed of two distinct rotations about the long axis, which resulted in the creation of unique collections of interacting residues consisting of hydrophobic, charged, and polar groups. Both interacting configurations

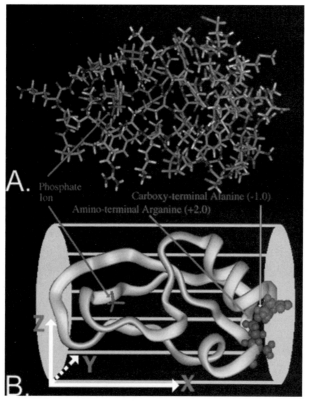

FIGURE 8.1   Standard wireframe (A) and ribbon (B) representations of the bovine pancreatic trypsin inhibitor (BPTI) protein from the "noninteracting system" at $t = 0$. In (B), the phosphate ion has been left in wireframe format, whereas the amino-terminal arginine (red) and carboxy-terminal alanine (purple) are shown in CPK format and have valence charges of $+2.0$ and -1.0, respectively. The generalized cylindrical shape is shown in (B), with the long axis considered as the X direction, whereas the short axis is taken to be in the Y direction. (See Color Plate 9)

contained approximately the same number of polar residues; the interacting residues of interacting system 1 were predominantly charged (46% vs 17% hydrophobic), whereas those of interacting system 2 were predominantly hydrophobic (38% vs 29% charged). This small number of initial orientations was expected to be representative of the possibilities inherent within the BPTI–MgO–water system but not representative of a true global energy search.

It was assumed that an aqueous environment was essential for a first approximation simulation of adsorption. In this work, all the energy calculations employed a group-based direct sum method with a 14.0-Å cutoff. Therefore, an aqueous environment can be approximated by a 15.0-Å layer of water. Based on the construction of the simulation box, water molecules at the vacuum–water boundary were not part of the virtual "solution," and we would expect the energy of these to deviate from their bulk values.

The energy of interaction between the protein molecule and the surface was evaluated as the difference between interacting protein-surface systems and a "noninteracting system," which we placed at a distance considered to be, functionally, at infinite separation. The reason for the quotation marks will be explained shortly. One conventional method of evaluating the energy of the "noninteracting system" is calculating the energy of a single protein molecule encased within a 15.0-Å layer of water and adding in the energy (calculated separately) for the surface system covered with a 15.0-Å layer of water. The energy of the interacting system is established by evaluating the combined protein–inorganic surface system collectively coated with a 15.0-Å layer of water. However, the disadvantage of this approach becomes clear when evaluating the energy of the water molecules themselves. First, the numbers of water molecules within the noninteracting and interacting systems will not, without deliberate intervention, be identical, and corrections to the total energy will need to be made based on a separately determined value for the energy of bulk water. More importantly, even if the numbers of water molecules were deliberately made equal, the water molecules within any interacting and noninteracting systems would have a considerable mismatch in the number of molecules in proximity to the surrounding vacuum. This water surface energy can be approximated but not without the introduction of a considerable uncertainty in the resulting interaction energy. Because we wish to simulate adsorption phenomena with known interaction energies on the order of hundreds of kilocalories per mole of protein, it would be much better to minimize this uncertainty.

Therefore, a "noninteracting system" was constructed within a single periodic system by placing the protein in a position *initially* considered sufficiently far (20 Å) from the surface that there could be no direct interaction between the biomolecule and the materials surface. The rest of the simulation box was filled with water molecules to within a minimum of 10Å above the uppermost segments of the protein molecule. As a result, a total of 7541 water molecules were included within the system. It is important to note at this point that the extended simulations reported here show that *this initial assumption appears to be incorrect.* At 20 Å, the results of extended simulations presented here indicate that the BPTI protein still interacts with the MgO surface. For continuity with previous work, we continue to refer to the protein 20 Å from the surface as the "noninteracting system" (hence the use of the quotation marks). In fact, at 1040 ps, the position of BPTI in the "noninteracting system" appears to be converging with one of the interacting systems. This is discussed in Section III.

The energies of the interacting systems were evaluated through construction of equivalent periodic systems containing equal numbers of water molecules, with the protein placed close to the surface. Of special concern was the structure of the interface between the MgO surface and BPTI protein molecule in these interacting systems. Water molecules must be included within this interface, but the initial number of water molecules is arbitrary. Because the protein molecule used to construct the complete system is encased within a 15-Å layer of water, the protein–MgO surface interface is created by elimination of the water layer up to the plane of interaction. As residues may extend quite a distance out into the water layer, the water can be removed only up to any such residues, leaving an amount of water within the interface that varies in accordance with the length of the residues in closest proximity to the MgO surface. The number of interfacial water molecules would be expected to adjust to a more "natural" level over extended simulation periods.

The resulting systems ("noninteracting" and "interacting") all contain a single, well-defined water–vacuum boundary that is very nearly identical in each case. This allows for straightforward comparison. The interaction energy is the difference between the total energies of the interacting and noninteracting systems. Binding (or negative interaction energy) occurs when an interacting system resides at a lower (i.e., more negative) energy than the noninteracting system.

The amorphous nature of the protein and water molecules requires special handling for the energy calculations. At the initial formation of the systems, we would expect the water molecules to be far from equilibrium. Yet, during molecular dynamics (MD) simulations, there is a need to simultaneously avoid protein "drift" without forcing excessive rigidity on the biomolecule. By fixing a few central α carbons (Tyr21, Phe22, Tyr23, Asn43, Asn44, Phe45), the water molecules and protein residues equilibrate over a period (here we used 40 ps of MD simulation) without the protein drifting away from its initial position and without forcing excessive rigidity on the protein molecule. We used 20,000 time steps at 2.0 fs per time step or 40 ps of MD simulation with direct temperature scaling. To avoid continuously "driving" the system, we adopted a very large temperature window (100 K), which allowed the system, to some extent, to find its own equilibrium temperature in the absence of temperature scaling after being initiated with random velocities according to a Boltzmann distribution at 320 K. The temperature quickly settled down to around 268 K. This preliminary, equilibration MD simulation is designed to remove local stresses only, and the fact that the system temperature rests at a level slightly below the physical freezing point of water is not a concern.

For a proper simulation of adsorption, especially adsorption energies, comparison between equilibrium noninteracting and interacting systems is required. However, there is no *a priori* means of determining just where the protein should be placed nor how many water molecules reside within the protein–surface interface. We would like to allow for the protein and the water molecules to establish these aspects independently. Yet, for a strictly MD simulation, the protein molecule must be constrained in some way; otherwise it is free to acquire a drift velocity.

Comparisons between interacting systems and a noninteracting system should occur between systems that are, in principle, representative of some type of quasi-global energy minimum. MD simulations, alone, do not provide such a description because of the very large number of system configurations available. MD simulations will likely settle into a local equilibrium, probably not far from the initial configuration. Because of the large number of water molecules, there will be many such local minima. This, in turn, will give rise to a complicated energy landscape. On the other hand, static lattice minimization (LM) is, itself, wholly unsuitable because the result is merely the lowest energy structure local to an initial configuration. In summary, a useful model must: (1) inhibit the acquisition of a drift velocity by the protein, (2) allow for an independent adjustment of water within the protein–surface interface, and (3) explore the lower-energy features of the "energy landscape," targeting "quasi-global" minima. The qualifier *quasi-* is used because it is possible to explore only a few of the enormous number of initial configurations available to the protein in the simulation box. Future work needs to address appropriate methods of sampling a significant cross section of this conformation space.

An MD-LM hybrid technique was developed by alternating 8-ps MD simulations, in the manner described previously (*without* fixed α carbons), with local, static LM. This procedure uses both MD and LM to exploit the advantages of each method while compensating for their respective weaknesses. The disadvantage of strict LM is alleviated by reshuffling the structures at each iteration. However, the technique is different from MD alone, because the simulations, at each iteration, are not at equilibrium and, thus, the average MD systems at each iteration are not mutually comparable. Within the procedure, the atoms are shuffled toward sites, near their initial sites, that lead to overall lower potential energy. It is important to recognize that the MD-LM procedure is an iterative approach. The final, lattice-minimized structures are singular, not representions of averaged properties as would be the case with MD. These structures can be visualized, systematically studied, and compared.

III. OUTPUT FROM EXTENDED SIMULATIONS

The behavior of the MD-LM procedure is demonstrated by the example of a system of 5328 water molecules with periodic boundary conditions 67.338 × 67.338 × 86.835 Å that has been equilibrated over a 40-ps MD simulation at 268 K and subsequently subjected to the MD-LM procedure for 40 iterations. Because this system consists entirely of water molecules, a change in the system energy can occur only via reorientation of the water molecules themselves. Examination of Fig. 8.2 reveals that the energy of the water system drops by 600 kcal/mol in the first 13 iterations (96 ps of MD simulation). It appears that the water is being jostled at each iteration toward lower-energy configurations in a more global manner than available to either MD or LM techniques alone.

Recognition of this fact is important because it indicates that a drop in system energy will be observed for all of the systems under consideration, whether interacting or not, simply as a result of the application of the MD-LM procedure on the water within each system. Therefore, a decrease in the system energy for the noninteracting system cannot be taken as *a priori* evidence of interaction.

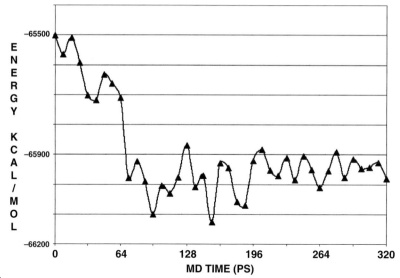

FIGURE 8.2   Graph of the energy of 5328 water molecules with periodic boundary conditions of 67.33800 × 67.33800 × 86.83450 Å versus the MD time in picoseconds, wherein a local minimization is applied to each system every 8 ps. An energy minimum is evident at the 152-ps mark, and the energy quickly reaches a point, thereafter, when further iterations yield little or no improvement in the calculated energies.

In addition, the systems obtained, at each iteration, are clearly in a state of transition and, therefore, not comparable to the next until we arrive at a point where further iterations cease to yield significant changes in the calculated potential energy (past the 200-ps mark). Nevertheless, it is clear that the configuration at 88 ps is more comparable with configurations at 80 or 96 ps than with one at, say, 24 or 208 ps. The point where further iterations yield little or no change in the calculated energy indicates that the minimization procedure is yielding consistent energy values. This is where equivalently handled systems can be adequately compared. This chapter presents results from such equivalently handled, extended simulations for three systems containing BPTI, MgO, and water. These systems represent three unique orientations of BPTI with respect to both distance from the MgO surface and rotation of the protein around the long (X) cylindrical axis (Fig. 8.1).

These systems, designated as "noninteracting system," system 1, and system 2, respectively, have been described in detail previously (Cormack et al., 2004). The aforementioned work identified energy minimum configurations obtained for these systems after 240 ps, or 30 MD-LM iterations. In this chapter we report on extended simulations, some out to 1040 ps or 130 MD-LM iterations. Total system energy versus MD-LM time for the "noninteracting system," system 1, and system 2 is represented graphically in Fig. 8.3.

A third-order polynomial fit to the data is included within Fig. 8.3.

Extending the MD-LM simulations to 1040 ps generated a number of important results. Three conclusions are directly related to the behavior of the so-called "noninteracting system":

1. In the extended MD-LM simulation, the system originally designated as "noninteracting" appears to approach a minimum energy configuration. As such, we must now conclude that this system interacts with

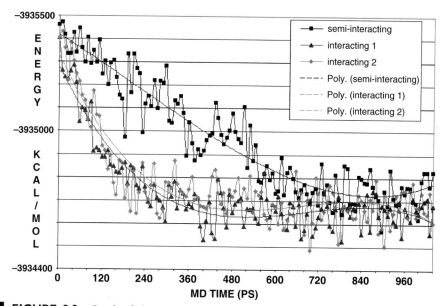

FIGURE 8.3 Graph of the energy of the "noninteracting system" and interacting systems 1 and 2 (labeled 1 and 2) versus the MD time in picoseconds, wherein a local minimization is applied to each system every 8 ps. In addition, third-order polynomials are fit to the data as a means of including a fuller range of data. The energy of the "noninteracting system" is very nearly equivalent to that of the interacting systems, making the calculation of an interaction energy impossible.

the MgO surface from a distance of 20 Å. For continuity with our previous work, we continue to designate this system as "noninteracting."

2. Given that the "noninteracting system" appears to interact, we no longer have a means of evaluating the interaction energies of system 1 or system 2. These energies were previously calculated as the difference between interacting and noninteracting systems.

3. As discussed above, the "noninteracting" system was constructed within a single periodic system by placing the protein what we considered sufficiently far from the surface (20 Å) so that there could be no direct force field interactions between the biomolecule and the materials' surface. Therefore, extended MD-LM simulation indicates that long-range, solvent-mediated interactions play a paramount role in the initial adsorption process.

Even in the absence of a "noninteracting" system with which to calculate binding energies, examination of the simulated linear and rotational motion of the BPTI molecule within the various systems yields a great deal of useful information. Because the BPTI–MgO–water systems are restarted with random velocities at each iteration, the total momentum must be initially zero or very nearly so. Therefore, any net displacement over multiple iterations should result from dispersion of the displacement values alone (i.e., the change in the protein molecule's position at each iteration should behave like a random walk) With this in mind, the mean squared displacement of the protein molecule ($<R^2>$, or the average squared distance traversed over many iterations) should be equivalent to the product of the number of iterations (n) and the average displacement per iteration ($<r^2>$) according to the relationship

$$<R^2> = n<r^2>. \tag{8.2}$$

Deviations from random motion would result in subsequent displacements being correlated such that Eq. (8.2) would be modified by multiplying the right side by a factor f, where f is the so-called correlation factor. Consequently, f would be approximately 1 for uncorrelated displacements (those behaving like a random walk) and larger than 1 for displacements that tend to proceed in one direction. Therefore, a correlation factor for the BPTI molecule displacement data that is found to be much greater than 1 clearly substantiates the existence of interaction between the BPTI molecule and its environment.

It is possible to evaluate the geometrical center of the protein molecule for the various systems, relative to the MgO surface, with respect to MD-LM simulation time, as has been done in Fig. 8.4.

A second-order polynomial fit to the position data is included. In all cases, the BPTI molecule is involved in highly correlated motion perpendicular to the MgO surface (Z direction), wherein the correlation factor for motion in the Z direction is determined to be:

- 3.1 for the "noninteracting" system (over the initial 600 ps of simulation),
- 3.4 for system 1 (over the initial 240 ps),
- 6.3 for system 2 (over the initial 560 ps of simulation).

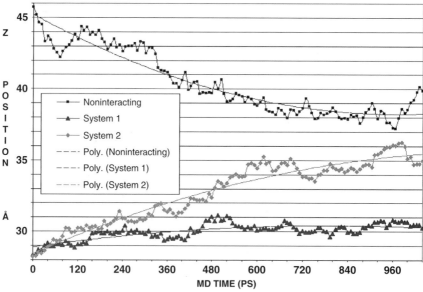

FIGURE 8.4   Graph of the motion of the geometrical centers, perpendicular to the MgO surface (taken to be the Z direction), for the "noninteracting system" and interacting systems 1 and 2 with respect to MD simulation time in picoseconds. Fit to the data are second-degree polynomials that indicate that the "noninteracting" system and interacting system 1 are at stable positions perpendicular to the MgO surface. System 2 is continuing to move away from the surface. The correlation factors for the various systems vary from 3.1 ("noninteracting" system over the initial 600 ps of simulation) to 3.4 (interacting system 1 over the initial 240 ps) to as high a value as 6.3 (as in the case of interacting system 2 over the initial 560 ps of simulation).

After the 600-ps point for the "noninteracting" system and system 2, and the 240-ps point for system 1, we observe that the fitted lines approach a slope of zero, thus indicating that BPTI is approaching a stable position normal to the MgO surface.

On the basis of the data in Figs. 8.3 and 8.4, it appears that the "noninteracting" system is in constant motion toward the MgO surface, with an accompanying decrease in energy. We tested this hypothesis by conducting an equivalent simulation using the "noninteracting" system minus the MgO surface. Figure 8.5 compares the geometrical centers of the BPTI–water system and the "noninteracting" system over a period of 600 ps relative to the Z axis of the protein. It is observed that the BPTI molecule behaves in an equivalent manner in both cases until about 400 ps of MD-LM simulation, whereupon the protein's position in the BPTI–water system reaches a stable point.

At the beginning of the simulation, the residues facing away from the MgO surface are about 10 Å from the water–vacuum interface, whereas the residues facing the MgO surface are about 20 Å away from it. When the MgO surface is removed from this "noninteracting" system, these distances remain the same. In the absence of the materials component, motion around 5 Å away from the water–vacuum interface would place the protein molecule at the middle position with respect to the surrounding water molecules, or at about the 32-Å point in Fig. 8.5. It is observed in Fig 8.5 that the protein molecule in the simple BPTI–water system does appear to oscillate around

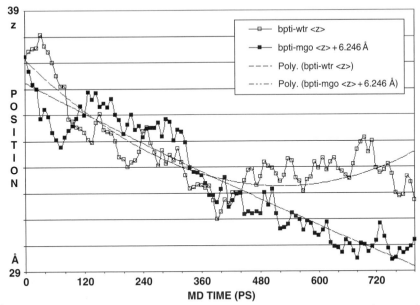

FIGURE 8.5 Graph of the motion of the geometrical centers over a simulation period of 800 ps, with respect to MD simulation time in picoseconds, perpendicular to the MgO surface for the "noninteracting" system (bpti–mgo) and a BPTI–water (bpti–wtr) system constructed from the original "noninteracting" system by removing the MgO slab. The position of the "noninteracting" system has been adjusted by a constant (6.246 Å) to create a match for the respective centers. Fit to the data are second-degree polynomials that indicate that the BPTI–water system has reached a stable position in the Z direction (what would be perpendicular to the MgO surface). The "noninteracting" system continues to move toward the surface, as is also readily apparent by inspection of Fig. 8.3.

this 32-Å point after a simulation period of about 400 ps. The initial motion of the protein molecule toward the middle of the surrounding water environment indicates that this motion can most likely be attributed to interaction with the water–vacuum interface. Initially, the same type of movement along the Z axis is observed for the BPTI molecule within the "noninteracting" system. However, it is clear, from Figs. 8.4 and 8.5, that the protein molecule within the "noninteracting" system continues to move past this midpoint toward the MgO surface. We suggest that this additional motion is the result of a *bona fide* interaction with the MgO surface, even though the surface remains in excess of the 14-Å cutoff.

After 560 ps of MD-LM simulation, system 1 appears to reach a stable position along the Z axis (Fig. 8.4). System 2 shows continuous motion away from the surface over the entire time course of the simulation (Fig. 8.4). It appears that, given sufficient simulation time, the positions of the BPTI molecule in the "noninteracting" system and system 2 converge. The fact that two very different configurations are converging clearly suggests that the BPTI molecule is moving toward a global energy minimum. This possibility will be investigated further.

In addition to motion observed perpendicular to the MgO surface, motion of the BPTI molecule occurs parallel with the MgO surface (in the X and Y directions). This parallel motion is illustrated in Figs. 8.5 and 8.6.

The displacement data in the X and Y directions (in the direction of the protein molecule's long and short axes, respectively) are very interesting because this motion is highly correlated. The "noninteracting" system is defined by correlation coefficients of 3.2 in the X direction over the initial 320 ps of MD-LM simulation and 2.5 in the Y direction over the entire

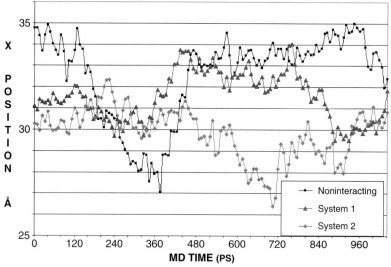

FIGURE 8.6 Graph of the motion of the geometrical centers, parallel to the MgO surface, in a direction along the short axis of the semicylindrical BPTI molecule (taken to be the Y direction), for the "noninteracting" system and interacting systems 1 and 2 with respect to MD simulation time in picoseconds. Correlated motion is apparent (particularly in the case of the noninteracting system, which is defined by a correlation coefficient of 3.2 over the initial 320 ps of MD-LM simulation), although of a more transient nature. Equally correlated displacements are observed for the other systems at various intervals of simulation.

1040 ps of MD-LM simulation. The major difference between observed motion in the Z direction and that in the X and Y directions is that the correlated displacements are more transient in nature and often switch back and forth over a period of hundreds of picoseconds of MD-LM simulation (particularly in the X direction). Correlated displacements in the Z direction are naturally related to interactions between the protein molecule and either the water–vacuum or water–MgO surface interface. However, there is no obvious mechanism by which correlated displacements may occur parallel to the MgO surface. One possibility is that the protein interacts with its own periodic images, mediated by the intervening water molecules (equivalent to a dog chasing its own tail *in silico*). Such interactions would be transient owing to the unstable configuration of the water molecules mediating this type of interaction. However, the distance over which such a postulated interaction must occur is about 28 Å for motion in the Y direction and approximately 43 Å for motion in the X direction. The fact that motion in the X direction is found to be more transient than that observed in the Y direction may be accounted for by the much larger distance over which such an interaction must develop. In fact, motion in the absence of an MgO surface, as is the case for the BPTI–water system (Fig. 8.7), shows an equivalent trend, wherein motion is highly correlated in the Y direction ($f = 3.0$ over the initial 320 ps of MD-LM simulation) and not at all in the X direction.

In addition to consideration of linear motion of the geometrical center of the BPTI molecule within the various systems, we must consider the manner in which rotational motion appears. For the "noninteracting" system and system 2, this motion is shown in Figs. 8.8a and b. Figure 8.8c shows

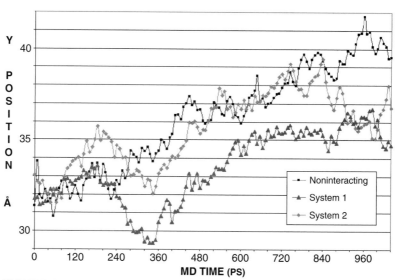

FIGURE 8.7 Graph of the motion of the geometrical centers, parallel to the MgO surface, in a direction along the long axis of the semicylindrical BPTI molecule (taken to be the X direction), for the "noninteracting" system and interacting systems 1 and 2 with respect to MD simulation time in picoseconds. Correlated motion is apparent (particularly in the case of the "noninteracting" system, which is defined by a correlation coefficient of 2.5 over the entire 1040 ps of MD–LM simulation). Equally correlated displacements are observed for the other systems at various intervals of simulation.

the superposition of the "noninteracting" system before and after rotation around its Y axis. System 1 (not shown) behaves in an equivalent manner.

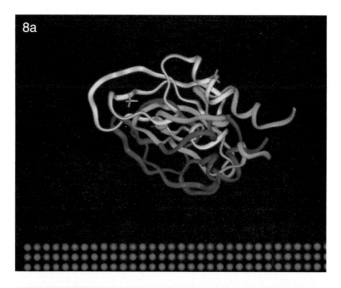

FIGURE 8.8 (a) Ribbon overlay, in the XZ plane of interacting system 2 at 1040 ps (blue) and the "noninteracting" system at 0 ps (yellow ribbon) versus 1040 ps (green ribbon). The "noninteracting" system and interacting system 2 are seen to merge toward a similar configuration with respect to the MgO surface. (b) Ribbon overlay, in the YZ plane, identical to (a) except that the BPTI systems have been rotated by 90°, system 2 is at 1040 ps (blue), and the "noninteracting" system is at 0 ps (yellow) versus 1040 ps (green). It is apparent that the "noninteracting" system has rotated by about 90° in the direction of interacting system 2. (c) Rotation of the "noninteracting" system around the Y axis results in movement of the protein's charged termini toward the MgO surface. The yellow ribbon represents $t = 0$; the green ribbon represents $t = 1040$ ps. The amino-terminal arginine (red CPK) and carboxy-terminal alanine (purple CPK) have valence charges of $+2.0$ and -1.0, respectively. As discussed in the text, extending the simulation to 1040 ps has caused us to reevaluate this system. It now appears that interaction occurs even though the protein was initially placed 20 Å from the MgO surface to be a "noninteracting" control. In both the "noninteracting" system and system 2, the protein molecule rotates about its short axis so as to orient its charged amino and carboxy termini toward the MgO surface. (See Color Plate 10)

FIGURE 8.8 (*Continued*)

▬▬ Table 8.2 **Distance between the MgO Surface and Charged Termini**

	Distance from MgO Surface (Å) of the amino- and carboxy-terminal BPTI residues	
	Arginine (+2.0)	Alanine (−1.0)
"Noninteracting" system	18	27
Interacting system 1	14	5
Interacting system 2	21	22

Again, we are confronted with evidence of an interaction between the BPTI molecule and the MgO surface, even though, as is evident from Table 8.2, the distances involved are well in excess of the 14-Å cutoff used in the simulation. Long-range, water-mediated interaction must be postulated to account for these data.

Examination of Fig. 8.8b also indicates that the protein molecule in the "noninteracting" system has rotated about its long axis in such a way as to increase its similarity to system 2 and has traversed approximately one-half the distance over the 1040 ps of MD-LM simulation. This simulated motion (shown as an overlay in Fig. 8.8c) is made more significant by the fact that no other system (including BPTI–water) shows rotation of BPTI about its long axis. It is reasonable to propose that the protein is moving iteratively toward a global minimum energy configuration with respect to the MgO surface and surrounding water molecules. If this conclusion is correct, the respective BPTI configurations of the "noninteracting" system and system 2 will approach each other asymptotically. After 1040 ps of MD-LM simulation, these two systems indeed appear to be converging.

In a previous publication (Cormack et al., 2004), another interacting system, denoted system 3, was included. In system 3, BPTI rotated approximately 30°

about its long axis in such a way as to make it nearly equivalent to system 1 (i.e., orienting the charged phosphate ion toward the MgO surface). In light of this fact, it is of interest that the "noninteracting" system would be observed to rotate in a manner opposite to this; even though a configuration like that of system 1 is much closer to its initial orientation. Coupled to this is the fact that the BPTI molecule in system 1 does not rotate about its long axis. As a result, the charged phosphate ion is oriented toward the MgO surface in all simulations (Figs. 8.8a–c). If it is assumed that the MD-LM technique minimizes energy in a global sense, it is possible that the initial temperature of 320 K is not sufficient to overcome the energy barrier that exists in system 1 as a result of the initial orientation of the phosphate ion toward the MgO surface.

It remains to be seen whether the MD-LM technique acts in the capacity of a global energy minimization scheme. The use of unconstrained MD simulation alone has shown that the protein molecule can be expected to undergo translational and rotational variation over as long a period as 5 ns (Agashe et al., 2005). Such information warrants extending the MD-LM timescale further to investigate the possibility of continued or additional motion of the protein molecule relative to the MgO surface in the various systems studied thus far.

By extending the number of iterations for systems 1 and 2, we may determine whether the BPTI configuration produced in system 2 continues to be energetically favorable relative to that produced in system 1. It will also be of interest to determine whether the positions of BPTI in the "noninteracting" system and system 2 continue to converge. Such a convergence from two radically different starting configurations would argue strongly for the existence of a global energy minimum that may point toward an atomistic structure for the adsorbed protein.

System 1 could be used as a means of optimizing the temperature initiated with each MD portion of the simulation. By starting system 1 with the phosphate ion aimed at the MgO surface, we may have inadvertently "trapped" it in a local energy minimum. An increase in temperature would increase the probability of surmounting such a local energy barrier and decrease the number of iterations required to reach a globally minimized state. However, an excessively high temperature would diminish the simulation's ability to discriminate between neighboring states of slightly lower energy, thus reducing the effectiveness of any given iteration. The appropriate balance between these alternatives remains to be achieved.

IV. CONCLUSION

The original intent of the MD-LM scheme was to provide some means of evaluating the energy of any initial system with a minimum introduction of arbitrary structural constraints while allowing for the eventual formation of an energy minimum configuration. The protein molecule was expected to remain fairly close to its original placement, and, as such, it was thought that this modeling strategy might not represent a global energy search. Extending the simulations out to 1040 ps or 130 MD-LM iterations has generated fairly conclusive evidence that the MD-LM procedure has the potential to identify global energy minima that, in turn, may represent an atomistic picture of the initial adsorption event.

ACKNOWLEDGMENTS

The author thanks Mr. R. J. Lewis for expert technical assistance and for useful input on the manuscript. The author also thanks Professor A. N. Cormack for useful discussions and comments. This work was supported, in part, by the Fierer Chair endowment.

SUGGESTED READING

The **RCSB** Protein Data Bank; Archive of experimentally-determined, biological macromolecule 3-D structures. www.rcsb.org/pdb/

Accelrys – Software for Pharmaceutical, Chemical, and Materials . . .

Supplier of bioinformatics software for gene and protein sequence analysis, and materials modeling software. www.accelrys.com

Goldstein lab at http://goldsteinlab.alfred.edu/projects.html. Click on <Projects> button for tutorial on biomaterials adsorption modeling.

REFERENCES

Agashe, M. Raut, V. Stuart S. J., and Latour, R. A. (2005) Molecular simulation to characterize the adsorption behavior of a fibrinogen γ-chain fragment. *Langmuir* **21:** 1103–1117.

Bockris, J. O'M.,and Reddy, A. K. N. (1970) *Modern Electrochemistry.* Plenum, New York.

Branden, C., and Tooze, J. (1991) *Introduction to Protein Structure.* Garland, New York.

Catlow, C. R. A. Freeman, C. M. Vessal, B. Tomlinson, S. M., and Leslie, M. (1991) Molecular dynamics studies of hydrocarbon diffusion in zeolites. *J. Chem. Soc. Faraday Trans.* **87:** 1947–1950.

Catlow, C. R. A., James, R., Mackrodt, W. C., and Stewart, R. F. (1982) Defect energetics in alpha-Al_2O_3 and rutile TiO_2. *Phys. Rev. B* **25:** 1006–1026.

Clare, A. G., Hall, M. M., Korwin-Edson, M. L., and Goldstein, A. H. (2003) Biomolecular characterization of glass surfaces *J. Phys. Condens. Matter* **15:** S2365–S2375.

Cleland, J. L. (1996) Introduction to protein engineering. In *Protein Engineering: Principles and Practice* (J. L. Cleland, and C. S. Craik, Eds.), pp. 1–32. Wiley–Liss: New York

Cormack, A. N., Lewis, R. J., and Goldstein, A. H. (2004) Computer simulation of protein to a material surface in aqueous solution. *J. Phys. Chem.* B **108:** 20408–20418.

Dauber-Osguthorpe, P. Roberts, V. A. Osguthorpe, D. J. Wolff, J. Genest, M., and Hagler, A. T. (1988) Structure and energetics of ligand binding to proteins: *E. coli* dihydrofolate reductase-trimethoprim, a drug-receptor system. *Proteins Struct. Funct. Genet.* **4:** 31–47.

De Leeuw, N. H. (1997) *Atomistic Simulation of the Structure and Stability of Hydrated Mineral Surfaces.* Doctoral thesis, University of Bath, Bath, UK.

Fernandez J. M.,and Li, H. (2004) Force-clamp spectroscopy monitors the folding trajectory of a single protein. *Science* **303:** 1674–1678.

Hagler, A. T. Osguthorpe, D. J. Dauber-Osguthorpe, P., and Hempel, J. C. (1985) Dynamics and conformational energetics of a peptide hormone: Vasopressin. *Science* **227:** 1309–1315.

Horbett, T. H. (1996) *Biomaterials Science: An Introduction to Materials in Medicine* (B. D. Ratner et al., Eds.), pp. 133–140. Academic Press, New York. [Note: Updated discussion in the second edition of this text (2004) does not differ significantly with respect to basic principles of protein adsorption to biomaterials discussed here.]

Kiselev, A. V. Lopatkin, A. A., and Shulga, A. A. (1985) Molecular statistical calculation of gas adsorption by silicalite. *Zeolites* **5:** 261–267.

Korwin-Edson, M. L., Clare, A. G., Hall, M. M., Goldstein, A. H. (2004) Bisospecificity of glass surfaces: Streptavidin attachment to silica. *J. Non-Cryst. Solids* **349:** 260–266.

Kuhlman, B. Dantas, G., Ireton, GC., Varani, G., Stoddard, B. L., and Baker, D. (2003) Design of a novel globular protein fold with atomic-level accuracy. *Science* **302:** 1364–1368.

Langel, W., and Parrinello, M. (1995) *Ab initio* molecular dynamics of H_2O adsorbed on solid MgO. *J. Chem. Phys.* **103**: 3240–3252.

Lennard-Jones, J. E. (1931) Cohesion. *Proc. Phys. Soc.* **43**: 461–482.

Marmier, A., Hoang, P. N., Picaud, M. S., Girardet, C., and Lynden-Bell, R. M. (1998) A molecular dynamics study of the structure of water layers adsorbed on MgO (100). *J. Chem. Phys.* **109**: 3245–3254.

McCarthy, M. I., Schenter, G. K., Scamehorn, C. A., and Nicholas, J. B. (1996) Structure and dynamics of the water/MgO interface. *J. Phys. Chem.* **100**: 16989–16995.

Ratner, B. D. (1996) Introduction. In *Biomaterials Science: An Introduction to Materials in Medicine* (B.D. Ratner, A. S. Hoffman, F. J. Schoen, J. E. Lemons, Eds.), p. 2. Academic Press, San Diego.

Sanders, M. J. (1984) *Computer Simulation of Framework Structured Minerals.* Doctoral thesis, University College, London.

Sastre, G., Catlow, C. R. A., Chica, A., and Corma, A. (2000) Molecular dynamics of C7 hydrocarbon diffusion in ITQ-2: The benefit of zeolite structures containing accessible pockets. *J. Phys. Chem. B* **104**: 416–422.

Scamehorn, C. A. Hess, A. C., and McCarthy, M. I. (1993) Correlation corrected periodic Hartree–Fock study of the interactions between water and the (001) magnesium oxide surface. *J.Chem. Phys.* **99**: 2786–2795.

West, J. K., Latour, J. K., Jr., Hench, L. L. (1997) Molecular modeling study of adsorption of poly-L-lysine onto silica glass. *J. Biomed. Mat. Res.* **37**: 585–591.

Westwood, A. D., Youngman, R. A., McCartney, M. R., Cormack, A. N., and Notis, M. R. (1995) Oxygen incorporation in aluminum nitride via extended defects: Part I. Refinement of the structural model for the planar inversion domain boundary. *J. Mat. Res.* **10**: 1270–1286.

9

MODEL CELL MEMBRANE SURFACES FOR MEASURING RECEPTOR–LIGAND INTERACTIONS

CRAIG D. BLANCHETTE, TIMOTHY V. RATTO, and MARJORIE L. LONGO

Biophysics Graduate Group and Department of Chemical Engineering and Materials University of California at Davis, Davis, California; Biophysical and Interfacial Science Group, Chemistry and Materials Science, Lawrence Livermore National Laboratory, Livermore, California, USA

In this chapter, practical and theoretical issues concerning measurements of receptor–ligand interactions at domain surfaces of model cell membranes are discussed. We begin with background on the existence and function of lipid domains (so-called "raft" structures) in cell membranes. We present supported lipid bilayers as a potential model system for studying lipid domains and receptor–ligand interactions at domain surfaces. Methods are discussed to form domains in supported lipid bilayers that allow them to adopt their "equilibrium" or nonequilibrim microstructure (especially with respect to domain size). This important development, in combination with variation in domain composition, makes it possible to vary a number of parameters that may modulate receptor–ligand interactions at domain surfaces, particularly in the case of multivalent binding. We present force-spectroscopy as a suitable method to quantify receptor–ligand interactions at domain surfaces of supported lipid bilayers, especially in the case of biological interactions involving a shear stress or force. The general principle and theory of dynamic force spectroscopy are set forth, followed by a practical discussion of how to obtain measurements for specific interactions using dynamic force spectroscopy. From a practical standpoint, it is still necessary to determine if the force to pull domain lipids out of a supported lipid bilayer is significantly different from the forces involved with receptor–ligand bond rupture, and the present understanding of the magnitude of these forces is discussed briefly. If it is assumed that forces of receptor–ligand bonds of interest are significantly larger than lipid pullout forces, multivalent receptor–ligand interactions at domain surfaces of supported lipid bilayers could be measured by varying force loading rates.

I. INTRODUCTION

Glycosphingolipids isolated from plasma membranes are generally detergent insoluble and are believed to exist in liquid ordered domains in the cell membrane. It is thought that one of the primary functions of glycosphingolipid domains or "rafts" is to enhance multivalent interactions with host pathogens including a wide range of viruses (HIV-1, influenza virus, murine polyoma virus), bacterial toxins (pertussis toxin, Shiga toxin, cholera toxin), and bacterial parasites (*Heliocobacter–pyroli, Bordetella pertussis, Pseudomonas aeruginosa*) (Connell et al., 1996; Fantini et al., 2000; Hirmo et al., 1997). Although several studies have suggested this function of rafts, very little research has been directed toward studying the effects of domain microstructure on domain–pathogen interactions. The next step in understanding how rafts function as pathogen binding sites is to use model membrane systems to determine how ligand spacing and fluidity affect the strength and interaction energy of domain–pathogen interactions.

Supported membranes offer several advantages in studying these binding events. Phase-separated domain formation in supported lipid bilayers can be readily controlled through thermal history and characterized through atomic force microscopy. The two-dimensional platform of supported membranes allows for the application of several techniques to quantify pathogen–domain binding, including total internal fluorescence microscopy for measuring macroscopic binding affinities and dynamic force spectroscopy for measuring single-molecule binding forces and interaction potentials. Applying these methods to supported membranes functionalized with domains of a particular glycosphingolipid and corresponding receptor protein will lead to greater insight into how plasma membrane rafts function as binding domains in the infective pathways of host pathogens including viruses, bacteria, and bacterial toxins.

II. LIPID DOMAINS IN LIVING CELL MEMBRANES

An issue of central importance in membrane biology is the existence and function of microdomains or rafts within the plane of the cellular membrane. Over the past decade, evidence has emerged indicating that cell membranes do not exist as a homogeneous lipid matrix, as was described in the fluid mosaic model (Singer, 1977), but rather certain lipid types (glycosphingolipids, cholesterol, and sphingomyelins) phase separate into microdomains or rafts (Hwang et al., 1998; Simons and Ikonen, 1997). Rafts are believed to serve several functions that include signaling, sorting, and trafficking through secretory and endocytic pathways, and acting as attachment platforms for host pathogens and their toxins (Fantini et al., 2000; Simons and Ikonen, 1997). It has also been shown that the lateral organization of glycosphingolipids in microdomains allows for selective partitioning of proteins involved in the signal transduction pathway (Simons and Ikonen, 1997). It is believed that rafts enriched in glycosphingolipids can act as cellular binding sites for several pathogens including a wide range of viruses (HIV-1, influenza virus, murine polyoma virus), bacterial toxins (pertussis toxin, Shiga toxin, cholera toxin), and bacterial parasites (*Heliocobacter pyroli, Bordetella pertussis, Pseudomonas aeruginosa*) (Connell et al., 1996; Fantini et al., 2000; Hirmo et al., 1997).

The mechanism driving lipid phase separation and raft formation is the hydrophobic mismatch between the acyl chains of lipids in two different phases. Without the addition of cholesterol or the application of an external pressure, lipids exist in two phases within the plane of a membrane or bilayer: the gel or crystalline state and the fluid or liquid crystalline state. The acyl chains of fluid-phase lipids exhibit relatively weak van der Waals interactions and freely diffuse through the two-dimensional plane of the membrane. In the gel phase, however, the lipid acyl chains exhibit strong van der Waals interactions and are locked in a crystalline structure. The tight acyl chain packing and interlipid interactions prevent lipid diffusion. Therefore, lipids in a gel state are relatively immobile in comparison to lipids in a fluid state. Without the influence of external pressure, the phase of a lipid is dependent primarily on temperature. The phase transition temperature (T_m) of a lipid is determined predominately by the length of the acyl chain. Longer acyl chains exhibit stronger van der Waals interactions and, therefore, display higher melting temperatures. As lipids transition from a fluid phase into a gel phase, the acyl chains stiffen and the lipids extend. As a result, within a two-phase bilayer, the gel-phase lipids protrude slightly higher than fluid-phase lipids, resulting in a hydrophobic mismatch. This causes gel-phase lipids to aggregate into domains, thereby reducing the gel–liquid interface.

Cellular membranes contain a wide range of lipids including lipids with long saturated hydrocarbon chains (glycosphingolipids and sphingolipids) that phase separate into domains or lipid "rafts" at physiological temperatures (Brown and Rose, 1992). Furthermore, rafts are believed to contain a high concentration of cholesterol. The presence of cholesterol has the effect of disrupting the hydrocarbon chain packing in the crystal or gel state and flattening out the phase transition temperature of lipids (Ohvo-Rekila et al., 2002; Ramstedt and Slotte, 2002). As a result of this phenomenon, rafts are believed to exist in a liquid ordered state, l_o (Ahmed et al., 1997; Silvius et al., 1996). Acyl chains of lipids in the l_o phase have properties that are intermediate between those of the gel and liquid crystalline states. They are extended and ordered, similar to the gel state, but contain high lateral mobility within the ordered domain, similar to the liquid crystalline phase.

Glycosphingolipid packing, spacing, and mobility should have dramatic impact on multivalent protein interactions with these lipids at raft surfaces. A predominate percentage of multivalent interactions occur at membrane surfaces and involve carbohydrate–carbohydrate binding, such as viral attachment and entry. Influenza viral attachment, for example, occurs by interactions between multiple trimers of hemagglutinin (HA, a lectin that is densely packed on the surface of the virus) and multiple moieties of N-acetylneuraminic acid. Because of its emerging importance, significant theoretical work has been directed at understanding the interactions of multivalent ligands with cell surface receptors. In particular, a large focus of the work has been directed at studying the effects of ligand density, mobility, and molecular arrangement (steric parameters) in bivalent interactions (Dembo and Goldstein, 1978; Hlavacek et al., 1999; Posner et al., 1995). This theoretical work has resulted in several possible mechanisms to describe multivalent interactions. One mechanism is based on the effects of entropy. After the initial ligand has bound, entropic effects for the second binding event may be reduced if the second ligand is optimally spaced. For very flexible systems, enhanced affinities can be explained through an increased local concentration of the second ligand after the initial binding of the first ligand. This phenomenon was determined to occur maximally at optimal ligand spacing (Kramer and Karpen, 1998). The feature common to all of these proposed mechanisms is the importance of ligand spacing, molecular orientation, and mobility to optimization of multivalent interactions. Little work has been done to study how the microstructure of a raft, that is, cholesterol concentration, glycosphingolipid density, and raft size, affects its ability to act as a binding platform for viruses, bacterial toxins, and bacterial parasites. One of the next steps in further understanding the function of rafts is to study the effects of such microstructural parameters on the binding events between rafts and multivalent receptor proteins.

III. SUPPORTED LIPID BILAYERS

A. Formation Methods

Over the past two decades a large number of studies have been conducted to examine lipid immiscibility in model membrane systems in an attempt to elucidate the nature of rafts in cellular membranes. One of the most widely used model membrane systems is the supported lipid bilayer (Crane and Tamm, 2004; Giocondi et al., 2004; Milhiet et al., 2002; Ratto and Longo, 2002; Rinia et al., 2001; Stottrup et al., 2004; Tokumasu et al., 2003; Yuan et al., 2002). A supported lipid bilayer is a bilayer supported on a hydrophilic substrate, with an ~1-nm water layer separating the bilayer from the substrate. Traditionally, two techniques have been used in forming supported lipid bilayers, Langmuir–Blodgett deposition and vesicle fusion. Langmuir–Blodgett deposition involves sequentially adding the two lipid layers of the bilayer from a lipid monolayer at the air–water interface. The first layer is deposited by slowly pulling a hydrophilic surface from the aqueous subphase across the monolayer. The second layer is deposited by slowly dropping the hydrophobic lipid monolayer from the air into the aqueous subphase. Vesicle fusion involves depositing a solution of small unilamellar vesicles (SUVs) onto a hydrophilic substrate. Bilayer formation is believed to occur through four steps: initial vesicle adsorption, fusion of adsorbed vesicle to form larger vesicles, vesicle rupture forming bilayer disks on the surface, and, finally, merging of bilayer disks to form a uniform two-dimensional supported lipid bilayer (Johnson et al., 2002; Keller et al., 2000; Reviakine and Brisson, 2000).

B. Characterization of Domains by Atomic Force Microscopy

A number of techniques, such as fluorescence microscopy and atomic force microscopy (AFM), can be readily applied for characterization of lipid-phase behavior and gel-phase domain formation in supported lipid bilayers. Because of its subnanometer resolution and ability to image in water, AFM has proven to be an invaluable technique for characterizing and imaging lipid-phase separation. Gel-phase lipids tend to protrude approximately 1–2 nm above fluid-phase lipids, which is well within the resolution of AFM. Phase-separated domains on the nanoscale have been imaged for a variety of lipid mixtures in supported lipid bilayers including 1,2-dilauroyl-*sn*-glycero-3-phosphocholine (DLPC):1,2-distearoyl-*sn*-glycero-3-phosphocholine (DSPC) (Ratto and Longo, 2002); 1,2-dipalmitoyl-*sn*-glycero-3-phosphocholine

(DPPC):DLPC:cholesterol (Tokumasu et al., 2003); 1-palmitoyl-2-oleoyl-*sn*-glycero-3-phosphocholine (POPC):sphingomyelin:cholesterol (Giocondi et al., 2004); 1,2-dioleoyl-*sn*-glycero-3-phosphocholine (DOPC):DPPC:cholesterol (Stottrup et al., 2004); and DOPC:sphingomyelin:GM1:cholesterol (Yuan et al., 2002). These studies have shown the existence of phase separation in multicomponent bilayers and the formation of domains enriched in gel-phase lipids. Interestingly, the domains remain stationary while the intervening fluid phase retains fluid-like mobility. Ratto and Longo (2002) demonstrated that symmetric DSPC gel-phase domains remained immobile for up to 3 h on a mica substrate. Tokumaso et al. (2002) noted that the phase behavior of DLPC:DPPC:cholesterol mixtures was quantitatively affected by the mica substrate. The immobility of these domains has been attributed largely to domain–substrate interactions, as this property is generally not seen in unsupported model membrane systems such as giant unilamellar vesicles (GUVs) (Feigenson and Buboltz, 2001; Veatch and Keller, 2003).

C. Controlling Domain Characteristics

Generally, nanometer-scale domains are reported for supported lipid bilayers. This propensity has been attributed to the effects of the substrate in supported lipid bilayers, and it has been suggested that the substrate–bilayer interactions impart nonequilibrium phase behavior. However, before the domains become immobile, it is possible to control the time given for domain aggregation and growth. Shorter times should result in "nonequilibrium" sizes and shapes, whereas much longer times should result in "equilibrium" sizes and shapes. The time for domain aggregation and growth can be controlled by the thermal history of the bilayer.

The exact size of gel-phase domain size on the nanoscale can be controlled through the technique of thermal quenching (Ratto and Longo, 2002) during vesicle fusion. The SUVs contain both the gel-phase and fluid-phase components and are heated during vesicle formation to above the phase transition temperature of the gel-phase component. The heated vesicles are deposited onto room temperature mica. Through the rate of cooling, gel-phase domain size can be varied between 30 nm and 1 μm, with faster cooling rates resulting in smaller domains. Figure 9.1a demonstrates the effects of cooling rates on domain size in the nanometer range. In this case, phosphatidylcholine lipids of differing acyl chain lengths, DSPC (18:0, $T_m = 55\,°C$) and DLPC (12:0, $T_m = -5\,°C$) were used to demonstrate the effect of quench rate on domain size. At room temperature DSPC exists in a crystalline or gel state and DLPC is in a liquid crystalline or fluid state. As a result, DSPC phase separates and forms domains within the fluid DLPC matrix.

A second method of forming supported lipid bilayers has been developed, slow thermal cooling vesicle fusion (Lin and Longo, 2004), in which gel-phase domain size can be controlled on the micron scale (Fig. 9.1b) Domain size on this scale has been reported predominately for GUVs, an equilibrium model membrane system (Feigenson and Buboltz, 2001; Veatch et al., 2003). Therefore, by the use of slow thermal cooling techniques, it is possible to overcome the nonequilibrium effects of the solid substrate, allowing for the formation of micron-scale equilibrium domain sizes. The process of forming micron-scale domains involves depositing a heated vesicle solution onto a

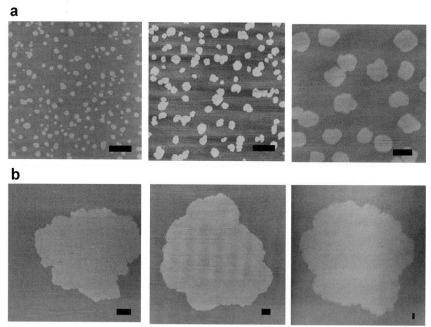

FIGURE 9.1   Controlling DSPC gel-phase domain size on the nanometer and micrometer scale in a DSPC:DLPC-supported lipid bilayer. (a) Thermal quenching rates were decreased from left to right, resulting in a gradual increase in DSPC gel-phase domain size from 50 nm at the fastest quenching rate to 1 μm at the slowest quenching rate. (b) Slow thermal cooling rates decreased from left to right, resulting in an increase in DSPC gel-phase domain size from 6 to 40 μm. Bar = 1 μm.

heated mica surface and then slowly cooling the bilayer in a temperature-controlled oven for 2–6 h. Domain size can be controlled by oven cooling rates where longer cooling times result in larger domains. See Fig. 9.1b for a demonstration using the DLPC:DSPC system. This process results in domain sizes ranging from 3 to 35 μm. At the longest cooling time, 5 h, gel-phase domains can become as large as 35 μm in diameter.

D. Imaging and Detection of Receptor–Ligand Interactions at Domain Surfaces

Specific binding between protein and lipid ligands can be visualized and correlated to the bilayer microstructure, using a combination of fluorescence microscopy and atomic force microscopy. Specific protein binding to functionalized supported lipid bilayers was explored using TRITC-conjugated neutravidin binding to a fluid-phase biotinylated dioleoylphosphatidyle-thanolamine (biotin–DOPE) and gel-phase biotinylated dipalmitoylphophatidyle-thanolamine (biotin–DPPE). Pure fluid-phase bilayers containing 5 mol% biotin–DOPE were incubated with TRITC–neutravidin and examined by optical fluorescence microscopy. The images in Fig. 9.2a show uniform fluorescence throughout the bilayer, indicating uniform binding. This was then repeated on bilayer containing phase-separated DLPC, DSPC, and 2 mol% biotin–DOPE, which partitions into the fluid DLPC phase, formed through slow thermal cooling vesicle fusion. This resulted in the formation of large DSPC domains (Fig. 9.2b). When incubated with TRITC–neutravidin, binding was directed to the fluid-phase DLPC regions with a fluorescence pattern that reflected the microstructure of the domains (Fig 9.2b). This was again

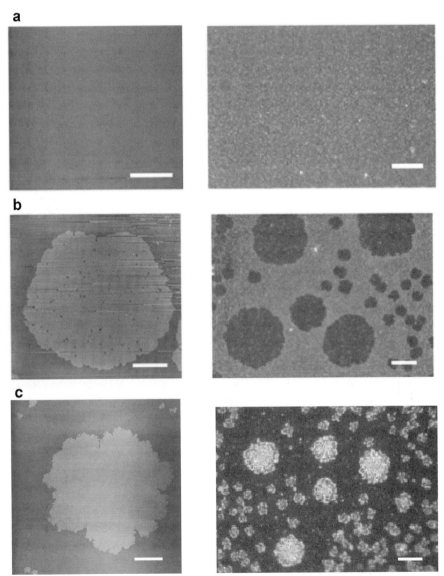

FIGURE 9.2 TRITC–neutravidin binding directed to fluid-phase biotin–DOPE and gel-phase biotin–DPPE. (a) On the left is an AFM image of a pure fluid bilayer of DLPC containing 5 mol% biotin–DOPE. On the right is an optical fluorescence microscopy image of the bilayer 30 min after incubation with TRITC–neutravidin. The uniform fluorescence indicates uniform binding across the bilayer. (b) On the left is an AFM image of a phase-separated bilayer containing DLPC, DSPC, and 2 mol% biotin–DOPE. The bilayer was formed through slow thermal cooling. On the right is an optical fluorescence microscopy image 30 min after incubation with TRITC–neutravidin. The regions of no fluorescence indicate there was no binding to gel-phase DSPC domains. (c) On the left is an AFM image of a phase-separated bilayer containing DLPC, DSPC, and 2 mol% biotin–DPPE. The bilayer was formed through slow thermal cooling. On the right is an optical fluorescence microscopy image 30 min after incubation with TRITC–neutravidin. The binding pattern strongly reflects selective binding to gel-phase DSPC domains. Bar = 5 μm (AFM) and 10 μm (fluorescence).

repeated with 2 mol% biotin–DPPE, which partitions into the gel phase. When the bilayer was incubated with TRITC–neutravidin, binding was directed to the gel-phase DSPC domains (Fig. 9.2c). These methods were recently extended to study lectin–glycolipid binding using glycolipid domains (with addition of cholesterol). These experiments have established the feasibility of using supported lipid bilayers as binding platforms in which membrane–protein interactions can be studied. The next step is to employ optical techniques such as total internal reflection fluorescence microscopy (TIRFM) to measure binding affinities and force spectroscopy to measure binding strength at the single-molecule level.

IV. DYNAMIC FORCE SPECTROSCOPY

A. General Principle and Basic Theory

Many molecular recognition events occurring at cellular membrane interfaces are initiated under a shear stress or force, such as virus– and bacteria–host attachment, antibody–antigen recognition, cell–cell adhesion, and lectin agglutination. Molecular binding interactions of these biological processes are commonly investigated through molecules free in solution or directly affixed to a solid support. In several of these processes, the ligand and/or protein are anchored via a flexible or semiflexible tether to a membrane, such as a cell, viral, or bacterial membrane. Although equilibrium measurements have provided important insight into molecular recognition events, they largely ignore the fact that under physiological conditions, these interactions are occurring at membrane interfaces, under nonequilibrium conditions, and within a limited interaction volume. It has been shown that equilibrium measurements in which at least one molecular entity is free in solution do not scale well with the binding properties of membrane-bound molecules restricted to the two-dimensional plane of the membrane (Chang and Hammer, 1999; Riper et al., 1998). *Ricinus communis* agglutinin (RCA), a known potent crosslinker of erythrocytes, and *Viscum album* agglutinin (VAA), a weak agglutinating lectin, were both shown to exhibit the same equilibrium off-rate, 1×10^{-3} s^{-1}, but under an applied force the rupture strengths were 1.5 times greater for RCA, 65.9 pN, than VAA, 43.5 pN (Dettmann et al., 2000). This demonstrates that binding strengths cannot be predicted based on zero-force dissociation kinetics and equilibrium thermodynamics for biological recognition events occurring under an applied force or shear stress. Therefore, biological processes that occur under an applied force, such as virus–host cell adhesion and agglutination of cells, may be better characterized using techniques that can measure the binding rupture force under different loading rates such as AFM (Binnig et al., 1986).

The ability of the AFM to measure recognition between biomolecular pairs was first demonstrated and quantified between biotin and avidin (Florin and Gaub, 1994; Lee et al., 1994). Subsequently several groups demonstrated recognition events between individual antibody–antigen pairs (Allen et al., 1997; Dammer et al., 1996; Hinterdorfer et al., 1996) and lectin– carbohydrate pairs (Dettmann et al., 2000; Fritz et al., 1998; Ratto et al., 2004). This technique has been further exploited to determine the energy landscapes of the protein–ligand interaction. On the basis of the reaction rate theory developed by Kramer (1940) and further modified by Hanggi et al. (1990), the unbinding force is directly related to the dissociation kinetics of the complex under an applied force and, therefore, depends on the loading rate (Evans and Ritchie, 1997). This relationship has been predicted theoretically (Bell, 1980; Strunz et al., 2000) and shown empirically (Dettmann et al., 2000; Evans, 1999; Merkel et al., 1999; Strunz et al., 2000) to be an exponential increase in dissociation rates with increasing applied forces. On the basis of these theories, the shape and structure of the energy landscape can be predicted from the dissociation kinetics under an applied force, using a technique called dynamic force spectroscopy (Evans and Ritchie, 1997; Evans, 1999).

The use of AFM for force measurements is based on the principal mechanism by which an atomic force microscope produces images, through tip–sample interactions. Initially, a functionalized atomic force microscope tip is brought into contact with the sample containing the ligand. On separation of the tip from the sample, the interacting force between the protein and ligand results in a downward deflection of the tip. A typical force curve is diagrammed in Fig. 9.3.

Initially the tip approaches the surfaces and makes contact (B), and on further approach, the tip is deflected upward. At the top of the approach cycle, the surface begins to retract until it stretches the length of the polyethyleneglycol spacer. (C) Further retraction causes a downward deflection of the tip until the bond ruptures, and the tip deflects to baseline. (D) The vertical displacement during bond rupture and cantilever spring constant can be used to calculate the rupture force through Hook's law, $F = k\Delta z$.

Energy landscapes of receptor–ligand bonds can be determined through continuous plots of rupture force expressed on a scale of $\log_e$ of loading rates. This correlation translates into sharp energy barriers traversed in the energy landscape along the force-driven pathway, as recently described in

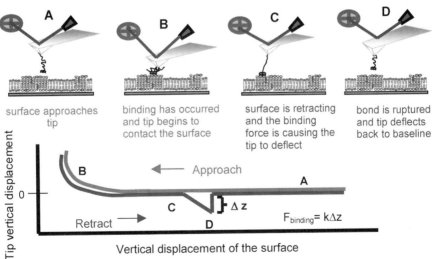

FIGURE 9.3 Measuring rupture forces with AFM. (A) Initially the surface begins to approach the tip. (B) The surface contacts the tip and deflects it upward, which is indicated by the increase in tip displacement. At some point in the approach cycle, the tethered protein binds ligand. (C) At the top of the approach cycle, the surface begins to retract until the PEG tether is fully extended. On further retraction, the tip is deflected downward due to protein–ligand binding. (D) The tip continues to be deflected downward until the bond is ruptured, at which point it will rapidly deflect back to baseline. The rupture force can be calculated by multiplying the calibrated cantilever spring constant by the distance the tip was deflected.

the study of energy landscapes for the biotin–avidin (Merkel et al., 1999), lectin–carbohdydrate (Dettmann et al., 2000), antibody–antigen (Schwesinger et al., 2000), and P-selectin complexes (Fritz et al., 1998). Without an applied force the transition from bound state to free state is described by the equilibrium dissociation constant. According to Kramer's (1940) theory, the equilibrium dissociation constant depends on an oscillation frequency (λ_0), an energy barrier (ΔU), and the inverse thermal energy ($\beta = 1/k_B T$):

$$\lambda = k_{off} = \lambda_0 e^{-\beta(\Delta U)}. \tag{9.1}$$

From this theory the ligand is seen as a point of gravity trying to cross the potential barrier, with frequency (λ_0), set by the height of an energy barrier (ΔU) and independent of the barrier shape. This behavior is altered in situations where there is an applied force. Externally applied forces reduce the energy barrier by the product of the applied force (F) and the width of the barrier (γ), where the width of the barrier is defined as the distance between the maximum and minimum of the potential well. The dissociation constant under an applied force is given by

$$\lambda(F) = \lambda_0 e^{-\beta(\Delta U - F \cdot \gamma)} = k_{off} e^{\beta F \gamma}. \tag{9.2}$$

To extract energy landscapes from Eq. (9.2), stochastic distributions must be used in which the stochastic process of bond breakage under a steady increase in force is analyzed. In this case, a probability density function for detachment between times t and $t + \Delta t$ during the loading process can describe the lifetime of the bond under force. As instantaneous force is set by time and loading rate, $F = r_f t$, where r_f, is the loading rate, the unbinding between forces F and $F + \Delta F$ is equivalent to the distribution of lifetimes:

$$P(F) = 1/r_f \, \lambda(F) e^{-1/r \int \gamma(F) dF}. \tag{9.3}$$

The peak of the distribution in the unbinding force occurs when the first derivative of the unbinding force distribution is zero. By combination of Eq. (9.2) with the first derivative of the unbinding force set to zero, a linear relationship between $\log_e$ of the loading rate, r_f, and rupture force, F, can be derivived:

$$F = \beta/\gamma \log_e(r_f) + \log_e(\gamma \cdot \lambda_0/\beta). \tag{9.4}$$

The y intercept and slope of this linear relationship provide the fitting parameters needed to calculate the energy barrier, ΔU, and barrier width, γ.

B. Measurements of Specific Interactions

One of the most challenging aspects of force spectroscopy experiments involves adding functionality to the atomic force microscope cantilever. The simplest method entails physically adsorbing the molecules of interest directly to the atomic force microscope tip, either by immersing the cantilever in a high-concentration solution containing the molecules or by bringing the tip down on a molecule laden-surface. This process has been used successfully for synthetic (Holland et al., 2003; Li et al., 1999a) and biological (Li et al., 1999b; Rief et al., 1997) polymers, as well as a variety of chemicals and proteins (Rief et al., 1997). One drawback of this method is that the coating on the tip is not well-characterized, making interpretation of the data relatively difficult due to issues surrounding the delamination of the coating. Perhaps even more serious is the fact that, as the molecules of interest are directly on the atomic force microscope tip, it is quite difficult to separate nonspecific tip–substrate interactions (van der Waals, hydration, depletion, etc.) from the specific interactions between the molecules of interest on the tip and the molecules of interest on the substrate. Although the problem of delamination has been addressed by chemically adsorbing the molecules of interest to the tip, either through silane linkages to silicon nitride cantilevers (Schwesinger et al., 2000) or thiol linkages to gold-coated levers (Friedsam et al., 2004), separating nonspecific from specific interactions remains a problem. Some researchers have attempted to avoid this problem by using statistical analysis, for example, accepting data only from tips that display interactions less than one time in 10 "touches" or contacts with the substrate. This is not a particularly rigid criterion, however, and the forces measured using molecules directly attached to atomic force microscope tips tend to have quite large standard deviations, larger than the natural stochastic deviation associated with bond strength.

An alternate method involves attaching the molecules of interest to the atomic force microscope tip through a polymer tether (Hinterdorfer et al., 2002; Riener et al., 2001; Riener et al., 2003). This has the advantage of moving the specific interactions away from the point of tip–sample contact, thus separating the nonspecific from the specific. Four sample force traces are shown in Fig. 9.4.

Other measures can be taken to ensure that the data from specific interactions can be identified. Blocking experiments can add confidence in the specificity of the interactions being measured. Blocking experiments, wherein the specific interaction between the tip and sample is blocked through the addition of one or the other of the binding molecules, abolish specific bond formation but tend to not affect nonspecific adhesion (for an example, see Fig. 9.5).

Figure 9.6 illustrates data taken from an experiment in which both the protein and ligand were attached with polymers to the tip and substrate, respectively (this was also the case in Fig. 9.5).

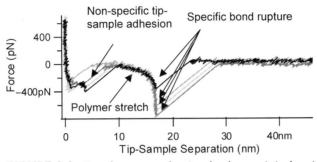

FIGURE 9.4 Four force traces showing the characteristic three interaction regions seen with polymer functionalized tips. In this case a thiol molecule was attached at the end of the polymer and used to measure the force required to rupture the thiol–gold bond. The four forces at which the bond ruptures reveals details about the stochastic nature of the thiol–gold bond.

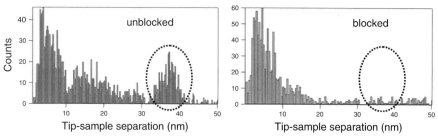

FIGURE 9.5 Histograms of adhesive interactions between a polymer-tethered protein and a ligand-functionalized substrate. When the buffer solution in the atomic force microscope liquid cell is exchanged with buffer containing free ligand, binding between the protein on the microscope tip and the ligand attached to the substrate is blocked, and the specific interactions seen from ~30 to 40 nm can no longer be seen. Note that the number of interactions at lengths less than 30 nm also decreases on addition of free ligand, implying that specific interactions may also occur at these lengths. These may be due to adherence of the tethered protein to the microscope tip or adherence of tethered ligands to the substrate prior to the formation of a bond.

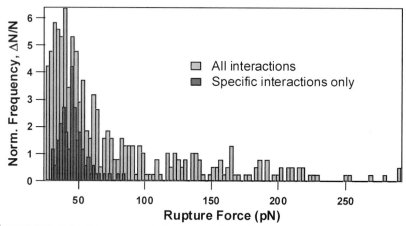

FIGURE 9.6 Histograms of the rupture force frequencies for all measured interactions between the tip and sample (gray) and for just the interactions taken from the specific cluster (black). Separating the specific bond rupture forces from the nonspecific interactions increases the accuracy of the measurement. Gaussian fits to both data sets reveal distributions of 76 ± 55 pN for all interactions and 45 ± 10 pN for the specific interactions alone.

V. MEASUREMENTS OF LIPID ANCHORING FORCES IN LIPID BILAYERS

It has been hypothesized that rafts can function as cellular binding sites for several pathogens including a wide range of viruses (HIV-1, influenza virus, murine polyma virus), bacterial toxins (pertussis toxin, Shiga toxin, cholera toxin), and bacterial parasites (*Heliocobacter pyroli, Bordetella pertussis, Pseudomonas aeruginosa*) (Connell et al., 1996; Fantini et al., 2000; Hirmo et al., 1997). These adhesion events generally occur at epithelial cell surfaces under a shear stress or force. One plausible reason that these events occur at rafts may be due to the increased energy or applied force required to pull lipids out of the membrane. Gel-phase lipids contain much higher interlipid van der Waals interactions relative to fluid-phase lipids. Therefore, lipids within gel-phase microdomains may act as better anchors during the biological process described earlier. In addition, cholesterol is known to modulate acyl chain packing and lateral fluidity of gel-phase lipids. Therefore, an additional role of cholesterol may be to control the interlipid van der Waals forces and lateral fluidity during recognition events under an external force either to cause the lipid to remain anchored in the membrane or to allow for lipid pullout.

Several techniques have been employed to measure the force required for lipid pullout, including the biomembrane force probe (BFP) (Evans and Ludwig, 2000), surface force apparatus (SFA) (Leckband et al., 1995), AFM (Desmeules et al., 2002), and molecular dynamics simulations (Marrink et al., 1998). With the BFP, the force to pull out biotin–polyethyleneglycol–distearoylphosphatidylethanolamine (biotin–PEG–DSPE) has been measured for giant vesicles consisting of pure DSPE (23 pN), SOPC (30 pN), and DSPC containing 40% cholesterol (41 pN) (Evans and Ludwig, 2000). Leckband et al. (1995), using SFA, calculated from adhesion energy measurements that the lipid pullout force from a DPPC-supported lipid bilayer is 70 pN. Desmeules et al (2002), using AFM, measured the adhesion forces between myristoylated recoverin and a supported DPPC bilayer in the absence and presence of calcium. The force to remove myristoyl from a supported DPPC bilayer was measured to be 48 pN. Marrink et al. (1998) conducted molecular dynamic simulations at very high loading rates (10^7–10^8 times faster than the experimental loading rates) and calculated lipid pullout forces for fluid-phase lipids to be greater than 200 pN. Despite the work done in measuring single-molecule lipid pullout, there exists a broad range in the measured forces. The broad range may be due to the differing techniques employed for the measurements in combination with the different lipid environments from which the lipids were pulled. In none of these studies has a comprehensive comparison been performed to determine the effects of lipid phase and cholesterol content within a given lipid mixture on the lipid pullout force. Before force spectroscopy experiments can be conducted to study membrane–protein interactions at supported lipid bilayer interfaces, it is necessary to determine in which force regime lipid pullout occurs and under which conditions membrane anchoring is optimal.

VI. CONCLUSIONS

One of the main functions of lipid domains, such as lipid rafts, in cell membranes may be to modulate multivalent carbohydrate–protein interactions in a variety of biological pathways. Currently, very little is known about how raft microstructure affects these interactions, and in general, the mechanisms and kinetics of multivalency are not well understood both experimentally and theoretically. As many of the interactions at membrane surfaces and rafts occur under external mechanical forces, techniques are needed to study binding behaviors under similar conditions. Dynamic force spectroscopy satisfies these conditions and yields important details of ligand–protein interactions under dynamic conditions. Varying force loading rates in combination with varying ligand domain composition and microstructure in supported lipid bilayers should provide important details about the molecular mechanism of multivalency for protein–membrane interactions in the future.

ACKNOWLEDGMENTS

M.L.L. acknowledges funding by the Center for Polymeric Interfaces and Macromolecular Assemblies, the Materials Research Institute at Lawrence Livermore National Laboratory, the Nanoscale Interdisciplinary Research Teams Program of the National Science Foundation, and a generous endowment from Joe and Essie Smith. CDB acknowledges funding from the National Institutes of Health Biotechnology Training Grant of the University of California at Davis. This work was performed under the auspices of the U.S. Department of Energy by the University of California/Lawrence Livermore National Laboratory under Contract W-7405-Eng-48.

SUGGESTED READING

Brown, D.A., and London, E. (1997) Structure of detergent-resistant membrane domains: Does phase separation occur in biological membranes? *Biochem. Biophys. Res. Commun.* **240**: 1–7.

Brown, D.A., and London, E. (1998) Functions of lipid rafts in biological membranes. *Annu. Rev. Cell Dev. Biol.* **14**: 111–136.

Kahya, N., Scherfeld, D., Bacia, K., and Schwille, P. (2004) Lipid domain formation and dynamics in giant unilamellar vesicles explored by fluorescence correlation spectroscopy. *J. Struct. Biol.* **147**: 77–89.

Thompson, T.E., and Tillack, T.W. (1985) Organization of glycosphingolipids in bilayers and plasma-membranes of mammalian-cells. *Annu. Rev. Biophys. Biophys. Chem.* **14**: 361–386.

Karlsson K.A. 1995. Microbial recognition of target-cell glycoconjugates. *Curr. Opin. Struct. Biol.* **5**: 622–635.

Noy, A., Vezenov, D.V., and Lieber, C.M. (1997) Chemical force microscopy. *Annu. Rev. Mater. Sci.* **27**: 381–421.

REFERENCES

Ahmed, S. N., Brown, D. A., and London, E. (1997) On the origin of sphingolipid/cholesterol-rich detergent-insoluble cell membranes: Physiological concentrations of cholesterol and sphingolipid induce formation of a detergent-insoluble, liquid-ordered lipid phase in model membranes. *Biochemistry* **36**: 10944–10953.

Allen, S., Chen, X. Y., Davies, J., Davies, M. C., Dawkes, A. C., Edwards, J. C., Roberts, C. J., Sefton, J., Tendler, S. J. B., and Williams, P. M. (1997) Detection of antigen–antibody binding events with the atomic force microscope. *Biochemistry* **36**: 7457–7463.

Bell, G. I. (1980) Theoretical-models for the specific adhesion of cells to cells or to surfaces. *Adv. Appl. Prob.* **12**: 566–567.

Binnig, G., Quate, C. F., and Gerber, C. (1986) Atomic force microscope. *Phys. Rev. Lett.* **56**: 930–933.

Brown, D. A., and Rose, J. K. (1992) Sorting of Gpi-anchored proteins to glycolipid-enriched membrane subdomains during transport to the apical cell-surface. *Cell* **68**: 533–544.

Chang, K. C., and Hammer, D. A. (1999) The forward rate of binding of surface-tethered reactants: Effect of relative motion between two surfaces. *Biophys. J.* **76**: 1280–1292.

Connell, H., Agace, W., Klemm, P., Schembri, M., Marild, S., and Svanborg, C. (1996) Type 1 fimbrial expression enhances *Escherichia coli* virulence for the urinary tract. *Proc. Natl. Acad. Sci. USA* **93**: 9827–9832.

Crane, J. M., and Tamm, L. K. (2004) Role of cholesterol in the formation and nature of lipid rafts in planar and spherical model membranes. *Biophys. J.* **86**: 2965–2979.

Dammer, U., Hegner, M., Anselmetti, D., Wagner, P., Dreier, M., Huber, W., and Guntherodt, H. J. (1996) Specific antigen/antibody interactions measured by force microscopy. *Biophys. J.* **70**: 2437–2441.

Dembo, M., and Goldstein, B. (1978) Theory of equilibrium binding of symmetric bivalent haptens to cell-surface antibody: Application to histamine-release from basophils. *J. Immunol.* **121**: 345–353.

Desmeules, P., Grandbois, M., Bondarenko, V. A., Yamazaki, A., and Salesse, C. (2002) Measurement of membrane binding between recoverin, a calcium-myristoyl switch protein, and lipid bilayers by AFM-based force spectroscopy. *Biophys. J.* **82**: 3343–3350.

Dettmann, W., Grandbois, M., Andre, S., Benoit, M., Wehle, A. K., Kaltner, H., Gabius, H. J., and Gaub, H. E. (2000) Differences in zero-force and force-driven kinetics of ligand dissociation from beta-galactoside-specific proteins (plant and animal lectins, immunoglobulin G) monitored by plasmon resonance and dynamic single molecule force microscopy. *Arch. Biochem. Biophys.* **383**: 157–170.

Evans, E., and Ludwig, F. (2000) Dynamic strengths of molecular anchoring and material cohesion in fluid biomembranes. *J. Phys. Condensed Matter* **12**: A315–A320.

Evans, E., and Ritchie, K. (1997) Dynamic strength of molecular adhesion bonds. *Biophys. J.* **72**: 1541–1555.

Evans, E. B. (1999) Looking inside molecular bonds at biological interfaces with dynamic force spectroscopy. *Biophys. Chem.* **82**: 83–97.

Fantini, J., Maresca, M., Hammache, D., Yahi, N., and Delezay, O. (2000) Glycosphingolipid (GSL) microdomains as attachment platforms for host pathogens and their toxins on intestinal epithelial cells: Activation of signal transduction pathways and perturbations of intestinal absorption and secretion. *Glycoconj. J.* **17**: 173–179.

Feigenson, G. W., and Buboltz, J. T. (2001) Ternary phase diagram of dipalmitoyl-PC/dilauroyl-PC/ cholesterol: Nanoscopic domain formation driven by cholesterol. *Biophys. J.* **80**: 2775–2788.

Florin, E. L., Moy, V. T., and Gaub, H. E. (1994) Adhesion forces between individual ligand–receptor pairs. *Science* **264**: 415–417.

Friedsam, C., Seitz, M., and Gaub, H. E. (2004) Investigation of polyelectrolyte desorption by single molecule force spectroscopy. *J. Phys. Condensed Matter* **16**: S2369–S2382.

Fritz, J., Katopodis, A. G., Kolbinger, F., and Anselmetti, D. (1998) Force-mediated kinetics of single P-selectin ligand complexes observed by atomic force microscopy. *Proc. Natl. Acad. Sci. USA* **95**: 12283–12288.

Giocondi, M. C., Milhiet, P. E., Dosset, P., and Le Grimellec, C. (2004) Use of cyclodextrin for AFM monitoring of model raft formation. *Biophys. J.* **86**: 861–869.

Hanggi, P., Talkner, P., and Borkovec, M. (1990) Reaction-rate theory: 50 Years after Kramers. *Rev. Mod Phys* **62**: 251–341.

Hinterdorfer, P., Baumgartner, W., Gruber, H. J., Schilcher, K., and Schindler, H. (1996) Detection and localization of individual antibody–antigen recognition events by atomic force microscopy. *Proc. Natl. Acad. Sci. USA* **93**: 3477–3481.

Hinterdorfer, P., Gruber, H. J., Kienberger, F., Kada, G., Riener, C., Borken, C., and Schindler, H. (2002) Surface attachment of ligands and receptors for molecular recognition force microscopy. *Colloids Surf. B* **23**: 115–123.

Hirmo, S., Kelm, S., Wadstrom, T., and Schauer, R. (1997) Lack of evidence for sialidase activity in *Helicobacter pylori*. *FEMS Immunol. Med. Microbiol.* **17**: 67–72.

Hlavacek, W. S., Posner, R. G., and Perelson, A. S. (1999) Steric effects on multivalent ligand-receptor binding: Exclusion of ligand sites by bound cell surface receptors. *Biophys. J* **76**: 3031–3043.

Holland, N. B., Hugel, T., Neuert, G., Cattani-Scholz, A., Renner, C., Oesterhelt, D., Moroder, L., Seitz, M., and Gaub, H. E. (2003) Single molecule force spectroscopy of azobenzene polymers: Switching elasticity of single photochromic macromolecules. *Macromolecules* **36**: 2015–2023.

Hwang, J., Gheber, L. A., Margolis, L., and Edidin, M. (1998) Domains in cell plasma membranes investigated by near-field scanning optical microscopy. *Biophys. J.* **74**: 2184–2190.

Johnson, J. M., Ha, T., Chu, S., and Boxer, S. G. (2002) Early steps of supported bilayer formation probed by single vesicle fluorescence assays. *Biophys. J.* **83**: 3371–3379.

Keller, C. A., Glasmastar, K., Zhdanov, V. P., and Kasemo, B. (2000) Formation of supported membranes from vesicles. *Phys. Rev. Lett.* **84**: 5443–5446.

Kramer, R. H., and Karpen, J. W. (1998) Spanning binding sites on allosteric proteins with polymer-linked ligand dimers. *Nature* **395**: 710–713.

Kramer, H. A. (1940) Brownian motion in a field of force and the diffusion model of chemical reactions. *Physica* **7**: 284–304.

Leckband, D., Muller, W., Schmitt, F. J., and Ringsdorf, H. (1995) Molecular mechanisms determining the strength of receptor-mediated intermembrane adhesion. *Biophys. J.* **69**: 1162–1169.

Lee, G.U., Kidwell, D. A., and Colton, R. J. (1994) Sensing discrete streptavidin biotin interactions with atomic-force microscopy. *Langmuir* **10**: 354–357.

Li, H., Rief, M., Oesterhelt, F., and Gaub, H. E. (1999a) Force spectroscopy on single xanthan molecules. *Appl. Phys. A* **68**: 407–410.

Li, H. B., Rief, M., Oesterhelt, F., Gaub, H. E., Zhang, X., and Shen, J. C. (1999b) Single-molecule force spectroscopy on polysaccharides by AFM – nanomechanical fingerprint of alpha-(1,4)-linked polysaccharides. *Chem. Phys. Lett.* **305**: 197–201.

Lin, W. C., and Longo, M. L. (2004) Lipid superposition and transmembrane movement in gel-liquid coexisting supported lipid bilayers, an AFM study. *Biophys. J.* **86**: 380A–380A.

Marrink, S. J., Berger, O., Tieleman, P., and Jahnig, F. (1998) Adhesion forces of lipids in a phospholipid membrane studied by molecular dynamics simulations. *Biophys. J.* **74**: 931–943.

Merkel, R., Nassoy, P., Leung, A., Ritchie, K., and Evans, E. (1999) Energy landscapes of receptor-ligand bonds explored with dynamic force spectroscopy. *Nature* **397**: 50–53.

Milhiet, P. E., Giocondi, M. C., and Le Grimellec, C. (2002) Cholesterol is not crucial for the existence of microdomains in kidney brush-border membrane models. *J. Biol. Chem.* **277**: 875–878.

Ohvo-Rekila, H., Ramstedt, B., Leppimaki, P., and Slotte, J. P. (2002) Cholesterol interactions with phospholipids in membranes. *Prog. Lipid Res.* **41**: 66–97.

Posner, R. G., Wofsy, C., and Goldstein, B. (1995) The kinetics of bivalent ligand bivalent receptor aggregation: Ring formation and the breakdown of the equivalent site approximation. *Math. Biosci.* **126**: 171–190.

Ramstedt, B., and Slotte, J. P. (2002) Membrane properties of sphingomyelins. *FEBS Lett.* **531**: 33–37.

Ratto, T. V., Langry, K. C., Rudd, R. E., Balhorn, R. L., Allen, M. J., and McElfresh, M. W. (2004) Force spectroscopy of the double-tethered concanavalin-A mannose bond. *Biophys. J.* **86**: 2430–2437.

Ratto, T. V., and Longo, M. L. (2002) Obstructed diffusion in phase-separated supported lipid bilayers: A combined atomic force microscopy and fluorescence recovery after photobleaching approach. *Biophys. J.* **83**: 3380–3392.

Reviakine, I., and Brisson, A. (2000) Formation of supported phospholipid bilayers from unilamellar vesicles investigated by atomic force microscopy. *Langmuir* **16**: 1806–1815.

Rief, M., Gautel, M., Oesterhelt, F., Fernandez, J. M., and Gaub, H. E. (1997) Reversible unfolding of individual titin immunoglobulin domains by AFM. *Science* **276**: 1109–1112.

Riener, C. K., Borken, C., Kienberger, F., Schindler, H., Hinterdorfer, P., and Gruber, H. J. (2001) A new collection of long, flexible crosslinkers for recognition force microscopy of single molecules. *Biophys. J.* **80**: 304a–305a.

Riener, C. K., Kienberger, F., Hahn, C. D., Buchinger, G. M., Egwim, I.O.C., Haselgrubler, T., Ebner, A., Romanin, C., Klampfl, C., Lackner, B., et al. (2003) Heterobifunctional crosslinkers for tethering single ligand molecules to scanning probes. *Anal. Chim. Acta* **497**: 101–114.

Rinia, H. A., Snel, M. M. E., van der Eerden, J. P. J. M., and de Kruijff, B. (2001) Visualizing detergent resistant domains in model membranes with atomic force microscopy. *FEBS Lett.* **501**: 92–96.

Riper, J. W., Swerlick, R. A., and Zhu, C. (1998) Determining force dependence of two-dimensional receptor–ligand binding affinity by centrifugation. *Biophys. J.* **74**: 492–513.

Schwesinger, F., Ros, R., Strunz, T., Anselmetti, D., Guntherodt, H. J., Honegger, A., Jermutus, L., Tiefenauer, L., and Pluckthun, A. (2000) Unbinding forces of single antibody–antigen complexes correlate with their thermal dissociation rates. *Proc. Natl. Acad. Sci. USA* **97**: 9972–9977.

Silvius, J. R., delGiudice, D., and Lafleur, M. (1996) Cholesterol at different bilayer concentrations can promote or antagonize lateral segregation of phospholipids of differing acyl chain length. *Biochemistry* **35**: 15198–15208.

Simons, K., and Ikonen, E. (1997) Functional rafts in cell membranes. *Nature* **387**: 569–572.

Singer, S. J. (1977) Citation classic: Fluid mosaic model of structure of cell-membranes. *Curr. Contents,* 13–13.

Stottrup, B. L., Veatch, S. L., and Keller, S. L. (2004) Nonequilibrium behavior in supported lipid membranes containing cholesterol. *Biophys. J.* **86**: 2942–2950.

Strunz, T., Oroszlan, K., Schumakovitch, I., Guntherodt, H. J., and Hegner, M. (2000) Model energy landscapes and the force-induced dissociation of ligand–receptor bonds. *Biophys. J.* **79**: 1206–1212.

Tokumasu, F., Jin, A. J., Feigenson, G. W., and Dvorak, J. A. (2003) Nanoscopic lipid domain dynamics revealed by atomic force microscopy. *Biophys. J.* **84**: 2609–2618.

Veatch, S. L., and Keller, S. L. (2003) Separation of liquid phases in giant vesicles of ternary mixtures of phospholipids and cholesterol. *Biophys. J.* **85**: 3074–3083.

Yuan, C. B., Furlong, J., Burgos, P., and Johnston, L. J. (2002) The size of lipid rafts: An atomic force microscopy study of ganglioside GM1 domains in sphingomyelin/DOPC/cholesterol membranes. *Biophys. J.* **82**: 2526–2535.

10
A FLOW CHAMBER FOR CAPILLARY NETWORKS: LEUKOCYTE ADHESION IN CAPILLARY-SIZED, LIGAND-COATED MICROPIPETTES

DAVID F. J. TEES, PRITHU SUNDD, and DOUGLAS J. GOETZ

Department of Physics and Astronomy and Department of Chemical Engineering Ohio University, Athens, Ohio, USA

Several mechanisms have been proposed to explain the arrest of leukocytes in capillaries. It has been hypothesized that cells become mechanically trapped in these small vessels or, alternatively, that leukocytes adhere to capillary endothelial cells via endothelial cell adhesion molecules (ECAMs) in a manner similar to that in venules. Recent work using glass microcapillary tubes coated with adhesion molecules supports the proposition that leukocyte arrest in capillaries involves both mechanical and adhesive forces and that the biochemical adhesive force is strongly modulated by mechanical forces that alter the area of contact between leukocytes and the endothelium. These *in vitro* microcapillary flow chambers allow the effects of vessel geometry, cell deformation, and ECAM chemistry to be studied in a well-controlled manner. The work promises to provide important insights into the process of leukocyte retention in capillaries in the same way that parallel-plate flow chamber assays helped to make possible a detailed understanding of leukocyte adhesion in the venular circulation.

I. INTRODUCTION

A. Physiological and Pathophysiological Adhesion in Capillaries

Mechanical/biochemical arrest of cells in capillaries has been observed in inflammation and sepsis for leukocytes (where it is a major cause of the sudden leukopenia that is observed in these conditions) (Andonegui et al., 2003; Erzurum et al., 1992; Gebb et al., 1995). This diminution of peripheral leukocytes is due in part to massive sequestration in capillaries of the lung. Arrest in lung alveolar capillaries is a necessary part of the response to infection or injury in the lung (Burns et al., 2003). There is also rapid leukopenia following injection (or presumably endogenous production during infection or inflammation) of chemotactic mediators such as interleukin-1 and complement fragment 5a (Doerschuk et al., 1990; Doyle et al., 1997; Kubo et al., 1999). Lung sequestration can also be produced by bacterial sepsis anywhere in the body that generates lipopolysaccharide (Andonegui et al., 2003).

B. Capillary Network Geometry

The capillary network in the lung and other organs has been extensively studied. Capillaries range from 2 to 8 μm in diameter (Doerschuk et al., 1993). Average neutrophil diameter has been variously estimated (by electron microscopy) at 6.8 μm (Doerschuk et al., 1993) and 8.5 μm (Shao et al., 1998). Most leukocytes must thus undergo considerable deformation on entering capillaries. There are two competing models for the geometry of alveolar capillary beds in the lung (Dawson and Linehan, 1997). One classic model treats the bed as a pair of parallel plates with posts (perhaps at high density) that run between the plates (Sobin and Fung, 1992). An alternate view is that the alveolar capillary bed is more tubelike (Guntheroth et al., 1982). An electron micrograph of a latex cast of the rat lung microcirculation that demonstrates the latter geometry is reproduced in Fig. 10.1.

In this latter tubule geometry, relatively straight capillary segments connect with other capillaries and branch off again. As shown schematically in

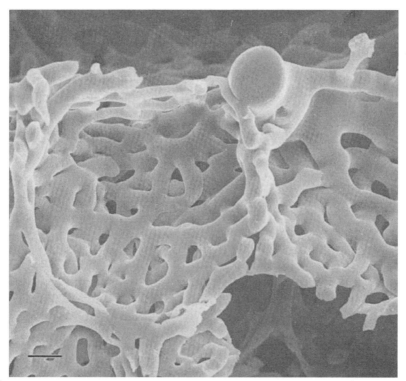

FIGURE 10.1   Scanning electron micrograph of a latex cast of the rat lung alveolar capillary bed. This shows that in this tissue, the geometry is tubular with many branching points. Bar = 10 μm. Reproduced, with permission, from Fig. 4 of Guntheroth et al. (1982).

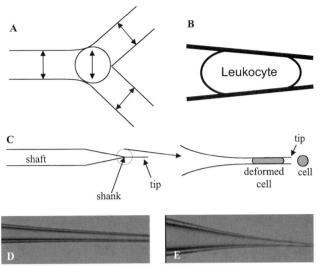

FIGURE 10.2 (A) Diagram showing the branching of a capillary into equal-sized branches. There is an expansion of diameter at the "vestibule," where the three vessels join (double-headed arrows are the same length in all vessels and also in the inscribed circle in the junction or "vestibule" region). (B) Diagram of a leukocyte in a tapering tube. The differing radii of the two end caps and the pressure on the tapering frustrum section can lead to cell arrest in the absence of adhesion molecules. (C) Micropipette nomenclature. A micropipette is a small tube. The shaft region is the same size as the glass capillary tubing used to form the micropipette (= 0.9 mm in this case). The shank section is the part that tapers down to the tip section. The tip section has a nearly constant inside diameter. Initially spherical cells introduced into the tip adopt a deformed capsule shape and are driven along the microvessel by a pressure difference between inside and outside. The dotted circle at the junction between the shank and the tip is shown enlarged. The micropipette tapers smoothly from the shank region into the relatively straight tip region. (D, E) Shank region of two micropipettes that have different taper rates. The width of the field of view for both images is 160 μm. (D) Wide-end inside diameter = 18 μm, narrow-end inside diameter = 5.3 μm. (E) Wide-end inside diameter = 30 μm, narrow-end inside diameter = 1.5 μm.

Fig. 2A, at the "vestibules" where vessels bifurcate, vessel diameters are significantly larger than in the straighter segments. The vestibules then branch and taper down again into a few exit vessels within a cell diameter or so. Mechanical arrest has been observed both in junction regions (where capillaries taper significantly) and in straight segments (where there is little tapering) (Gebb et al., 1995). The degree of tapering in capillary networks varies, but it is estimated that diameter can change from 8 to 4 μm within one or two cell diameters (8–16 μm) (Bathe et al., 2002; Guntheroth et al., 1982).

C. Mechanisms for Cell Arrest in Capillaries

Two major mechanisms have been proposed to explain leukocyte retention in capillaries. The current consensus has been summarized as follows: initial arrest of leukocytes in capillaries is through mechanical trapping, but retention and extravasation are dependent on interactions between adhesion molecules and on cell activation (Burns et al., 2003).

1. Mechanical Trapping

Mechanical trapping is a term used to describe any arrest mechanism that does not involve adhesion. More specifically it is cell stasis in the presence of a pressure gradient, mediated, presumably, by geometrical factors in the vessel. For a cell to be stationary inside a microvessel, there has to be a balance of forces acting on it. Cells are driven along capillaries by a pressure difference between the cell's leading and trailing edges. To a first approximation, if the cell almost completely fills a straight capillary with radius R_{cap}, the force F_{drive} on the cell is given by the simple relation $F_{drive} = \Delta P(\pi R_{cap}^2)$, where ΔP is the pressure difference across the cell. A more detailed analysis of the force on the capsule shape shows that the force also depends on the thickness of the lubrication layer between cell and tube surface. If this layer is small, however (and experiments on neutrophils in small tubes by Shao et al. (1998) show that it is), the simple analysis holds extremely well (Shao and Hochmuth, 1996). It is not clear, however, how the cell can arrest in a straight tube (at least if the tube is rigid).

If the leading and trailing edges of the cell in the vessel have different radii of curvature (as shown in Fig. 10.2B), then by the Law of Laplace[1] there will be an additional pressure difference that opposes or facilitates the motion of the cell depending on the direction of the taper (Tran-Son-Tay et al., 1994; Yeung and Evans, 1989; Zhelev et al., 1994). If this end-cap curvature difference is large enough, the force differential due to the end caps can balance the circulatory pressure difference across the cell as a whole and inhibit cell motion (Tran-Son-Tay et al., 1994). A recent study of the effect of vessel geometry on a finite-element mechanical model of a leukocyte squeezing through a capillary shows that the situation in a continuously tapering vessel is even more complicated than what would be expected from the Law of Laplace (Bathe et al., 2002). As described in the previous section, capillaries do change size with some tapering at the vestibules (Fig. 10.2A) between straight segments in the complex vessel networks that are seen in lung alveolar capillaries (Bathe et al., 2002; Huang et al., 2001).

2. Biochemical Adhesion

The idea that adhesion should play a role in leukocyte sequestration and plugging in capillaries is suggested by the role that it clearly plays in post-capillary venules (Abbassi et al., 1993; Alon et al., 1996; Borges et al., 1997; Finger et al., 1996; Lawrence et al., 1995; Lawrence and Springer, 1991; Lawrence and Springer, 1993; Ramos et al., 1997; Springer et al., 2002). In venules, the endothelial cells (ECs) that line blood vessels express receptor adhesion molecules that interact with ligands expressed on the leukocyte. E-selectin and P-selectin are carbohydrate-binding proteins that are expressed

---

[1]A pressure difference, ΔP, across a membrane held along its edge causes the membrane to bulge toward the low-pressure side. The force balance between pressure on the membrane and surface tension, τ, at the membrane edge yields the Law of Laplace. For a spherical cell membrane with radius of curvature R, this law can be written $\Delta P = 2\tau/R$. The Law of Laplace has been used to find the critical pressure for aspiration of a cell into a micropipette. For a cell partially aspirated into a micropipette (see, e.g., Fig. 10.5B), the segment of the cell that is outside the micropipette has a radius approximately equal to that of an unaspirated cell ($R = R_{cell}$). This shape is maintained by the difference between the cell internal pressure, P_{cell} and the pressure outside the micropipette, P_{out}. The curved membrane segment inside the micropipette (which has radius of curvature $R = R_{pipette}$) is maintained by the difference between P_{cell} and the pressure inside the micropipette, P_{in}. The Law of Laplace for these two curved surfaces can be used to eliminate P_{cell} and obtain $P_{out} - P_{in} = 2\tau(1/R_{pipette} - 1/R_{cell})$. The pressure difference across the cell must be larger than this critical value for the cell to be completely aspirated.

on stimulated ECs. L-selectin is expressed on resting leukocytes (it is rapidly shed following cell activation) (Kansas, 1996). All three selectins bind to the tetrasaccharide sialyl Lewis^x (sLe^x), and this sugar moiety is ubiquitous on leukocyte and EC surfaces (Varki, 1997). On leukocytes, there is a glycoprotein ligand for P- and E-selectin called PSGL-1 that is constitutively expressed. The β_2-integrins LFA-1 and Mac-1 are expressed on leukocytes and bind to the EC immunoglobulin superfamily members ICAM-1 and ICAM-2. Leukocyte adhesion to postcapillary venules in flow involves a series of steps: initial tethering, transient rolling, firm arrest, activation, and transendothelial migration (Springer, 1994). The role of many of these molecules has been clarified using *in vitro* parallel-plate flow chamber assays. In venules, the selectins are responsible primarily for tethering and transient attachment to ECs at physiological shear rates, whereas integrins are responsible for the firm attachment step.

For capillaries, the receptor–ligand situation is not as clear. Leukocytes most likely express the same adhesive ligands (L-selectin, PSGL-1, and inactive forms of the β_2-integrins) whether traveling through capillaries or venules (Burns et al., 2003). Then there arises the question of whether lung capillary ECs express the complementary receptors that bind to these ligands. Studies of lung ECs show that there are Weibel–Palade bodies in lung capillary ECs (Burns et al., 2003). Studies have been done using antibodies or other antagonists to try to block adhesion. It has been shown that there is some effect of fucoidin (a selectin inhibitor) on sequestration (Kubo et al., 1999; Kuebler et al., 1997). Antibodies to β_2-integrins also caused cell release, but antibodies to P-selectin had little effect (Kubo et al., 1999). There is also no observed change in sequestration in P/E-selectin knockout mice or in β_2-integrin knockout mice (Andonegui et al., 2003). β_2-Integrins and ICAM-1 have been shown to play a role in adhesion in some circumstances, however (Doerschuk et al., 1990). Clearly, there are conflicting data on the role of EC adhesion molecules (ECAMs) in arrest in capillaries.

The low levels of adhesion molecules that may be present could still play a role in leukocyte arrest if the forces from elastic shape recovery of leukocytes either make larger contact areas between cell and vessel wall or force the leukocyte surfaces into closer contact with capillary ECs than hydrodynamic forces could push them in venules (Burns et al., 2003). Increased extent and intimacy of contact could increase the rate for bond formation. Adhesion molecules could then generate a force that opposes the force from the pressure drop across the cell in a straight tube. Even small numbers of receptor–ligand bonds, when exposed over a large surface, could generate sufficient dynamic friction to cause cell arrest (King and Hammer, 2001).

This observation motivates a hypothesis: Biochemical adhesive forces due to ECAMs facilitate cell arrest at higher pressures than would be possible if the ECAMs were not present, and much lower surface densities of ECAMs are needed to mediate firm adhesion in capillaries than in postcapillary venules.

Recognition of the limitations of intravital microscopy, combined with an appreciation of the impact of the parallel-plate flow chamber assay on our understanding of adhesion in postcapillary venules, motivated the development of an *in vitro* adhesion assay that allows systematic study of the biochemical and mechanical mechanisms involved in leukocyte adhesion in the capillary network.

II. EXPERIMENT AND STIMULATION OF LEUKOCYTE ARREST IN CAPILLARIES

A. Previous Experiments on Leukocyte Motion in Small Tubes

As described earlier, there have been considerable *in vivo* studies, using intravital microscopy and electron microscopy of tissue sections, on leukocyte adhesion in the capillary network of the lungs (Andonegui et al., 2003; Burns et al., 2003; Doerschuk et al., 1990; Gebb et al., 1995). Although these *in vivo* studies have resulted in significant advances in our understanding, much is left unanswered regarding the mechanisms of biochemical and mechanical arrest in capillaries. This lack of understanding stems, in part, from the lack of control with *in vivo* assays. Specifically, although one can examine vessels of different diameters *in vivo* and probes can be used to narrow vessels and change the flow rate, it is difficult to systematically study the relationship between the parameters that are important for arrest (e.g., vessel diameter and geometry, pressure drop, ECAM chemistry). It is thus difficult to understand the relationship between biochemical forces and mechanical forces that mediate leukocyte adhesion in capillaries.

Because *in vivo* experiments cannot easily control the geometry of the network, the pressure difference across the cell, or the expression of adhesion molecules on the endothelial surface, there have been *in vitro* experiments that do control some of these factors. Many of these experiments have used micropipette aspiration. In a typical micropipette aspiration experiment, a pressure difference between micropipette interior and the bulk fluid outside the micropipette causes a single cell to be aspirated into a glass micropipette with inner diameter (ID) smaller than the cell size. As it is aspirated, the cell deforms, and flows into the micropipette. If the aspiration pressure is above a critical value (needed to overcome the additional pressure from the Law of Laplace), the cell is completely aspirated and forms a capsule shape inside the micropipette. It then travels down the tube at a velocity that depends on the pressure difference across the cell.

Classic micropipette aspiration work from Cokelet's laboratory (Fenton et al., 1985) examined entrance times for neutrophils and lymphocytes into micropipettes. They found that the time for complete aspiration of a cell depends on cell diameter, D_c and micropipette diameter, D_p, but not differential pressure (over the range 200–400 Pa). They found a correlation between log (entrance time) and the ratio of D_c/D_p (which is greater than 1 in the capillary geometry because the average diameter of a capillary is somewhat smaller than the diameter of a leukocyte). Similar experiments have been repeated for HL-60 cells, with similar results (see Fig. 10.3).

There have been several studies in which the motion of cells inside micropipettes has been studied in the presence of a chemotactic gradient (Usami et al., 1992; You et al., 1999) without attempting to look at adhesion to the vessel wall. A study that is more explicitly targeted to the motion of leukocytes in capillaries, however, is the elegant work from the Hochmuth laboratory on the hydrodynamic motion of leukocytes in capillary-sized micropipettes in the absence of adhesion molecules (Shao and Hochmuth, 1997). Neutrophil velocity as a function of pressure gradient was measured and found to be a function of micropipette size. Their results showed that the motion was well described using lubrication theory for the thin layer of fluid between cell and micropipette wall. No mechanical arrest was observed, but there was some pausing at extremely low pressures which was attributed to nonspecific adhesion.

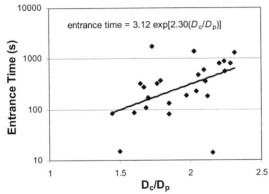

FIGURE 10.3 Scatterplot of HL-60 cell entrance time into micropipettes as a function of D_c/D_p. The time for a cell to be drawn completely into a micropipette is measured for aspiration pressures of 150–200 Pa. These data are consistent with similar results for leukocytes (Fenton et al., 1985).

B. Micropipette Studies of Leukocyte Deformability

A considerable body of work exists on the use of micropipette aspiration to measure leukocyte mechanical properties (Evans and Rawicz, 1990; Evans, 1989; Hochmuth, 2000; Tsai et al., 1993; Yeung and Evans, 1989). Evans' laboratory has demonstrated that partial cell aspiration can be used to measure the bending modulus and area expansion modulus of the membranes of leukocytes and many other cell types (Evans and Rawicz, 1990; Longo et al., 1997). In addition, the time for shape recovery following complete aspiration can be used as a measure of the cell's viscoelastic properties (Yeung and Evans, 1989).

C. Simulations of Cell Motion in Capillary-Sized Tubes and Constrictions

One of the first studies to simulate the dynamics of cell motion inside a tapering micropipette was that of Tran-Son-Tay et al. (1994). A leukocyte was modeled as a liquid drop with three layers: surface, cortex, and liquid core. The flow pattern inside the cell was calculated as it moved along a tapering tube. The pressures on the curved end caps were calculated from the Law of Laplace, but for the full force balance of the cell, the force due to the tapering frustrum section between the end caps, produced by a normal force due to the tube walls, was also calculated. The authors found a stepwise numerical solution for the velocity, pressure, and shape of the cell as it moved down the tube. The radial component of velocity inside the cell was found to be effectively zero, and this allowed a simpler theory to be used to fit experimental data for cell length versus time in tapering tubes (Bagge et al., 1977). The cell slowed down and extended along the tube as the tube narrowed, but it is not clear whether the cell ever stopped. The authors proposed that better-defined taper experiments were needed.

Bathe et al. (2002) describe how capillary geometry and changes in leukocyte mechanical properties (produced, e.g., by cell-activating agents) affect transit time into and through capillary segments. The entry time estimates were based on the Cokelet and Evans experiments described earlier and used a Maxwell model of the neutrophil (Skalak et al., 1990), with parameters measured from indentation experiments from the late 1980s (Worthen et al., 1989).

The capillary segments were modeled as having a stenosis with a constant radius of curvature. The vessel wall was taken to be rigid with a 100-nm-thick glycocalyx that the cell cannot penetrate, but the fluid can. They found a relationship between cell transit time and radius of curvature of the constriction and also found that when neutrophil compressibility is decreased (as happens with fMLP stimulation to activate the cell), the transit time is increased, but cells do not arrest. The authors proposed that further micropipette studies of the mechanical properties of fMLP-activated neutrophils were needed, along with experiments to validate the theory.

A follow-up study from the same laboratory simulated leukocytes flowing through networks of small tubes and examined how pressure fluctuations due to passage of cells through nearby vessels could lead to temporary stasis for other cells (Huang et al., 2001).

D. Experiments with Cells in Adhesion Molecule-Coated Micropipettes

Use of the parallel-plate flow chamber (i.e., an assay that recreates the mechanical (shear) force in venules) has significantly advanced our mechanistic understanding of adhesion in postcapillary venules. A similar assay for the capillary microcirculation could potentially shed light on the role of capillary geometry and adhesion molecule expression under well-controlled conditions. The lumen (i.e., the height) in a typical parallel plate-flow chamber assay is nearly 20-fold larger than the diameter of a leukocyte. Thus, the parallel-plate flow chamber does not capture important mechanical forces that are present during leukocyte adhesion in the capillaries. Although it is conceivable that one could conduct assays in which the height of the flow chamber was reduced to less than 8 μm, it would be difficult to achieve reproducible chamber heights. In addition, the leukocyte would still be constrained only by the upper and lower surfaces of the chamber and not surrounded by the vessel wall, as leukocytes are in the capillary network of the lung.

The hypothesis at the end of Section I proposed that leukocyte retention in capillaries involves both mechanical and adhesive forces. Although the micropipette studies described in Sections II.B and II.C have revealed much about the mechanical properties of leukocytes, it appears that no one has performed adhesion experiments using micropipettes coated with adhesion molecules to explore the combined effect of mechanical and biochemical forces on leukocyte arrest in capillaries. To explore the hypothesis, an adhesion assay has been developed in which individual leukocytes are aspirated into micropipettes that can be coated with ECAMs. This adhesion assay builds on earlier work from the Hochmuth laboratory on the motion of leukocytes inside straight micropipettes that had not been coated with adhesion molecules (Shao and Hochmuth, 1997), and examines the motion of freshly isolated individual leukocytes in ECAM-coated micropipettes.

I. Description of the Assay

A. Fabrication and Characterization of Micropipettes

Figure 10.2C is a diagram of a micropipette with the parts labeled for reference. Micropipettes with tip inner aperture diameters, D_p, ranging from 2 to 10 μm and with varying taper geometries were fabricated using a vertical micropipette puller (David Kopf Instruments; Model 730) from 0.9-mm-OD glass capillary tubes (Friedrich & Dimmock, Millville, NJ, USA). Micropipette tips were smoothed using a microforge (MF-200; World Precision Instruments), and the D_p values of micropipette tips were obtained from the distance that calibrated glass microneedles could be inserted into the tips.

B. Cells

Two cell types have been used for micropipette experiments.

1. HL-60 cells: Undifferentiated HL-60 premyelocytic leukemia cells were obtained from American Type Culture Collection. Cells were grown in RPMI-1640 medium (Cambrex, Walkersville, MD, USA) supplemented with 10% fetal bovine serum and penicillin/streptomycin as described previously (Goetz et al., 1996). On the day of an experiment, an aliquot of cell suspension was transferred to a centrifuge tube, which was kept at 37 °C in a water bath for up to 4 h until use. Samples of cells were drawn from this tube, washed three times (by microcentrifugation at 380g for 2 min) in plain RPMI-1640 medium, and finally suspended in HBSS+ (Hanks' balanced salt solution containing Ca^{2+} and Mg^{2+}) to which 1% BSA was added. The median size of the cell population was 13.4 μm, with 90% of the cells within the diameter range 12.4 μm $< D_c <$ 14.6 μm.

2. Neutrophils: Because the micropipette approach uses only one cell at a time, collection of large amounts of blood by venipuncture, as in the neutrophil isolation protocol, is not an absolute requirement. Other studies have shown that collection of small samples by fingerstick provides more than enough leukocytes for an experiment following minimal processing (Moazzam et al., 1997; Shao and Hochmuth, 1997; Spillmann et al., 2004). Fresh capillary whole blood samples were collected by fingerstick from healthy volunteers into heparinized microhematocrit tubes. Neutrophils were isolated by adding an equal volume of Mono-Poly Resolving Medium (M-PRM; MP Biomedicals, Irvine, CA, USA) to the hematocrit tube. The hematocrit tube was centrifuged for 20 min at 380g, and the neutrophil layer was cut out of the tube and diluted into an Eppendorf tube containing 2 ml of HBSS + 1% bovine serum albumin, pH 7.4. The few remaining red cells were easily distinguished from the neutrophils and were used to adjust the manometer.

C. Coating of Micropipettes with ECAM

To date the ECAM used for micropipette experiments has been a soluble form of P-selectin. The assay could be easily modified by using other adhesion molecules (e.g., ICAM-1). This molecule was dissolved in 0.1 M sodium bicarbonate (pH 9.2) and adsorbed to the inner surface of micropipettes in a specially designed chamber (shown in Fig. 10.4A) consisting of a modified micropipette holder (MPH-1, E.W. Wright, Guilford, CT, USA) glued to an Eppendorf tube. One percent bovine serum albumin (BSA; A7030, Sigma, St. Louis, MO, USA) in HBSS+ was used as a negative control. Micropipette tips were cleaned and sterilized by aspirating 100% ethanol into the tip with a syringe. The tips were rinsed three times with Dulbecco's phosphate-buffered saline without Ca^{2+} and Mg^{2+} (DPBS-) by syringe aspiration and expulsion. Solutions containing 1% BSA or P-selectin were aspirated into the tips of glass micropipettes. Micropipette tips were incubated with protein at room temperature for 2 h, then washed three times before being blocked with DPBS+ (DPBS with Ca^{2+} and Mg^{2+}) containing 1% BSA to prevent nonspecific adhesion.

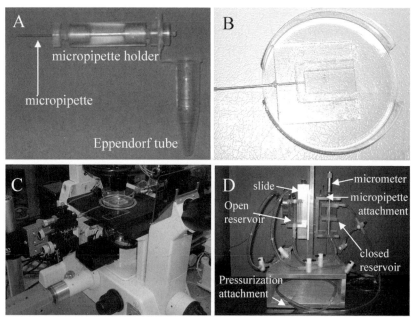

FIGURE 10.4 (A) Side view of the Eppendorf/holder system used for coating micropipettes with adhesion molecules. Micropipettes are inserted into micropipette holders that have been glued to the top of an Eppendorf tube. For incubations, the Eppendorf tube is placed in a rack with the micropipette vertically above it. ECAM solutions are placed in the Eppendorf tube, the tube is closed, and the micropipette is immersed. ECAM solution is aspirated into the micropipette tips with a syringe (not shown). (B) The viewing chamber for micropipette assays consists of a rectangular cavity glued to a tissue culture dish and covered with a vinyl microscope slide. The chamber is filled through a channel connected to a syringe by a length of tubing. (C) The chamber is placed on the heated stage of an inverted microscope. (D) Filled micropipettes are connected (via a holder labeled "micropipette attachment" (MM1–6, Technical Products International, St. Louis, MO, USA)) to a custom-built manometer so that controlled aspiration pressures can be applied. The reference pressure is set by an open reservoir that can be moved up and down with a slide so that there is no hydrostatic pressure difference between the manometer and the microscope stage. Precise pressure differences from 0 to 200 Pa can then be established with a micrometer that moves a closed reservoir. The closed reservoir can be pressurized to rapidly aspirate or remove cells.

D. Micromanipulation of Cells

HL-60 cells or neutrophils suspended in HBSS+ containing 1% BSA were placed in a viewing chamber for use in the micropipette assay. The viewing chamber (shown in Fig. 10.4B) was assembled by attaching a custom-made lexan plate to a plastic tissue culture dish using silicone sealant, and then attaching a vinyl microscope slide (Rinzl, Fisher Scientific, Hanover Park, IL, USA) to the top of the lexan plate. One end of the chamber was left open so that micropipettes could be inserted. The viewing chamber was filled and drained through a tube in the end of the viewing chamber. As shown in Fig. 10.4C, the viewing chamber was placed on an inverted microscope (Nikon Eclipse TE 300, Fryer Co., Huntley, IL, USA) with a heated stage to allow experiments to be performed at 37 °C. Micropipettes that had been coated with 1% BSA or soluble P-selectin were filled by aspirating a small amount of HBSS+ into the tips using a gastight syringe and a length of tubing. The main bodies of the micropipettes were then directly filled with HBSS+ using a 9.8-cm-long, 28-gauge Microfil needle (WPI, Sarasota, FL, USA) and a syringe. Filled micropipettes were attached to a custom-built manometer (see Fig. 10.4D) that applied controlled aspiration pressures ranging from 0 to 200 Pa. The resolution of the manometer was ±1 Pa in the range 0–200 Pa. This range of pressures spans the 10- to 50-Pa region that is thought to be acting in the lung capillary network (Bathe et al., 2002; Huang et al., 2001). For reference, a normal diastolic blood pressure (across the whole cardiovascular system) is 80 mm Hg or 10.6 kPa. Other groups have used both lower ranges, 0–6 Pa (Shao and Hochmuth, 1997) and higher ranges, 200–400 Pa (Fenton et al., 1985).

To ensure that the manometer was properly calibrated, red blood cells (RBCs) were used. For neutrophil experiments, sufficient RBCs are always left over from neutrophil purification. For HL-60 cell experiments, small quantities of blood were collected by fingerstick, and samples of RBCs were added (after washing) to the HL-60 cell suspension. Individual RBCs were aspirated into the micropipette before aspiration of each leukocyte to allow the pressure reference level of the manometer (i.e., the height for which $\Delta P = 0$ between micropipette interior and atmospheric pressure) to be adjusted. When $\Delta P = 0$, a RBC will not move in the micropipette. The RBC was then expelled and a neutrophil or HL-60 cell was chosen and aspirated. Cell motion was monitored with a CCD video camera (Cohu, Fryer Co., Huntley, IL, USA) attached to the side viewing port of the microscope. Video images were displayed on a video monitor and recorded on a VCR for later analysis.

E. Aspiration Procedure

Figure 10.5 shows a sequence of frames from the aspiration of an HL-60 cellinto a 6.0-μm-ID micropipette. In Fig. 10.5A, the micropipette tip is

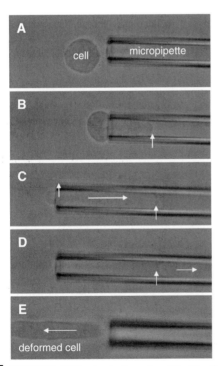

FIGURE 10.5 Aspiration of an HL-60 cell. (A) An HL-60 cell before aspiration into a 6.0-μm micropipette. The neutrophil is initially spheroidal. The micropipette tip is placed next to the cell, which is then completely aspirated into the micropipette (B, C). The edges of the cell in the micropipette are indicated by arrows. Following aspiration, the cell moves freely down the micropipette (D). If the cell is expelled from the micropipette, it retains its deformed shape for several minutes.

positioned next to the cell. A large negative aspiration pressure is applied to the cell to aspirate it rapidly. Aspiration pressure is defined as $\Delta P = P_{in} - P_{out}$, where P_{in} is the pressure inside the micropipette and P_{out} is the pressure outside in the viewing chamber. For cell aspiration, where $P_{in} < P_{out}$, ΔP will thus be negative. To speed the experiments, large aspiration forces were used to bring the cells in, followed by the application of small negative pressures (a similar procedure was used by Shao and Hochmuth, 1997). Complete aspiration of the HL-60 cell takes several seconds. At first (Fig. 10.5B), a tonguelike section of the cell is aspirated into the micropipette. Several seconds later (Fig. 10.5C), the cell has been almost completely aspirated and has deformed into a capsule shape. If smaller aspiration pressures are used, the cell will move into the micropipette more slowly (up to several minutes for the pressures of 29.4 and 58.8 Pa used in the experiments to be described). Cells that had been completely aspirated usually did not move for a short time following application of negative pressure. Once a cell began moving, however, it moved at a constant velocity inside the micropipette under the influence of the preset pressure gradient (Figs. 10.5C, D). If the cell was expelled from the micropipette by a positive pressure, several minutes were required for the cell to return to its original spherical shape from the capsule shape it assumed inside the micropipette (Fig. 10.5E). Shape recovery experiments have been performed for HL-60 cells. Figure 10.6A is a sample trace of HL-60 cell shape recovery as a function of time, and Fig. 10.6B is a graph of recovery time versus D_c/D_p following aspiration into 7- to 8-μm-diameter micropipettes.

F. Analysis of Cell Motion in the Micropipette

Recorded video images of cell motion inside the micropipette were captured using a frame grabber board (PCI-1409, National Instruments, Austin, TX, USA) and Labview IMAQ image acquisition software (National Instruments). Images were analyzed frame by frame to find the instantaneous cell position as a function of time. The position of the leading edge of the cell was quantified on the captured images using a video caliper tool in Labview (National Instruments). The pause time was found by examining the displacement-versus-time curves and measuring the lengths of pauses. If a cell did not move within 5 min of the start of a pause, then it was scored as an arrested cell and assigned a pause time of 300 s.

2. Micropipette Aspiration Experiments

Adhesion of HL-60 cells in micropipettes coated with soluble P-selectin has also been studied. It was found that there was much stronger adhesion to P-selectin-coated micropipettes than to micropipettes coated with BSA. The pause times were longer on P-selectin than on BSA (and, indeed, many of the cells were stationary in the micropipette for more than 10 min), and the fraction of cells that exhibited any pausing at all was much higher on P-selectin than on BSA. The firm adhesion seen in the micropipette could be contrasted with parallel-plate flow chamber assays that had HL-60 cells moving over soluble P-selectin–coated glass surfaces. In the parallel-plate flow chamber assay, only rolling adhesion was observed, even though the experiments were done with the same cells, reagents, and coating procedures, and the estimated minimum forces on cells in the micropipette were 1.5 times larger than the largest force used in the parallel-plate flow chamber assay. Adhesion in the micropipette could be inhibited with EDTA, but,

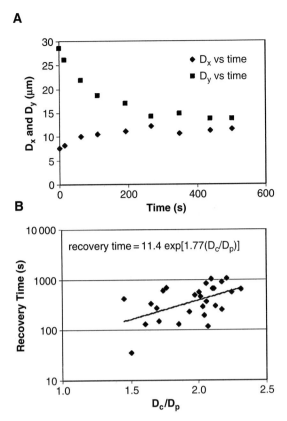

FIGURE 10.6 Plots of HL-60 cell recovery time following complete aspiration into a micropipette. (A) Sample trace of the shape recovery as a function of time for a typical cell *i*. D_x is the length of the minor axis, and D_y is the length of the major axis. (B) Scatterplot of recovery time versus deformation ratio, D_c/D_p. Recovery time is defined as the time for the relative difference between major and minor axes $(D_y - D_x)/D_{\text{pre-aspiration}}$ to reach $<15\%$ (the major/minor axis difference that included 100% of HL-60 cells before aspiration).

unlike in the parallel-plate flow chamber, it proved impossible to block the adhesion with antibodies.

We have found, then, that the percentage of interacting HL-60 cells that are arrested in the micropipette assay is much larger than that in the parallel-plate flow chamber, even though the forces on an interacting cell are larger in the micropipette than in the flow chamber.

Similar experiments have also been performed with neutrophils. As shown in Fig. 10.7, average pause times for neutrophils in soluble-P-selectin-coated micropipettes were significantly longer (165 ± 34 s, mean $\pm$ SE) than pause times in BSA-coated micropipettes (25 ± 18 s, mean $\pm$ SE). Figure 10.8 shows that the fraction of cells that paused in P-selectin-coated micropipettes (55%) was significantly higher than the fraction that paused in BSA-coated micropipettes (10%).

3. Characterization of Deformation

There are at least two ways to characterize the deformation of cells. The first method follows Fenton et al. (1985) in defining a "deformation ratio," DR, as the diameter of the cell, D_c, divided by the diameter of the micropipette, D_p.

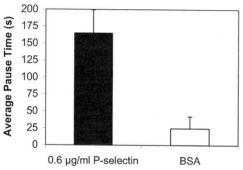

FIGURE 10.7 Many neutrophils paused (firmly adhered) following aspiration into micropipettes. Average pause times in 6.60 ± 0.05-μm-ID (mean ± SD) micropipettes coated with either 0.6 μg/ml soluble P-selectin or 1% BSA are compared. The cells were all subjected to an applied pressure difference of 29.4 Pa. Error bars represent SEM.

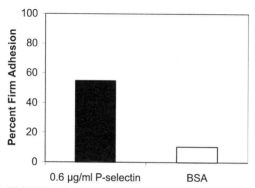

FIGURE 10.8 Percentage of aspirated human neutrophils that arrested in the micropipette tip following aspiration into 6.60 ± 0.05-μm-ID (mean ± SD) micropipettes coated with either 0.6 μg/ml soluble P-selectin or 1% BSA. The cells were all subjected to an applied pressure difference of 29.4 Pa.

When DR is greater than 1, the leukocyte is larger than the micropipette. When DR is less than 1, the leukocyte is smaller than the vessel (as in the parallel-plate flow chamber). The second method uses the contact area between cell and micropipette surface. As the cell deforms to a capsule shape, the area in contact with the micropipette increases. It might be expected, in light of the hypothesis that mechanical forces modulate adhesive forces, that pause time in micropipettes depends on one or both of these measures of deformation.

Recent work from the Waugh laboratory using a two-micropipette impingement assay has demonstrated that when a neutrophil is forced against a smaller spherical microbead coated with ECAMs, the bond formation rate increases due to an increase in the contact area between cell and microbead (Lomakina et al., 2004; Spillmann et al., 2004) (see also Chapter 5). The formation rate is also affected when the contact stress changes. Although the experiments described in this chapter used a different geometry, the leukocyte surface area exposed for binding in the micropipette is much larger than that in the parallel-plate flow chamber, and the contact stress is also likely to be different. The Waugh results would suggest that this should lead to more bond formation and, hence, to more durable arrest.

To test for similar correlations in the micropipette data, the pause time for neutrophils following aspiration into micropipettes coated with 0.6 μg/ml P-selectin is plotted against deformation ratio DR (Fig. 10.9A) and contact area (Fig. 10.9B).

Individual neutrophils in this system either were nonadherent (in which case they were assigned a pause time of zero) or remained adherent for the entire 5 min over which they were observed (hence the 300-s pause times). Although adhesion was "all or nothing" in this way, the average DR for those cells that paused was significantly larger than that for cells that did not pause. This is consistent with the hypothesis that deformation can modulate adhesive forces. The average contact area for the cells that paused was significantly larger than that for cells that did not pause. This is consistent with the Waugh laboratory data on the dependence of bond formation on contact area.

In a parallel-plate flow chamber, gravity and hydrodynamic interactions with other cells force leukocytes into adhesive contact with the vessel wall. For

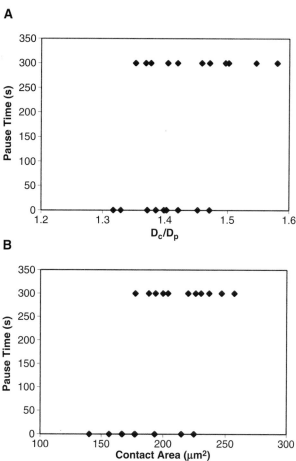

FIGURE 10.9 Individual neutrophils either were nonadherent (in which case they were assigned a pause time of zero) or remained adherent for the entire 5 min over which they were observed. (A) Scatterplot of pause time as a function of the deformation ratio, $DR = D_c/D_p$, for human neutrophils following aspiration into micropipettes coated with 0.6 μg/ml soluble P-selectin. (B) Scatterplot of pause time as a function of cell–micropipette contact area for human neutrophils following aspiration into micropipettes coated with 0.6 μg/ml soluble P-selectin. The cells were all subjected to an applied pressure difference of 29.4 Pa.

leukocytes moving in venules or flow chambers, the cell area that is in contact with the wall is approximately one-sixth to one-half of the cell surface (depending on whether the cell is considered to be a cube or a pancake). In the micropipette, by contrast, the elastic shape recovery of the cell forces a large proportion of the cell area into close contact with the vessel wall. A cell that is deformed inside a micropipette forms a long cylinder with diameter equivalent to the ID of the micropipette. When an 8-μm-diameter cell is aspirated into a 4-μm-ID micropipette, the resulting cylindrical section is 18.7 μm long and 82% of its surface is in contact with the micropipette (assuming equal volume before and after aspiration (Shao and Hochmuth, 1997)).

This large increase in contact area, plus the possibility that elastic shape recovery can bring the cell and vessel wall into more intimate contact than in venules, should mean that lower densities of adhesion molecules are required for adhesion in capillaries than for adhesion in venules.

III. PROSPECTS FOR FUTURE RESEARCH

One direction for future research is to look at different micropipette geometries to determine the conditions required for mechanical trapping. Micropipettes with different taper geometries can be pulled (as shown in Fig. 10.2E), thus setting the stage for a systematic investigation into the role of taper geometry in leukocyte arrest. The taper geometry can be characterized by the taper rate, $T = (D_i - D_f)/d$, where D_i is the initial micropipette diameter (larger), D_f is the final diameter, and d is the distance between the initial and final inside diameters along the micropipette bore. For reference, for the micropipettes shown in Figs. 2E and F, $T = 0.079$ (C) and $T = 0.18$ (D). Given that an isolated leukocyte is $\sim$8 μm, we can set D_i equal to 8 μm. An examination of the literature regarding the geometry of capillary beds indicates that the smallest D_f is 2 μm (Doerschuk et al., 1993) and a reasonable value for d is one to two leukocyte diameters, $\sim$8–16 μm (Bathe et al., 2002; Guntheroth et al., 1982). Capillaries with $D_f > 7$ μm are of little interest, so we set the highest useful D_f equal to 7 μm. With these numbers, the physiologically important range of taper rate, T, runs from 0.0625 to 0.75. Thus, to explore the physiologically relevant ranges of taper, we should use micropipettes with D_f between 7 and 2 μm, with d values of 8 and 16 μm. This would result in a range of taper rates from 0.0625 to 0.75.

From this data set, the relationship between aspiration pressure, taper rate, and velocity can be determined. Importantly, it can be determined if there exists a physiologically relevant parameter set that leads to mechanical arrest of the leukocyte and, if this does exist, the extent of taper needed to achieve mechanical arrest.

The only ECAM used so far has been P-selectin, but ICAM-1, E-selectin, and sLe^x (a ligand for L-selectin) are also of great interest. It has been observed that if the BSA used to inhibit nonspecific adhesion is not heat-treated before use, cells can adhere firmly to one side of the tube (creating a gap through which RBCs can pass). The origin of this interesting behavior still needs to be clarified.

Another important area of study is the possibility of activation of the leukocytes. This can be done artificially (e.g., by treating them with phorbol myristate acetate or fMLP for 10 min) prior to or during the adhesion assay.

The resulting adhesive behavior can be compared with any mechanical or adhesive activation observed in regular experiments. This line of inquiry is motivated by the observation that leukocyte activation causes a change in the mechanical (Worthen et al., 1989) as well as the biochemical adhesive (e.g., activation of integrins and shedding of L-selectin) (Griffin et al., 1990; Lawrence and Springer, 1991) properties of the leukocytes.

IV. CONCLUSIONS

The questions surrounding the role of capillary geometry and biochemical binding in mechanical/biochemical arrest in capillaries demand a method for clarifying the arrest mechanism in a well-defined geometrical and fluid mechanical environment. An assay that uses micropipettes to simulate the capillary should be a perfect complement to the well-established parallel-plate flow chamber assay, which has proved indispensable in unraveling the mechanics of leukocyte adhesion in postcapillary venules. It is anticipated that development of a "flow chamber" assay for adhesion in capillaries will prove equally valuable.

ACKNOWLEDGMENTS

The authors acknowledge Dr. Raymond T. Camphausen (Wyeth Research, Cambridge, MA, USA) for generously supplying the P-selectin used in this research. This work was supported by an award from the American Heart Association (D.F.J.T.) and by National Institutes of Health Grant GM057640 (D.J.G.). D.J.G. is an Established Investigator of the American Heart Association.

SUGGESTED READING

Apart from the many excellent references cited in the chapter, a number of articles and reviews have proved especially useful as starting points from which to approach the field.

For the biology of lung capillary networks, several excellent recent reviews have proved extremely helpful. Burns et al. (2003) summarizes the state of knowledge of receptor expression and theories for mechanisms of arrest. Similar ground is covered in a slightly older review by Doerschuk (2001).

For cell movement in micropipettes, Shao and Hochmuth (1997) describe the motion of leukocytes in micropipettes in the absence of adhesion molecules (with a good development of the relevant theory). Bathe et al. (2002) summarize the simulation work on the motion of leukocytes in capillaries and make their own contribution.

The web site *Cell Adhesion under Fluid Flow (http://www.ent.ohiou.edu/~adhesion/)* maintained by D.J.G. is a good starting point for this topic, with video and links to other web sites on the subject.

REFERENCES

Alon, R., Fuhlbrigge, R. C., Finger, E. B., and Springer, T. A. (1996) Interactions through L-selectin between leukocytes and adherent leukocytes nucleate rolling adhesions on selectins and VCAM-1 in shear flow. *J. Cell Biol.* **135:** 849–865.

Andonegui, G., Bonder, C. S., Green, F., Mullaly, S. C., Zbytnuik, L., Raharjo, E., and Kubes, P. (2003) Endothelium-derived toll-like receptor-4 is the key molecule in LPS-induced neutrophil sequestration into lungs. *J. Clin. Invest.* **111**: 1011–1020.

Bagge, U., Skalak, R., and Attefors, R. (1977) Granulocyte rheology: Experimental studies in an in-vitro microflow system. *Adv. Microcirc.* **7**: 29–48.

Bathe, M., Shirai, A., Doerschuk, C. M., and Kamm, R. D. (2002). Neutrophil transit times through pulmonary capillaries: The effects of capillary geometry and fMLP-stimulation. *Biophys. J.* **83**: 1917–1933.

Borges, E., Eytner, R., Moll, T., Steegmaier, M., Campbell, M A., Ley, K., Mossmann, H., and Vestweber, D. (1997) The P-selectin glycoprotein ligand-1 is important for recruitment of neutrophils into inflamed mouse peritoneum. *Blood* **90**: 1934–1942.

Burns, A. R., Smith, C. W., and Walker, D. C. (2003) Unique structural features that influence neutrophil emigration into the lung. *Physiol. Rev.* **83**: 309–336.

Dawson, C. A., and Linehan, J. H. (1997) Dynamics of blood flow and pressure–flow relationship. In *The Lung: Scientific Foundations* (R. G. Crystal and J. B. West, Eds.), 2nd ed., pp. 1503–1522. Lippincott–Raven, Philadelphia.

Doerschuk, C. M. (2001) Mechanisms of leukocyte sequestration in inflamed lungs. *Microcirculation* **8**: 71–88.

Doerschuk, C. M., Beyers, N., Coxson, H. O., Wiggs, B., and Hogg, J. C. (1993) Comparison of neutrophil and capillary diameters and their relation to neutrophil sequestration in the lung. *J. Appl. Physiol.* **74**: 3040–3045.

Doerschuk, C. M., Winn, R. K., Coxson, H. O., and Harlan, J. M. (1990) CD18-dependent and -independent mechanisms of neutrophil emigration in the pulmonary and systemic microcirculation of rabbits. *J. Immunol.* **144**: 2327–2333.

Doyle, N. A., Bhagwan, S. D., Meek, B. B., Kutkoski, G. J., Steeber, D. A., Tedder, T. F., and Doerschuk, C. M. (1997) Neutrophil margination, sequestration, and emigration in the lungs of L-selectin-deficient mice. *J. Clin. Invest.* **99**: 526–533.

Erzurum, S. C., Downey, G. P., Doherty, D. E., Schwab, B., III, Elson, E. L., and Worthen, G. S. (1992) Mechanisms of lipopolysaccharide-induced neutrophil retention: Relative contributions of adhesive and cellular mechanical properties. *J. Immunol.* **149**: 154–162.

Evans, E., and Rawicz, W. (1990) Entropy-driven tension and bending elasticity in condensed-fluid membranes. *Phys. Rev. Lett* **64**: 2094–2097.

Evans, E. A. (1989) Structure and deformation properties of red blood cells: Concepts and quantitative methods. *Methods Enzymol.* **173**: 3–35.

Fenton, B. M., Wilson, D. W., and Cokelet, G. R. (1985) Analysis of the effects of measured white blood cell entrance times on hemodynamics in a computer model of a microvascular bed. *Pfluegers Arch.* **403**: 396–401.

Finger, E. B., Puri, K. D., Alon, R., Lawrence, M. B., von Andrian, U. H., and Springer, T. A. (1996). Adhesion through L-selectin requires a threshold hydrodynamic shear. *Nature (London)* **379**: 266–269.

Gebb, S. A., Graham, J. A., Hanger, C. C., Godbey, P. S., Capen, R. L., Doerschuk, C. M., and Wagner, W. W., Jr. (1995) Sites of leukocyte sequestration in the pulmonary microcirculation. *J. Appl. Physiol.* **79**: 493–497.

Goetz, D. J., Ding, H., Atkinson, W. J., Vachino, G., Camphausen, R. T., Cumming, D. A., and Luscinskas, F. W. (1996) A human colon carcinoma cell line exhibits adhesive interactions with P-selectin under fluid flow via a PSGL-1-independent mechanism. *Am. J. Pathol.* **149**: 1661–1673.

Griffin, J. D., Spertini, O., Ernst, T. J., Belvin, M. P., Levine, H. B., Kanakura, Y., and Tedder, T. F. (1990) Granulocyte–macrophage colony-stimulating factor and other cytokines regulate surface expression of the leukocyte adhesion molecule-1 on human neutrophils, monocytes, and their precursors. *J. Immunol.* **145**: 576–584.

Guntheroth, W. G., Luchtel, D. L., and Kawabori, I. (1982) Pulmonary microcirculation: Tubules rather than sheet and post. *J. Appl. Physiol.* **53**: 510–515.

Hochmuth, R. M. (2000) Micropipette aspiration of living cells. *Journal of Biomechanics* **33**: 15–22.

Huang, Y., Doerschuk, C. M., and Kamm, R. D. (2001) Computational modeling of RBC and neutrophil transit through the pulmonary capillaries. *J. Appl. Physiol.* **90**: 545–564.

Kansas, G. S. (1996) Selectins and their ligands: Current concepts and controversies. *Blood* **88**: 3259–3287.

King, M. R., and Hammer, D. A. (2001) Multiparticle adhesive dynamics: Interactions between stably rolling cells. *Biophys. J.* **81**: 799–813.

Kubo, H., Doyle, N. A., Graham, L., Bhagwan, S. D., Quinlan, W. M., and Doerschuk, C. M. (1999) L- and P-selectin and CD11/CD18 in intracapillary neutrophil sequestration in rabbit lungs. *Am. J. Resp. Crit. Care Med.* **159**: 267–274.

Kuebler, W. M., Kuhnle, G. E. H., Groh, J., and Goetz, A. E. (1997) Contribution of selectins to leucocyte sequestration in pulmonary microvessels by intravital microscopy in rabbits. *J. Physiol. London* **501**: 375–386.

Lawrence, M. B., Berg, E. L., Butcher, E. C., and Springer, T. A. (1995) Rolling of lymphocytes and neutrophils on peripheral node addressin and subsequent arrest on ICAM-1 in shear flow. *Eur. J. Immunol.* **25**: 1025–1031.

Lawrence, M. B., and Springer, T. A. (1991) Leukocytes roll on a selectin at physiological flow rates: Distinction from and prerequisite for adhesion through integrins. *Cell* **65**: 859–874.

Lawrence, M. B., and Springer, T. A. (1993) Neutrophils roll on E-selectin. *J. Immunol.* **151**: 6338–6346.

Lomakina, E. B., Spillmann, C. M., King, M. R., and Waugh, R. E. (2004) Rheological analysis and measurement of neutrophil indentation. *Biophys. J.* **87**: 4246–4258.

Longo, M. L., Waring, A. J., and Hammer, D. A. (1997) Interaction of the influenza hemagglutinin fusion peptide with lipid bilayers: Area expansion and permeation. *Biophys. J.* **73**: 1430–1439.

Moazzam, F., DeLano, F. A., Zweifach, B. W., and Schmid-Schönbein, G. W. (1997) The leukocyte response to fluid stress. *Proc. Natl. Acad. Sci. USA* **94**: 5338–5343.

Ramos, C. L., Kunkel, E. J., Lawrence, M. B., Jung, U., Vestweber, D., Bosse, R., McIntyre, K. W., Gillooly, K. M., Norton, C. R., Wolitzky, B. A., et al. (1997) Differential effect of E-selectin antibodies on neutrophil rolling and recruitment to inflammatory sites. *Blood* **89**: 3009–3018.

Shao, J.-Y., and Hochmuth, R. M. (1996) Micropipette suction for measuring piconewton forces of adhesion and tether formation from neutrophil membranes. *Biophys. J.* **71**: 2892–2901.

Shao, J. Y., and Hochmuth, R. M. (1997) The resistance to flow of individual human neutrophils in glass capillary tubes with diameters between 4.65 and 7.75 μm. *Microcirculation* **4**: 61–74.

Skalak, R., Dong, C., and Zhu, C. (1990) Passive deformations and active motions of leukocytes. *J. Biomech. Eng* **112**: 295–302.

Sobin, S. S., and Fung, Y. C. (1992) Response to challenge to the Sobin–Fung approach to the study of pulmonary microcirculation. *Chest* **101**: 1135–1143.

Spillmann, C. M., Lomakina, E., and Waugh, R. E. (2004). Neutrophil adhesive contact dependence on impingement force. *Biophys. J.* **87**: 4237–4245.

Springer, T. A. (1994) Traffic signals for lymphocyte recirculation and leukocyte emigration: The multistep paradigm. *Cell* **76**: 301–314.

Springer, T. A., Chen, S., and Alon, R. (2002) Measurement of selectin tether bond lifetimes. *Biophys. J.* **83**: 2318–2323.

Tran-Son-Tay, R., Kirk, T. F., III, Zhelev, D. V., and Hochmuth, R. M. (1994) Numerical simulation of the flow of highly viscous drops down a tapered tube,. *J. Biomech. Eng* **116**: 172–176.

Tsai, M. A., Frank, R. S., and Waugh, R. E. (1993) Passive mechanical behavior of human neutrophils: Power-law fluid. *Biophys. J.* **65**: 2078–2088.

Usami, S., Wung, S. L., Skierczynski, B. A., Skalak, R., and Chien, S. (1992) Locomotion forces generated by a polymorphonuclear leukocyte. *Biophys. J.* **63**: 1663–1666.

Varki, A. (1997) Selectin ligands: Will the real ones please stand up? *J. Clin. Invest.* **99**: 158–162.

Worthen, G. S., Schwab, B., Elson, E. L., and Downey, G. P. (1989) Mechanics of stimulated neutrophils: Cell stiffening induces retention in capillaries. *Science* **245**: 183–186.

Yeung, A., and Evans, E. (1989) Cortical shell–liquid core model for passive flow of liquid-like spherical cells into micropipettes. *Biophys. J.* **56**: 139–149.

You, J., Mastro, A. M., and Dong, C. (1999) Applications of the dual-micropipette technique to the measurement of tumor cell locomotion. *Exp. Cell Res.* **248**: 160–171.

Zhelev, D. V., Needham, D., and Hochmuth, R. M. (1994). Role of the membrane cortex in neutrophil deformation in small pipets. *Biophys. J.* **67**: 696–705.

IV

CELL ADHESION TO SURFACES UNDER FLOW

11

ADHESION OF FLOWING NEUTROPHILS TO MODEL VESSEL SURFACES: CONSTRAINT AND REGULATION BY THE LOCAL HEMODYNAMIC ENVIRONMENT

GERARD B. NASH and G. ED RAINGER

Center for Cardiovascular Sciences, The Medical School, University of Birmingham, Birmingham, UK

Recruitment of circulating neutrophils to the endothelium that lines blood vessels is essential for protective, innate, immune responses. Capture from flow, activation, and stabilization of adhesion and onward migration of neutrophils all operate under constraints applied by the local hemodynamics. Even though the neutrophils circulate at high speed on a molecular scale, efficient capture occurs through specially adapted, rapidly binding selectin receptors. The rheological properties of the blood influence both the transport of the neutrophils to the vessel wall and the efficiency of this attachment. The responses of the selectins to forces arising from viscous drag on the cells and the deformation of bound cells determine the rate at which the neutrophils roll along the wall, effectively scanning the surface for proadhesive signals. Ability to respond to these signals by activating more long-lived integrin-mediated bonds in less than a second ensures precise localization of captured neutrophils. Subsequent rate, direction, and efficiency of migration over and through the endothelial surface are also conditioned by forces impinging on the emigrating neutrophils. This multistage process of recruitment is an important example of a biological response that is conditioned by its physical environment and that requires use of biomechanical theory and experimentation for its understanding. Moreover, the responses of neutrophils, and indeed of vascular endothelial cells, to forces applied to them offer a paradigm for the emerging concept of cellular mechanotransduction.

I. INTRODUCTION

Neutrophilic granulocytes must adhere to the wall of blood vessels and migrate into tissue to carry out their protective functions. Intravital observations of the microcirculation in animals and studies using flow models *in vitro* have revealed a series of events that underlie neutrophil recruitment (see Suggested Reading for reviews). The process is illustrated in Fig. 11.1.

Initially, margination of neutrophils by centrally flowing red cells brings them into contact with the vessel wall. If the endothelial cells have been exposed to inflammatory mediators (such as the cytokines tumor necrosis factor α (TNF) and interleukin-1β), they express receptors of the selectin family. These receptors have high forward rate constants for binding their carbohydrate ligands borne by specific neutrophil glycoproteins. The result is initial bond formation or capture (sometimes referred to as *tethering*). These bonds also have rapid reverse rate constants and so are short-lived. Consequently the neutrophil may roll along the surface, making and breaking bonds, but moving much more slowly than free-flowing cells. Stabilization of adhesion typically requires formation of slower-acting but longer-lived bonds between members of the β₂-integrin family on the neutrophils and cognate endothelial receptors such as intercellular adhesion molecule 1 (ICAM-1). The integrins themselves are not generally in an "active" state on unstimulated neutrophils, and so an activating stimulus (such as the cytokine interleukin-8) must be presented to the neutrophils by the endothelial surface. Neutrophils can respond rapidly to these receptor-mediated signals and may become immobilized soon after contact with the endothelium or roll for some time before stopping. The same or similar stimuli cause the neutrophils to spread on the surface, engage their motile apparatus. and migrate first over and then through the endothelial monolayer.

The phases of initial attachment and migration over the endothelial cells occur in the presence of flowing blood and are constrained by hemodynamic factors. First, the patterns of flow and rheological properties of the blood influence margination of leukocytes in the bloodstream and contact with the vessel wall. When contact is made, the probability of forming adhesive bonds and the survival of these bonds depend on local velocity of flow and shear forces applied to the captured cells. The rate of response of captured, rolling cells to activating stimuli, and their rolling velocities, determine how precisely they become localized in the tissue. In addition, forces exerted on the adherent cells can influence their migratory behavior. As hemodynamics varies between

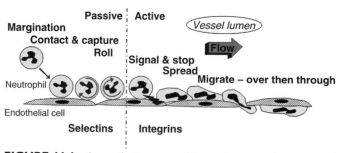

FIGURE 11.1 Stages during neutrophil recruitment to the wall of an inflamed blood vessel. The central flow of red cells marginates neutrophils toward the wall (see Fig. 11.3), where capture is mediated through fast-acting selectin receptors on the surface of the endothelial cells. Subsequently, breaking and formation of bonds under the action of shear stress cause rolling adhesion, which stops when a signal is transduced from the surface of the endothelial cells to the neutrophils. The signal causes activation of neutrophil integrins, which bind firmly to endothelial counterligand(s). They also cause the neutrophils to spread on the surface, and migrate over and then through it, while continually modulating their integrin-mediated anchorage.

regions of the vascular tree, between organs and in disease, these physico-chemical constraints are not only fundamental determinants of protective immunity, but can also influence localization or progression of inflammatory disease. Failure or disruption in the control of leukocyte recruitment is potentially pathogenic, for instance, during inappropriate or nonresolving inflammatory responses. Thus, effects of hemodynamic variables on adhesion and migration are important, and have been widely studied.

Here we consider how capture of flowing neutrophils by selectin-coated surfaces depends on the local characteristics of flow and the rheological properties of the blood, and how the kinetics of the subsequent activation and migration of neutrophils are influenced by the local physical environment. The work described revolves mainly around studies using simple receptor-coated surfaces, or surfaces coated with cultured endothelial cells, where systematic experimental variation in physical variables is possible. Comparisons to major phenomena observed *in vivo* are made where possible. Overall, these studies exemplify application of biomechanical and cell engineering approaches to studies of cell–surface interactions that underlie important physiological, and sometimes pathogenic, processes.

II. CAPTURE OF FLOWING NEUTROPHILS ON SELECTIN-COATED SURFACES

In general, a flowing neutrophil making contact with a surface that bears P-selectin or E-selectin may be captured by the formation of initial bonds between the selectins and sulfated, sialylated, fucosylated glycoprotein ligands on the cell (Kansas, 1996). The neutrophil instantaneously becomes immobilized. However, the bond(s) is typically short-lived, and the cell may rapidly come free and rejoin the flow stream. More likely, if there is a sufficient density of local receptors, the cell pivots on its initial bond due to torque imposed by flow, and makes further contacts and bonds downstream of that bond. Then, as the original bonds break, the cell "peels" from the rear and goes on to roll (or more accurately staggers) in the direction of flow as bonds are repeatedly formed and reversed (Hammer and Apte, 1992; Tozeren and Ley, 1992). The main question addressed here is: What hemodynamic and rheological factors determine the efficiency of capture? In addition, we consider what determines the fate of the cells (whether they roll or detach) and how fast they roll forward immediately after capture. In all cases, the density of receptors is important, because it influences the probability of capture and of subsequent formation of new bonds and, hence, rate of rolling.

A. Effects of Shear Rate and Shear Stress on Selectin-Mediated Capture in Suspensions of Isolated Neutrophils

1. Theoretical Considerations

The establishment of contact between perfused, isolated neutrophils and a selectin-coated surface depends on sedimentation. If adhesive capture is efficient, such delivery will be rate limiting for adhesion. The rate of delivery by sedimentation along a perfused horizontal vessel and its effect on levels of adhesion have been formally analyzed by Munn and colleagues (1994). For instance, using horizontal coated tubes, we find negligible adhesion to the upper surface; and if the tube is oriented vertically, adhesion to any surface is negligible.

After neutrophils reach the surface, their attachment and subsequent behavior depend on the wall shear rate (γ_w; local velocity gradient) and shear stress (product of shear rate and fluid viscosity) (see Fig. 11.2).

The shear rate determines the velocity of the cell. The velocity of a spherical cell grazing the surface is theoretically strongly dependent on the size of the gap between the cell and wall (Goldman et al., 1967). However, experimentally, the velocity of cells (which inevitably are not perfectly smooth) at the surface is approximately equal to $r\gamma_w$, where the cell radius, r, is much smaller than the vessel height (Tissot et al., 1992). In adhesion studies, this cell velocity is ~1 mm/s, but it should be borne in mind that the motion of a receptor on the cell surface is not simply translation along the surface. Because the cell is rotating in the shear flow, a receptor on the cell surface slows its velocity component parallel to the wall as it approaches, goes though a minimum at the point of nearest contact, and then speeds up again as it moves away from the wall (effectively a simple harmonic motion imposed on the steady forward motion of the cell). Once a bond is formed and the cell becomes briefly stationary, then the local flow over the cell exerts a shear force (F) on it. This force depends on the local fluid shear stress ($\tau = \eta^*\gamma_w$, where η is the fluid viscosity) and the cell radius, and is given by $F = 32^*\tau^*r^2$ (Goldman et al., 1967). The force accelerates the already rapid reverse rate constant of the selectin bonds and effectively promotes breakage (Hammer and Apte, 1992). Such a simple analysis appears to predict that

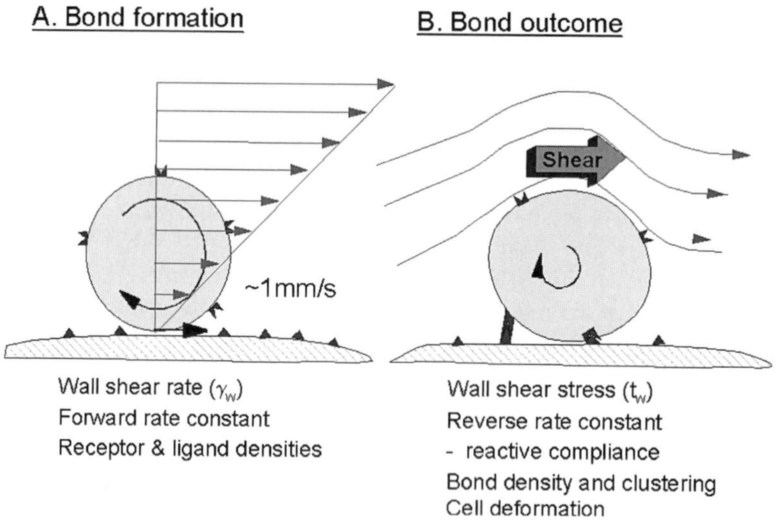

A. Bond formation

Wall shear rate (γ_w)
Forward rate constant
Receptor & ligand densities

B. Bond outcome

Wall shear stress (t_w)
Reverse rate constant
- reactive compliance
Bond density and clustering
Cell deformation

FIGURE 11.2 Schematic illustration of critical factors influencing (A) bond formation when neutrophils approach a selectin-coated surface and (B) bond outcome. Bond formation depends on the cell velocity (~1 mm/s, which in turn depends on the local wall shear rate), on the forward rate constant of the bonds, and on the density of receptors and ligands. Bond outcome depends on the force applied to the adherent cells (which depends on the local wall shear stress), on the reverse rate constant of the bond and its response to applied force (reactive compliance), on the distribution of bonds (as clustering opposes peeling), and on cell deformation. Typically the bonds are extended at the rear of the captured cells and peeling follows, with the rate of forward motion (or rolling velocity) depending on the force and receptor density. Deformation enables greater area contact and, thus, bond formation, and reduces viscous drag on the cell.

efficiency of capture is critically affected by the shear rate, whereas survival of bonds and velocity of rolling depend on the shear stress. However, an added complication is the finding that at very low levels of loading, applied force may actually stabilize selectin-mediated bonds (catch bonds), although at higher loading, the dissociation rate again increases (slip bonds) (Marshall et al., 2003). In any case, the number of cells observed to be adherent at any time represents a complex balance between the capture and disruption processes. Experimentally, because the shear rate and stress are linearly related for simple fluids, it may be difficult to prove their relative importance (see below).

Detailed theoretical analyses of effects of shear forces on adhesion have tended to center on the evolution of bonds after formation and, hence, the fate of cells and their rolling velocity (e.g., Zhu, 2000, for review), rather than delivery to the surface and initial capture. Such studies indicate that probabilistic analysis of the behavior of discrete bonds is required to understand the effects of shear on adhesive behavior (Cozens-Roberts et al., 1990). Existence of rolling or static attachment was critically dependent on the forward and reverse rate constants of the bonds, the sensitivity of the reverse rate constant to applied force (Hammer and Apte, 1992; Tozeren and Ley, 1992), and the dispersed nature of bonds (Ward et al., 1994), because clustering of bonds allows sharing of stress, limits peeling, and thus stabilizes attachment. More recent analyses have extended these approaches to predict cell behavior based on variations in densities and rate constants for receptor–ligand pairs, and to evaluate how mixtures of receptors with different characteristics influence stabilization of adhesion (Bhatia et al., 2003). Although these analyses typically predict increasing rolling velocity with increasing shear stress and decreasing receptor density, analyses also need to take into account the fact that leukocytes are viscoelastic solids, which deform under the action of fluid shear stress when attached to a surface. Increasing cell deformation with increasing shear stress tends to stabilize adhesion by increasing the cell–surface contact area and lowering the viscous drag experienced for a given shear stress. Analysis of these processes predicts that although rolling velocity increases with increasing shear stress, it tends to reach a plateau as cells deform (Lei et al., 1999). At the other extreme, at very low receptor density, single bonds may form, but cells quickly detach according to the lifetime of the bond. Analysis of such events may yield values for reverse rate constants, for example, of P-selectin bonds (Alon et al., 1995). However, recent studies indicate that such detachment rates are critically influenced by energy dissipation in localized cell deformation and in breakage of ligand–cytoskeleton links, as well as being dependent on the intrinsic kinetics of bond fracture (King et al., 2005).

Capture *per se*, or initial bond formation for moving cells, has been modeled less frequently. Chang and Hammer (1999) considered the effects of movement of cells over a surface, and concluded that as velocity increases, the likelihood of formation of bonds depends on a balance between the increasing receptor–ligand encounter rate and decreasing duration of collisions. It was predicted that the absolute capture rate would initially increase with shear rate (because there would be more collisions and effectively more cells delivered), but that total cell binding (as a percentage of cells delivered) would decrease with increasing shear rate. The possibility of a threshold phenomenon also arises from this analysis, where rate of capture is very low as flow increases from zero, until adequate collisions are generated. Subsequent extension of these analyses has shown that the cells that first attach to a surface can disturb the paths of unattached cells flowing close to the wall in such a way as to promote their secondary capture (King and Hammer, 2001) (see Chapter 12).

2. Experimental Studies

The first studies of adhesion of flowing neutrophils to purified selectin were carried out by Lawrence and Springer (1991), who studied binding to P-selectin inserted into lipid layers deposited on a glass slide. They found that the number of adherent cells decreased monotonically with increasing wall shear stress or rate, and was negligible at a shear stress of ~0.35 Pa (shear rate ~350 s^{-1}). The adherent cells typically rolled, with a velocity of ~10 μm/s, which increased with increasing shear stress. Coating with higher densities of immobilized receptor caused the number of captured neutrophils to increase and the rolling velocity to decrease. We observed similar shear dependence of adhesion of flowing neutrophils to P-selectin presented by activated platelets, but with relatively weak dependence of rolling velocity (Buttrum et al., 1993). Adhesion to P-selectin expressed by histamine-stimulated endothelial cells was also observed over a similar range of shear stresses, but there was strong shear-dependence of rolling velocity in the range ~5–30 μm/s (Jones et al., 1993). E-selectin expressed on the surface of a transfected cell line supported capture and rolling adhesion at a shear stress of ~0.2 Pa (Abbassi et al., 1993). In a direct comparison, purified E- and P-selectin supported rolling adhesion with similar rolling velocity and shear dependence, but an immobilized ligand for L-selectin supported much more rapid rolling (Puri et al., 1997). In these studies, "tethering" or initial capture was also quantified, separately from the level of adhesion that built up, and again, capture decreased to a negligible level for all three receptors at around 0.4 Pa. At the other extreme, neutrophil binding through L-selectin appears to have a "shear threshold" below which attachment is not detected (Finger et al., 1996). A less well-defined loss of binding to P-selectin may occur at low stress, when the receptor is present at low surface density (Lawrence et al., 1997). These phenomena may relate to the ability of selectin–ligand interactions to act as catch bonds, the effectiveness of which initially increases as stress increases from zero.

In principle, the number of bound cells at any time depends on the capture rate and the stability of the attachment. Thus, changes in the number bound may reflect changes in cell delivery rate, in collision time and hence capture efficiency, and in rates of dissociation of formed bonds which can theoretically lead to detachment from the surface. In fact, experience shows that on surfaces coated with high density of selectin receptors, rolling cells are rarely observed to detach and the number bound increases steadily with time (e.g., Buttrum et al., 1993). Moreover, shear stress can be increased greatly after adhesion is established with little evident wash-off. Multiple bonds have clearly formed, and although these peel continuously and more rapidly at high stress, the cells also roll forward and form new bonds more rapidly. Indeed, from observation of cells rolling at high stress, it is apparent that as they deform, they roll more smoothly and less erratically than at low stress. The fluctuations in velocity observed at lower stress evidence the stochastic formation and breakage of bonds (Goetz et al., 1994). Taking this further, on surfaces sparsely coated with selectins, binding does not build up with time. In this case, the rates of neutrophil detachment (or durations of individual attachments) after initial capture have been used to estimate bond lifetimes. It is assumed that the shortest-lived attachments represent single bonds. Alon et al. (1995) estimated lifetimes for P-selectin-mediated bonds of the order of 1 s, which decreased about threefold over the range of shear stresses at which capture was possible from flow. Interestingly, when neutrophils were perfused over minimally-stimulated human umbilical vein endothelial cells (HUVECs), we detected short-lived attachments supported by P-selectin that had a characteristic lifetime of ~0.5 s. Reverse rate constants for neutrophils adhering via E-selectin, P-selectin, and L-selectin have since been compared (Smith et al., 1999), and although they were similar in an unstressed state (~ 3 s^{-1}), the response to shear was greater for L-selectin than the other selectins, which may explain the higher velocity of rolling observed with this receptor (Puri et al., 1997). As noted earlier, doubt exists regarding whether these cell detachment–based estimates are true reflections of the intrinsic reverse rate constants for the receptor–ligand pairs (King et al., 2005). Indeed, direct comparisons of detachment kinetics of neutrophils and rigid spheres coated with P-selectin glycoproten ligand 1 (PSGL-1) indicate that cell deformation and formation of elongated tethers prolong attachment in flow models, beyond that which would be supported by direct load bonds (Park et al., 2002).

In an attempt to separate contributions of shear rate and stress in capture and transient adhesion on selectins, Chen and Springer (2001) studied capture rates and duration of attachments over a range of shear rates, while medium viscosity was also increased. The finding that capture rate was optimal at a similar shear rate for each viscosity was taken as evidence that capture was regulated by kinetics rather than stress, even though the capture rates decreased as viscosity increased for any given shear rate. The data were corrected for changes in sedimentation caused by changes in viscosity, but not for variations in the flux of neutrophils. The bell-shaped response to increasing shear rate may thus have reflected a balance between increasing collision frequency and decreasing duration of collision, as suggested by Chang and Hammer (1999). The same study showed that increasing stress caused increasing rate of cell dissociation from the surface, regardless of whether the increase in stress arose from an increase in shear rate or an increase in medium viscosity (Chen and Springer, 2001).

Several of these early studies were carried out at room temperature. Because temperature might be expected to influence forward and reverse rate constants for binding, we compared adhesion and rolling velocity on P-selectin presented by platelets over the range 10 to 37 °C for a range of increasing flow rates (Nash et al., 2001). We found that at lower temperature, for a given shear rate, rolling velocity decreased strongly, but numbers adherent were less affected. As lowering temperature also increased the fluid viscosity, shear stress effectively increased at the lower temperature. In fact, when adhesion was plotted as a function of wall shear stress (rather than rate), adhesion was greater at the lowest temperature. Although all adhesion data were corrected for the numbers of cells perfused, the colder cells were flowing more slowly for a given shear stress, and this may have promoted attachment. Potential changes in delivery because of changes in sedimentation should have been negligible because although slower perfusion allows more time for cells to sediment as they flow through the chamber, a proportional decrease in sedimentation rate occurs because of the increased fluid viscosity.

To attempt to unravel the relative effects of shear rate and stress in these experiments, we increased viscosity at 37 °C using dextran to reproduce the viscosities at the lower temperatures. Here we found that when shear rate and viscosity are identical, adhesion is equal at 37 and 26 °C, but greater at 10 °C. This is somewhat unexpected, but suggests that the level of adhesion attained is dependent on both forward and reverse rate constants, as both are likely to decrease at the lower temperature.

Most of the mechanistic studies referred to earlier used model surfaces other than endothelial cells. *In vitro*, endothelial cells stimulated with interleukin-1 (IL-1) or tumor necrosis factor α (TNF-α) may present mixtures of selectins or selectin ligands to flowing neutrophils. Nevertheless, in our hands, binding of neutrophils to IL-1-treated HUVECs decreases with increasing shear stress and is negligible above 0.2 Pa, as is the case with purified P- or E-selectin. Rolling velocities and stability of adhesion vary depending on the degree of endothelial stimulation (e.g., dose of TNF-α), and short-lived or erratic rolling can be observed in such models (Goetz et al., 1994; Rainger et al., 1997b). Intravital studies also indicate that rolling adhesion is supported by P-selectin, E-selectin, and L-selectin in inflamed microvessels in mice (e.g., Jung and Ley, 1999). Although the rolling velocities observed are sometimes higher than those measured *in vitro*, this depends on the stimulus applied to the microvasculature, and comparable values are observed, for example, with TNF-α as an *in vivo* agonist (Jung et al., 1998; Norman et al., 1998). Adhesion can also generally be detected *in vivo* at higher shear stress than *in vitro*, but this is in the presence of blood, which increases fluid viscosity and may have other effects on the attachment process (see below). Free-flow leukocyte velocities observed in microvessels (~ mm/s; Atherton and Born, 1973) are comparable to those in flow systems used for studies of adhesion *in vitro*.

B. Selectin-Mediated Capture from Flowing Blood

In principle, the presence of red cells and plasma proteins in blood increases viscosity and, hence, shear stress for any given shear rate. However, red cells also distort the velocity profile in tube flow, because of wall exclusion effects and a tendency for inward migration that is increased at low shear rate by the phenomenon of red cell aggregation. Because of this distortion and the non-Newtonian nature of the blood (for which the viscosity is lower, the higher the shear rate), it is difficult to define the actual shear rate or stress at the wall where adhesion occurs under any conditions (see Fig. 11.3).

Moreover, the red cells influence the distribution of neutrophils in the flow stream and, hence, contact with the wall, through a process termed *margination*, which is again promoted by red cell aggregation and tendency for inward migration. Collisions with red cells can also be thought of as effectively increasing the diffusion rate of neutrophils in the stream. In fact, there have been few studies of the effects of the rheological behavior of the blood on adhesion of neutrophils, although margination and, presumptively, wall contact have been shown to increase at slower flow and greater red cell aggregation (Goldsmith and Spain, 1984; Nobis et al., 1985).

Munn et al. (1996) showed that addition of red cells to lymphocyte suspension promoted adhesion to endothelial cells even at low hematocrit, but increasing adhesion with increasing hematocrit did not appear to arise from

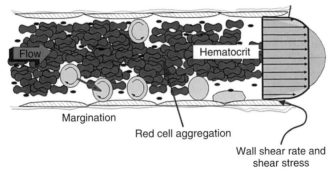

FIGURE 11.3   Flow of blood in a microvessel. Central flow of red cells, promoted by red cell aggregation, assists margination of neutrophils. Increasing hematocrit and aggregation increase the blunting of the flow velocity profile (plug flow) and increase the wall shear rate. The wall shear stress cannot be computed: increased shear rate may be offset by reduction in red cell concentration (and effective viscosity) near the wall.

increased margination of the cells. As noted above, we found that in vertical tubes, adhesion of isolated neutrophils to P-selectin was negligible (Abbitt and Nash, 2001). However, in native whole blood, many adhered at equal volumetric flow rates. The efficiency of adhesion decreased with increasing flow rate, and the dependence of adhesion on the "apparent" shear rate (calculated assuming a parabolic flow profile) was remarkably close to that for isolated cell suspensions. However, due to flow blunting, the actual wall shear rate, and also shear stress, was higher during adhesion in the presence of blood than in the isolated suspensions (see later text). In addition, it was notable that with increasing shear rate or stress, the rolling velocity was remakably constant (Abbitt and Nash, 2001).

Taking this further, we artificially manipulated the hematocrit of blood from 10 to 50% by adding autologous red cells or plasma (Abbitt and Nash, 2003). Increasing hematocrit to 30% increased leukocyte (predominantly neutrophil) adhesion, after which a plateau was reached. There was little evidence of any increase in margination (flux of neutrophils close to the wall), but it was evident that for constant volumetric flow rate, the velocity of cells near the wall was greater, the higher the hematocrit. Thus there was progressive flow blunting, and hence, increasing shear rate near the wall. Shear stress reached values comparable to those at which adhesion is observed *in vivo* (up to ~1 Pa) (Abbitt and Nash, 2001, 2003). Thus, adhesion increased even though the neutrophils were traveling faster and shear stress was increasing. Again, rolling velocity was remarkably constant with increasing hematocrit and hence shear stress. These results imply that red cells potentiate and stabilize neutrophil adhesion, in addition to any effect via margination, perhaps through operation of a force normal to the wall (Munn et al., 1996).

C. Effects of Flow Disturbance on Selectin-Mediated Capture

Adhesion in the microcirculation during inflammation occurs preferentially in postcapillary venules, where flow appears laminar. On the arterial side of the circulation, vessels typically have shear rates and stresses above those limiting for selectin-mediated adhesion. However, at bends and bifurcations in large arteries flow disturbances occur, with establishment of vortices and regions where flow has relatively low shear stress. These regions are associated with development of atheroma, and because of the immense pathogenic importance of this inflammatory process, effects of disturbed flow on leukocyte adhesion have been studied. Early studies in abruptly expanding cylindrical vessels with low wall shear stress showed promotion of adhesion of monocytes to nonspecific surfaces downstream of the expansion (Pritchard et al., 1995). We evaluated adhesion of isolated neutrophils perfused over a backward-facing step, where the surface was coated with P-selectin (Skilbeck et al., 2001). A stable vortex was established downstream of the step, and a reattachment point could be defined where the streamlines in the main jet returned to the wall. There was increased adhesion to P-selectin on either side of the reattachment point, but capture died away downstream of the reattachment point as parabolic flow redeveloped and the shear rate returned to ~ 400 s^{-1}. Extending this to studies with whole blood, we again found that adhesion occurred on either side of the reattachment point, but was negligible downstream (Skilbeck et al., 2004). Interestingly, the position of the reattachment point was quite accurately predicted by computational fluid dynamics for the isolated neutrophil suspension, but not for blood, when a Newtonian viscosity equal to that measured in a bulk viscometer was used for calculations. Again, patterns of flow and local shear stress and rate are hard to define in flowing blood. However, the conclusion is clear: binding to selectins can occur if flow is disturbed in regions where shear stresses would otherwise be too high for attachment.

III. KINETICS OF NEUTROPHIL ACTIVATION AND MIGRATION UNDER FLOW

A. Receptors Supporting Neutrophil Recruitment on Endothelial Cells

I. Importance of Selectins in Capture

The recruitment of neutrophils from flow by endothelial cells (ECs) appears to be dominated by selectin receptors and their counterligands. (Ley, 1996) Intravital experiments on the inflamed microvasculature of mice show that deletion of the genes for endothelial selectins (E-selectin and P-selectin) or leukocyte selectin (L-selectin), singly or in combination, reduces the levels of neutrophil adhesion to the vessel wall and increases the rolling velocities in microvessels, indicating that all three receptors may be involved in leukocyte capture in these models (e.g., Jung and Ley, 1999). Importantly, the inflamed microvasculature of genetically modified mice lacking all three selectins does not support visible adhesion of neutrophils (Jung and Ley, 1999). However, selectins may not be required for attachment in specific organs, such as the lungs and liver, where local fluid shear stresses are low and neutrophils squeeze through narrow capillary segments or sinuses (e.g., Doerschuk, 2000). This may relate to the early finding *in vitro* that flowing activated neutrophils can bind directly to endothelial cells through β2-integrins, if the wall shear stress is below ~ 0.05 Pa (Lawrence et al., 1990). *In vitro*, HUVECs have been most commonly used for flow-based studies of neutrophil recruitment. HUVECs constitutively express a granular depot of P-selectin (in the Weibel–Palade bodies), which can be mobilized within minutes by agents such as histamine and thrombin (Jones et al., 1993). HUVECs also upregulate E-selectin and P-selectin more slowly (over hours) in response to inflammatory cytokines. Although both E-selectin and P-selectin can support the capture of flowing neutrophils to HUVECs, to date counterligands for L-selectin have not been described on HUVECs or on endothelial cells in the inflamed microvasculature of mice.

2. Receptors Supporting Neutrophil Migration

Interactions between selectins and their counterligands localize neutrophils at the surface of activated ECs, but cannot support subsequent spreading, locomotion, and transendothelial migration. In human models, these processes appear to be supported solely by integrins of the $\beta 2$ family (CD18), which must first be "activated" by stimuli presented by endothelial cells (Springer, 1995). Four α units are known to associate with $\beta 2$-integrin: α_L, α_M, α_X, and α_d, designated CD11a, CD11b, CD11c, and CD11d, respectively. Circulating neutrophils constitutively express CD11a/CD18, CD11b/CD18, and CD11c/CD18. A substantial internal pool of CD11b and CD18 is also present in cytoplasmic granules of neutrophils and can be rapidly mobilized on chemotactic stimulation. Neutrophils do not express CD11d/CD18. Antibody blockade studies show that CD18 is essential for immobilization of rolling neutrophils on ECs on receipt of an activating stimulus (Bahra et al., 1998). Blockade of different α-subunits indicates that both CD11a/CD18 and CD11b/CD18 are used, perhaps in sequence, during neutrophil migration and diapedesis (Smith et al., 1989).

B. Kinetics of Neutrophil Activation and Immobilization by Integrins

In principle, to achieve precise localization of neutrophils at a site of infection, the transformation from unstable or rolling adhesion to stable immobilization should occur through a locally presented activating signal that is able to rapidly activate integrin-mediated adhesion. Several pathways that have been proposed to achieve this goal are described here, along with a paradigm for enabling a response to be effectively restricted to the region around an inflammatory locus.

I. Neutrophil Activation by Selectin-Mediated Signals

The primary function of selectin receptors and their ligands is to localize neutrophils to ECs of inflamed tissue. However, the facts that selectin expression is restricted to loci of inflammation and that the bonds formed between selectins and their ligands are highly specific have led to the idea that these interactions could provide a stimulus leading to localized neutrophil activation. Experimentally, neutrophils can be activated by crosslinking of surface-borne L-selectin or PSGL-1 with antibodies or soluble multimeric ligands. Such strategies have been reported to promote activation of $\beta 2$-integrins (Simon et al., 1999) and increase $\beta 2$-integrin–mediated adhesion (Blanks et al., 1998). However, to date no equivalent data have been generated for neutrophils rolling on surface-presented selectins under physiological regimes of shear stress. In our experience, neutrophils can roll on surfaces coated with P-selectin or E-selectin for prolonged periods without undergoing integrin activation, as evidenced by conversion to stationary adhesion. It is therefore a matter of debate whether selectin–ligand interactions play a significant role in the physiological process of neutrophil activation.

2. Neutrophil Activation through G-Protein-Coupled Receptors

The transition from rolling adhesion to integrin-mediated adhesion typically occurs when EC-borne chemoattractants bind neutrophil G-protein–linked receptors, which activate neutrophil β2-integrins. The expression of these activating agents is induced on exposure of ECs to inflammatory mediators. For example, thrombin, which upregulates the expression of P-selectin in minutes, can also induce rapid expression of the lipid-derived activator platelet-activating factor (PAF) (Macconi et al., 1995). Stimulation of ECs with inflammatory cytokines (e.g., TNF-α and IL-1β) promotes the transcriptional upregulation of chemokines such as IL-8, as well as the production of PAF (Kuijpers et al., 1992).

In the only study to date that has investigated the dynamics of neutrophil activation on ECs in response to surface-presented agonists, it was shown that neutrophils could respond to chemotactic stimulation very rapidly (<1 s) and become immobilized within one to two cell diameters from their point of first contact with the EC monolayer (Rainger et al., 1997b). Neutrophils bound transiently to low levels of P-selectin induced on ECs by treatment with hydrogen peroxide, and were activated after a mean rolling duration of only 280 ms when IL-8 was co-immobilized on the EC surface. When PAF was co-immobilized on the EC surface, neutrophils were activated after a mean rolling duration of 840 ms. Thus, it is clear that on tethering, neutrophils can respond very rapidly to surface-presented activating stimuli and can be localized with great precision to inflamed ECs. More recent studies have confirmed that IL-8 co-immobilized on plastic with P-selectin and ICAM-1 could cause rapid immobilization of flowing neutrophils, that is, within 200 μm of the point of first tethering via P-selectin (DiVietro et al., 2001). In addition, when neutrophil agonists such the bacterial formyl peptide fMLP and PAF are perfused over rolling neutrophils, immobilization follows within seconds (Sheikh and Nash, 1996; Rainger et al., 1997b). The neutrophils subsequently spread on the surface over ~30 s and commence migrating over the next few minutes.

In general, these responses appear to be all or none, with rapid transformation from the "resting" unstimulated state to an activated one. However, it has been proposed that rolling neutrophils might progressively reduce their rolling velocity as they integrate activating signals from the EC surface, eventually reaching a critical threshold for activation of their β2-integrins and immobilization (reviewed in Jung et al., 1998). This model of neutrophil activation proposes the requirement for a prolonged period of "transit" on the EC surface, during which numerous suboptimal activating signals are integrated by the rolling cells. Data in favor of this model have been derived from observations in intravital studies where the identity of the activating signal(s) is unknown (Jung et al., 1998). *In vitro* observations of neutrophils binding to cytokine-stimulated ECs yield evidence of mixed responses (Luu et al., 1999). For instance, when flowing neutrophils were captured by TNF-α–stimulated HUVECs, ~40% become activated and immobilized almost immediately (<1 s). Another 40% rolled for an extended period (averaging ~30 s) before becoming immobilized, while a subpopulation of ~20% of neutrophils rolled indefinitely. The activating signal was dominantly mediated through chemokine receptors on the neutrophils.

3. A Paradigm for Precise Localization of Neutrophils

Although neutrophils clearly have the ability to change their adhesive behavior rapidly, there is little direct evidence on how neutrophils can be targeted precisely to the locus of inflammation *in vivo*. Any inflammatory stimulus is likely to be diffuse on the scale of the microvasculature. We have previously postulated that a tissue gradient of an inflammatory cytokine such as TNF-α might maximize the efficiency of neutrophil recruitment by establishing a graded EC response depending on the local activity of that cytokine (Bahra et al., 1998) (Fig. 11.4).

We know that ECs exposed to low levels of cytokine support predominantly a rolling form of adhesion. Selectin expression (and coexpression) increase with increasing dose, and rolling velocity decreases. At the same time, increasing dose is associated with increasing efficiency of induction of stable adhesion and migration, presumably through more efficient contact with increasing levels of chemokine(s). Thus, ECs at the margins of cytokine influence would be minimally activated so that neutrophils captured at these

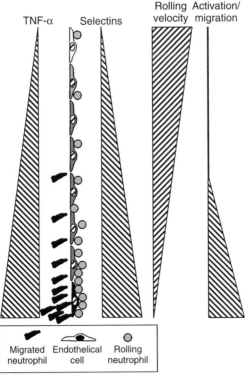

FIGURE 11.4 Schematic representation of how graded responses of endothelial cells to a cytokine diffusing from an inflammatory locus could lead to efficient local recruitment of neutrophils from the microcirculation. Flowing toward the locus, TNF-α concentration would increase, as would endothelial cell activation. At low concentration, rapid rolling adhesion would occur and cells would roll toward the locus. As TNF-α concentration increased, selectin expression would be higher, rolling would slows, and neutrophil activation by EC-borne agents would start. At high TNF-α concentration, both newly captured and already rolling cells would be efficiently activated. On the downstrean side, the opposite trends would be observed, and at the outer limits, neutrophils would roll away and detach.

peripheral sites would roll rapidly toward the focus of inflammation. Here they would be exposed to greater densities of selectin expression, leading to progressively reduced rolling velocities, and eventually to sites where surface-borne signals could mediate their activation and transendothelial migration. Neutrophils making contact with the vessel wall at sites of high cytokine activity would be exposed to ECs simultaneously expressing a high density of rolling receptors and activating stimuli and thus be rapidly immobilized close to their initial site of tethering. In downstream areas, neutrophils might be captured but roll out of the inflamed area. In this way, sites close to the inflammatory locus not only recruit neutrophils that bind directly to the vessel, but also neutrophils that roll in from more distant sites, and unnecessary and potentially damaging recruitment is avoided elsewhere. Thus, the kinetics of bond formation and rolling adhesion, and of activation responses and immobilization, may combine to allow efficient local accumulation of neutrophils.

C. Regulation of Neutrophil Migration by the Physical Environment

The processes that regulate $\beta2$-integrin–mediated neutrophil migration on ECs are poorly understood. However, there is evidence indicating that neutrophils receive cues from the physical environment, which are important for regulating their direction and velocity of locomotion, and possibly also the rate of penetration of the endothelial monolayer.

1. Regulation of the Direction of Neutrophil Migration by Hemodynamic Forces

When neutrophils have migrated into the extravascular compartment, the direction of their migration is regulated by gradients of chemotactic agents within the tissues. However, in the vascular compartment, gradients of soluble agents are disallowed by blood flow. It is possible that gradients of chemotactic stimuli or of adhesive receptors (haptotactic gradients) could be present on the surface of ECs, although evidence of such molecular organization on EC membranes is lacking. In flow-based *in vitro* models, neutrophils may be captured at, and appear to preferentially become activated and transmigrate at, the junctions between ECs (Gopalan et al., 2000). In addition, the forces generated by the flow of blood provide an alternative directional cue for neutrophil migration within the vascular compartment. The first demonstrations that neutrophils (or, indeed, any leukocytes) could sense flow-generated forces in a manner that regulated their migratory effort were conducted on substrates of immobilized recombinant adhesion receptors (CD31 or fibronectin) and on platelet monolayers (Rainger et al., 1999). On both of these substrates, activated neutrophils moved preferentially in the direction of flow. However, their migratory effort was directionally randomized by turning the flow off or by inhibiting the function of the neutrophil $\alpha v\beta3$-integrin. Subsequently, it was found that neutrophils on the apical surface of TNF-α–stimulated ECs also migrated preferentially in the direction of flow (Luu et al., 2003). However, antibodies that blocked the function of $\alpha v\beta3$-integrin had no effect on the direction of migration. Rather, directional migration was disturbed in the presence of antibodies that blocked homotypic interactions between CD31 found on the surfaces of neutrophils and that found on the surfaces of ECs. As might be expected, neutrophils that had transmigrated into the sub-EC environment, and thus could not sense flow-generated shear forces, showed a random pattern of migration. Thus, neutrophils can sense flow-generated shear stress by at least two distinct mechanisms, via αvβ3-integrin-mediated signaling or by homotypic CD31–CD31 signaling, and they use these signals to coordinate their migratory effort with direction of flow.

2. Regulation of Neutrophil Migration Velocity by the Physicochemical Environment

The velocity at which activated neutrophils migrate can also be regulated by interactions with surface receptors. Studies in slow-moving transfected cell lines have given rise to the concept that migration speed is essentially regulated by cell–substrate adhesive strength. However, neutrophils migrate rapidly (~10 μm/min) through continuously regulated adhesion via their β2-integins, although they can also use other integrins to achieve complex regulation of their speed (see, e.g., Hendey et al., 1996). In flow models we found rather different kinetics of migration, observed immediately after capture (Rainger et al., 1997a). Neutrophils captured by P-selectin on surfaces coated with immobilized, purified, recombinant adhesion receptors or monolayers of activated platelets demonstrated a marked phase of acceleration over a period of 5–7 min after their activation, before reaching a maximal migration velocity. The final velocity of migration was increased by accessory signals derived from ligation of CD31 if it was co-immobilized at increasing concentrations on the protein substrate. Antibody inhibition of the homotypic interactions between neutrophil CD31 and platelet CD31 slowed migration on that surface. On the other hand, addition of ICAM-1, which increased the levels of stable adhesion, did not accelerate the migration. Thus, in flow models, migration velocity of activated neutrophils is regulated by the adhesive substrate, but in a more receptor-specific fashion than through adhesive strength, which can itself be actively regulated.

Neutrophils migrating on or under ECs also show evidence of a complex pattern of velocity regulation (Luu et al., 2003). Neutrophils migrating on the apical surface of TNF-α–stimulated ECs migrated with a velocity of approximately 6 μm/min. This was insensitive to CD31 blockade and may represent a basal rate of neutrophil migration that has not been increased by accessory signals. However, when the neutrophils were activated with the bacterial formyl peptide fMLP, they migrated on the apical surface of the ECs with a velocity of approximately 14 μm/min. This increase in the terminal migration velocity was sensitive to CD31 blockade, indicating that the neutrophils have been "sensitized" to CD31-mediated acceleratory signals. Neutrophils that transmigrated through the EC monolayer also showed a marked increase in their migration velocities, without requirement for exogenous activating agents. This acceleration was again sensitive to CD31 blockade.

3. Flow-Mediated Regulation of Rate or Efficiency of Neutrophil Transmigration through Endothelium

The foregoing shows that the rate and direction of neutrophil migration are sensitive to applied forces. There is also some suggestion that such responses translate into an effect on efficiency of migration through the endothelial monolayer. Kitayama et al. (2000) found that the kinetics of penetration of neutrophils through TNF-α–treated HUVECs was more rapid in a flow assay where neutrophils were perfused over the endothelium than in a static assay in which they were simply allowed to settle onto it. The final proportions of adherent cells transmigrating were, however, similar. Studies in our own laboratory support this finding. Following individual neutrophils, we observed that the time from capture, to immobilization, to migration over and then through the monolayer averaged about 4 min and that for larger populations of cells, transmigration was complete between 5 and 10 min after delivery of a bolus of cells (Luu et al., 1999). However, in recent unpublished observations on static EC cultures, transmigration was not complete for about 20 min. Other recent studies have agreed that presence of flow has little effect on the proportion of neutrophils transmigrating though TNF-α–treated HUVECs (Cinamon et al., 2004). Yet it was found that if an exogenous agent (PAF) was bound onto the surface of HUVECs that had not been highly activated, then migration was more efficient in the presence of flow than its absence.

IV. CONCLUSIONS

We have discussed how the application of flow-generated forces to neutrophils and the local physical environment combine to regulate critical steps in the process of neutrophil recruitment. The concepts that shear rates and stresses influence rate-limiting steps such as selectin-mediated capture and subsequent rolling are well established. The role of blood rheology in the delivery of neutrophils and in bond stabilization has been less extensively studied. In addition, the ability of the hemodynamic environment to provide cues guiding or regulating neutrophil migration has attracted relatively little attention. These areas appear worthy of further investigation, for example, through detailed modeling of transport of neutrophils in the particulate flow and of the effects of interactions between flowing cells on contact efficiency and bond survival. The effects of flow on neutrophil migration offer an example of mechanotransduction that has been much more widely studied in the context of endothelial cell responses. Here, we have not addressed the important role of hemodynamic forces in regulating the phenotype or inflammatory responses of the ECs of the vasculature. In the cases of both neutrophils and endothelial cells, there is much to be discovered about the force transduction mechanisms, downstream signaling, and the mechanisms by which these signals are integrated with others, such as those from chemoattractants or cytokines.

The fact that neutrophil behavior is sensitive to the effects of flow at all stages of the recruitment process that occur in the vascular compartment indicates that hemodynamic factors have played an important part in shaping the molecular and biochemical evolution of leukocyte recruitment. Similar factors also influence lymphocyte and monocyte adhesion, although these have not been studied quite so extensively as neutrophils. The great significance of these processes in the functioning of the immune system suggests that further mechanistic studies are required to understand both physiological protective responses and their malfunction in diverse pathologies.

ACKNOWLEDGMENTS

Work in the laboratories of G.B.N. and G.E.R. is supported by a Programme Grant from the British Heart Foundation and a British Heart Foundation Non-Clinical Senior Lectureship Award to G.E.R.

SUGGESTED READING

Cines, D.B., Pollak, E.S., Buck, C.A., Loscalzo, J., Zimmerman, G.A., McEver, R.P., Pober, J.S., Wick, T.M., Konkle, B.A., Schwartz, B.S., Barnathan, E.S., McCrae, K.R., Hug, B.A., Schmidt, A.M. and Stern, D.M. (1998) Endothelial cells in physiology and in the pathophysiology of vascular disorders. *Blood* **91**: 3527–3561.

Wide-ranging review including the mechanisms of leukocyte adhesion to endothelial cells and the consequences of these and other EC-regulated processes.

Goldsmith, H.L. and Turitto, V.T. (1986). Rheological aspects of thrombosis and haemostasis: Basic principles and applications. ICTH-Report: Subcommittee on Rheology of the International Committee on Thrombosis and Haemostasis. *Throms. Haemost.* **55**: 415–435.

Clear and extended review of the hemodynamic and rheological factors that affect the distribution of cells in flowing blood, their velocities, and the forces acting on them.

Orsello, C.E., Lauffenburger, D.A., and Hammer, D.A. (2001) Molecular properties in cell adhesion: A physical and engineering perspective. *Trends Biotechnol.* **19**: 310–316.

Brief but very useful description of the essential physical properties of cells and of bonds that affect dynamic cellular adhesive processes.

REFERENCES

Abbassi, O., Kishimoto, T. K., McIntire, L. V., Anderson, D. C., and Smith, C. W. (1993) E-selectin supports neutrophil rolling in vitro under conditions of flow. *J. Clin. Invest.* **92**: 2719–2730.

Abbitt, K. B., and Nash, G. B. (2001) Characteristics of leucocyte adhesion directly observed in flowing whole blood *in vitro*. *Br. J. Haematol.* **112**: 55–63.

Abbitt, K. B., and Nash, G. B. (2003) Rheological properties of the blood influencing selectin-mediated adhesion of flowing leukocytes. *Am. J. Physiol.* **285**: H229–H240.

Alon, R., Hammer, D. A., and Springer, T. A. (1995) Lifetime of the P-selectin-carbohydrate bond and its response to tensile force in hydrodynamic flow [published erratum appears in Nature 1995;376:86]. *Nature* **374**: 539–542.

Atherton, A., and Born, G. V. (1973) Relationship between the velocity of rolling granulocytes and that of the blood flow in venules. *J. Physiol.* **233**: 157–165.

Bahra, P., Rainger, G. E., Wautier, J. L., Luu, N.-T., and Nash, G. B. (1998) Each step during transendothelial migration of flowing neutrophils is regulated by the stimulatory concentration of tumor necrosis factor-alpha. *Cell. Adv. Commun.* **6**: 491–501.

Bhatia, S. K., King, M. R., and Hammer, D. A. (2003) The state diagram for cell adhesion mediated by two receptors. *Biophys. J.* **84**: 2671–2690.

Blanks, J. E., Moll, T., Eytner, R., and Vestweber, D. (1998) Stimulation of P-selectin glycoprotein ligand-1 on mouse neutrophils activates beta 2-integrin mediated cell attachment to ICAM-1. *Eur. J. Immunol.* **28**: 433–443.

Buttrum, S. M., Hatton, R., and Nash, G. B. (1993) Selectin-mediated rolling of neutrophils on immobilized platelets. *Blood* **82**: 1165–1174.

Chang, K. C., and Hammer, D. A. (1999) The forward rate of binding of surface-tethered reactants: Effect of relative motion between two surfaces. *Biophys. J.* **76**: 1280–1292.

Chen, S., and Springer, T. A. (2001) Selectin receptor–ligand bonds: Formation limited by shear rate and dissociation governed by the Bell model. *Proc. Natl. Acad. Sci. USA* **98**: 950–955.

Cinamon, G., Shinder, V., Shamri, R., and Alon, R. (2004) Chemoattractant signals and beta 2 integrin occupancy at apical endothelial contacts combine with shear stress signals to promote transendothelial neutrophil migration. *J. Immunol.* **173**: 7282–7291.

Cozens-Roberts, C., Lauffenburger, D. A., and Quinn, J. A. (1990) Receptor-mediated cell attachment and detachment kinetics: I. Probabilistic model and analysis. *Biophys. J.* **58**: 841–856.

DiVietro, J. A., Smith, M. J., Smith, B. R., Petruzzelli, L., Larson, R. S., and Lawrence, M. B. (2001) Immobilized IL-8 triggers progressive activation of neutrophils rolling *in vitro* on P-selectin and intercellular adhesion molecule-1. *J. Immunol.* **167**: 4017–4025.

Doerschuk, C. M. (2000) Leukocyte trafficking in alveoli and airway passages. *Resp. Res* **1**: 136–140.

Finger, E. B., Puri, K. D., Alon, R., Lawrence, M. B., von Andrian, U. H., and Springer, T .A. (1996) Adhesion through L-selectin requires a threshold hydrodynamic shear. *Nature* **379**: 266–269.

Goetz DJ, el-Sabban ME, Pauli B. U., and Hammer DA. (1994) Dynamics of neutrophil rolling over stimulated endothelium *in vitro. Biophys. J.* **66**: 2202–2209.

Goldman, A. J., Cox, R.G., and Brenner, H. (1967) Slow viscous motion of a sphere parallel to a plane wall: II – Couette flow. *Chem. Eng. Sci.* **22**: 653–660.

Goldsmith, H. L. and Spain, S. (1984) Margination of leukocytes in blood flow through small tubes. *Microvasc. Res.* **27**: 204–222.

Gopalan, P. K., Burns, A. R., Simon, S. I., Sparks, S., McIntire, L. V., and Smith, C. W. (2000) Preferential sites for stationary adhesion of neutrophils to cytokine-stimulated HUVEC under flow conditions. *J. Leukoc. Biol.* **68**: 47–57.

Hammer, D. A., and Apte, S. M. (1992) Simulation of cell rolling and adhesion on surfaces in shear flow: General results and analysis of selectin-mediated neutrophil adhesion. *Biophys. J.* **63**: 35–57.

Hendey, B., Lawson, M., Marcantonio, E. E., and Maxfield, F. R. (1996). Intracellular calcium and calcineurin regulate neutrophil motility on vitronectin through a receptor identified by antibodies to integrins alphav and beta3. *Blood* **87**: 2038–2048.

Jones, D. A., Abbassi, O., McIntire, L. V., McEver, R. P., and Smith, C. W. (1993) P-selectin mediates neutrophil rolling on histamine-stimulated endothelial cells. *Biophys. J.* **65**: 1560–1569.

Jung, U., and Ley, K. (1999) Mice lacking two or all three selectins demonstrate overlapping and distinct functions for each selectin. *J. Immunol.* **162**: 6755–6762.

Jung, U., Norman, K. E., Scharffetter-Kochanek, K., Beaudet, A. L., and Ley, K. (1998) Transit time of leukocytes rolling through venules controls cytokine-induced inflammatory cell recruitment *in vivo. J. Clin. Invest.* **102**: 1526–1533.

Kansas, G. S. (1996) Selectins and their ligands: Current concepts and controversies. *Blood* **88**: 3259–3287.

King, M. R., and Hammer, D. A. (2001) Multiparticle adhesive dynamics: Hydrodynamic recruitment of rolling leukocytes. *Proc. Natl. Acad. Sci. USA* **98**: 14919–14924.

King, M. R., Heinrich, V., Evans, E., and Hammer, D. A. (2005) Nano-to-micro scale dynamics of P-selectin detachment from leukocyte interfaces: III. Numerical simulation of tethering under flow. *Biophys. J.* **88**: 1676–1683.

Kitayama, J., Hidemura, A., Saito, H., and Nagawa, H. (2000) Shear stress affects migration behavior of polymorphonuclear cells arrested on endothelium. *Cell. Immunol.* **203**: 39–46.

Kuijpers, T. W., Hakkert, B. C., Hart, M. H., and Roos, D. (1992) Neutrophil migration across monolayers of cytokine-prestimulated endothelial cells: A role for platelet-activating factor and IL-8. *J. Cell Biol.* **117**: 565–572.

Lawrence, M. B, Kansas, G. S, Kunkel, E. J., and Ley, K. (1997) Threshold levels of fluid shear promote leukocyte adhesion through selectins (CD62L,P,E) *J. Cell Biol.* **136**: 717–727.

Lawrence, M. B., Smith, C. W., Eskin, S. G., and McIntire, L. V. (1990) Effect of venous shear stress on CD18-mediated neutrophil adhesion to cultured endothelium. *Blood* **75**: 227–237.

Lawrence, M. B., and Springer, T. A. (1991). Leukocytes roll on a selectin at physiologic flow rates: Distinction from and prerequisite for adhesion through integrins. *Cell* **65**: 859–873.

Lei, X., Lawrence, M. B., and Dong, C. (1999) Influence of cell deformation on leukocyte rolling adhesion in shear flow. *J. Biomech. Eng.* **121**: 636–643.

Ley, K. (1996) Molecular mechanisms of leukocyte recruitment in the inflammatory process. *Cardiovasc. Res.* **32**: 733–742.

Luu, N. T., Rainger, G. E., Buckley, C. D., and Nash, G. B. (2003) CD31 regulates direction and rate of neutrophil migration over and under endothelial cells. *J. Vasc. Res.* **40**: 467–479.

Luu, N. T., Rainger, G. E., and Nash, G. B. (1999) Kinetics of the different steps during neutrophil migration through cultured endothelial monolayers treated with tumor necrosis factor-alpha. *J. Vasc. Res.* **36**: 477–485.

Macconi, D., Foppolo, M., Paris, S., Noris, M., Aiello, S., Remuzzi, G., and Remuzzi, A. (1995) PAF mediates neutrophil adhesion to thrombin or TNF-stimulated endothelial cells under shear stress. *Am. J. Physiol.* **269**: 42–47.

Marshall, B. T., Long, M., Piper, J. W., Yago, T., McEver, R. P., and Zhu, C. (2003) Direct observation of catch bonds involving cell-adhesion molecules. *Nature* **423**: 190–193.

Munn, L. L., Melder, R. J., and Jain, R. K. (1994) Analysis of cell flux in the parallel-plate flow chamber: Implications for cell capture studies. *Biophys. J.* **67**: 889–895.

Munn, L. L., Melder, R. J., and Jain, R. K. (1996) Role of erythrocytes in leukocyte–endothelial interactions: Mathematical model and experimental validation. *Biophys. J.* **71**: 466–478.

Nash, G. B., Abbitt, K., Tate, K., Jetha, K. J., and Egginton, S. (2001) Changes in the mechanical and adhesive behavior of neutrophils on cooling: potential modifiers of microvascular perfusion. *Eur. J. Physiol.* **442**: 762–770.

Nobis, U., Pries, A. R., Cokelet, G. R., and Gaehtgens, P. (1985) Radial distribution of white cells during blood flow in small tubes. *Microvasc. Res.* **29**: 295–304.

Norman, K. E., Anderson, G. P., Kolb, H. C., Ley, K., and Ernst, B. (1998) Sialyl Lewis(x) (sLe(x)) and an sLe(x) mimetic, CGP69669A, disrupt E-selectin-dependent leukocyte rolling *in vivo*. *Blood* **91**: 475–483.

Park, E. Y., Smith, M. J., Stropp, E.S., Snapp, K. R., DiVietro, J. A., Walker, W. F., Schmidtke, D. W., Diamond, S. L., and Lawrence, M. B. (2002) Comparison of PSGL-1 microbead and neutrophil rolling: microvillus elongation stabilizes P-selectin bond clusters. *Biophys. J.* **82**: 1835–1847.

Pritchard, W. F., Davies, P. F., Derafshi, Z., Polacek, D. C., Tsa, R., Dull, R. O., Jones, S. A., and Giddens, D. P. (1995) Effects of wall shear stress and fluid recirculation on the localization of circulating monocytes in a three-dimensional flow model. *J. Biomech.* **28**: 1459–1469.

Puri, K. D., Finger, E. B., and Springer, T. A. (1997) The faster kinetics of L-selectin than of E-selectin and P-selectin rolling at comparable binding strength. *J. Immunol.* **158**: 405–413.

Rainger, G. E., Buckley, C., Simmons, D. L., and Nash, G. B. (1997a) Cross-talk between cell adhesion molecules regulates the migration velocity of neutrophils. *Curr. Biol.* **7**: 316–325.

Rainger, G. E., Buckley, C. D., Simmons, D. L., and Nash, G. B. (1999) Neutrophils sense flow-generated stress and direct their migration through alphaVbeta3-integrin. *Am. J. Physiol* **276**: 858–864.

Rainger, G. E., Fisher, A. C., and Nash, G. B. (1997b) Neutrophil rolling is rapidly transformed to stationary adhesion by IL-8 or PAF presented on endothelial surfaces. *Am. J. Physiol.* **272**: H114–H122.

Sheikh, S., and Nash, G. B. (1996) Continuous activation and deactivation of integrin CD11b/CD18 during de novo expression enables rolling neutrophils to immobilize on platelets. *Blood* **87**: 5040–5050.

Simon, S. I., Cherapanov, V., Nadra, I., Waddell, T. K., Seo, S. M., Wang, Q., Doerschuk, C. M., and Downey, G. P. (1999) Signaling functions of L-selectin in neutrophils: Alterations in the cytoskeleton and colocalization with CD18. *J. Immunol.* **163**: 2891–2901.

Skilbeck, C., Westwood, S. M., Walker, P. G., David, T., and Nash, G. B. (2001) Dependence of adhesive behavior of neutrophils on local fluid dynamics in a region with recirculating flow. *Biorheology* **3**: 213–227.

Skilbeck, C. A., Walker, P. G., David, T., and Nash, G. B. (2004) Disturbed flow promotes deposition of leucocytes from flowing whole blood in a model of a damaged vessel wall. *Br. J. Haematol.* **126**: 418–427.

Smith, C. W., Marlin, S. D., Rothlein, R., Toman, C., and Anderson, D. C. (1989) Cooperative interactions of LFA-1 and Mac-1 with intercellular adhesion molecule-1 in facilitating adherence and transendothelial migration of human neutrophils *in vitro*. *J. Clin. Invest.* **83**: 2008–2017.

Smith, M. J., Berg, E. L., and Lawrence, M. B. (1999) A direct comparison of selectin-mediated transient, adhesive events using high temporal resolution. *Biophys. J.* **77**: 3371–3383.

Springer, T. A. (1995) Traffic signals on endothelium for lymphocyte recirculation and leukocyte emigration. *Annu. Rev. Physiol.* **57**: 827–872.

Tissot, O., Pierres, A., and Bongrand, P. (1992) Motion of cells sedimenting on a solid surface in laminar shear flow. *Biophys. J.* **58**: 641–652.

Tozeren, A., and Ley, K. (1992) How do selectins mediate leukocyte rolling in venules? *Biophys. J.* **63**: 700–709.

Ward, M. D., Dembo, M., and Hammer, D. A. (1994) Kinetics of cell detachment: Peeling of discrete receptor clusters. *Biophys. J.* **67**: 2522–2534.

Zhu, C. (2000) Kinetics and mechanics of cell adhesion. *J. Biomech.* **33**: 23–33.

12

HYDRODYNAMIC INTERACTIONS BETWEEN CELLS ON REACTIVE SURFACES

DOOYOUNG LEE and MICHAEL R. KING

Department of Chemical Engineering and Department of Biomedical Engineering, University of Rochester, Rochester, New York, USA

Leukocyte recruitment to sites of inflammation is initiated by adhesive tethering and rolling on the activated vascular wall under shear flow. Although the binding kinetics between adhesion molecules is primarily responsible for leukocyte recruitment, we have found that hydrodynamic interactions between cells near adhesive surfaces affect leukocyte behavior in the initial recruiting step onto the vascular endothelial wall. Previous experimental evidence has shown that cell–cell physical adhesion is one multicellular mechanism that accelerates cell recruitment to an adhesive surface. Our laboratory has shown, however, that hydrodynamic collisions between cells induce the capture of free-stream cells to a ligand-presenting surface under shear flow, verified by monitoring leukocytes in postcapillary venules, as well as by using carbohydrate-coated microspheres or neutrophils in flow chambers. We also demonstrate that neighboring cells roll more slowly as two rolling cells become closer due to cell–cell and cell–surface hydrodynamic interactions, not only *in vitro* but also *in vivo*. To explain and probe these phenomena, we developed the multiparticle adhesive dynamics (MAD) theoretical simulation that fuses an adhesive dynamics model with a boundary integral calculation of suspension flow. Our simulation results show good agreement with experiments measuring binary cell capture and pair rolling velocity. In this chapter, therefore, we compile evidence that a distinct hydrodynamic mechanism of cell–surface interaction in blood flow plays a significant role in the inflammation cascade.

I. INTRODUCTION

The adhesion of leukocytes to the luminal surface of the microvasculature plays an important role in the inflammatory response and lymphocyte homing to lymphatic tissues. The initial step in the leukocyte adhesion cascade involves capture and rolling of leukocytes on the receptor-bearing endothelial cell wall (Springer, 1994; Vestweber and Blanks, 1999; McEver, 2002). The process known as "capture" or "tethering" represents the first contact of leukocytes with the activated wall after cells have entered a position close to the wall due to a hydrodynamic mechanism termed *margination*. The next key step in these adhesive interactions is rolling, in which the adhesion of cells to the surface slows, but does not stop, the motion of a cell under hydrodynamic flow (Lawrence and Springer, 1991). Rolling is caused by the coordinated formation and dissociation of receptor–ligand bonds at the front and back of the cell, respectively. P-selectin on endothelium is considered the primary adhesion molecule for capture and the initiation of rolling (Lawrence and Springer, 1991). The ligand on leukocytes for P-selectin is P-selectin glycoprotein ligand-1 (PSGL-1) (Moore et al., 1995). In addition, many studies have suggested that the binding between L-selectin on leukocytes and its ligand on surfaces also plays an important part in both processes (Alon et al., 1996; Finger et al., 1996; Dwir et al., 2003). In addition to the primary binding of cells with the vessel wall, some researchers have suggested that secondary leukocyte–leukocyte tethering enhances the rate of leukocyte accumulation or rolling on the walls of blood vessels (Alon et al., 1996; Walcheck et al., 1996; St. Hill et al., 2003). Transient tether formation between an already adherent cell and one freely suspended in the blood, mediated by the bonding of L-selectin with L-selectin ligand, causes the freely flowing cells to be captured by the surface and to roll adhesively.

Hydrodynamic shear force induced by blood flow affects the behavior of both leukocytes and endothelial cells. Leukocytes in circulation are reported to minimize pseudopod formation (Moazzam et al., 1997), and when exposed to low shear stresses during adhesion to endothelial cells, they often retract pseudopodia and reduce their attachments (Coughlin and Schmid-Schonbein, 2004). In addition, shear force is imposed directly on the endothelium and modulates endothelial structure and function, as well as endothelial cell gene expression. For instance, flow-conditioned endothelial cells influence leukocyte adhesion by expressing ICAM-1 and E-selectin together (Burns and DePaola, 2005) or by decreasing inflammatory stimuli (Sheikh et al., 2003). However, little information is available from either simulations or experiments on how hydrodynamic interactions between cells or between cells and the vessel wall affect cell adhesion during the inflammation process.

The purpose of this chapter is to explore the hydrodynamic interactions of cells with reactive surfaces as revealed by numerical simulations of multiple particles and corresponding experiments *in vivo* as well as *in vitro*. The multiparticle adhesive dynamics (MAD) simulation is the first attempt at joining a precise calculation of multiparticle fluid flow with a realistic model for specific cell–cell adhesion. Comparison of results from the simulation and experiments demonstrates that previously unidentified biophysical mechanisms controlling dynamic cell adhesion under flow exist, and frequently occur in both *in vitro* and *in vivo* experimental systems.

II. MULTIPARTICLE ADHESIVE DYNAMICS

The MAD algorithm is based on adhesive dynamics and includes the effects of multiple particles (i.e., cells). The adhesive dynamics (AD) algorithm has been thoroughly described in several articles (Hammer and Apte, 1992; Chang and Hammer, 2000; Chang et al., 2000). Essentially, a large number of adhesive receptors and ligands are randomly distributed on the surfaces of a rigid sphere and a planar wall, with individual molecules modeled as linear springs. The motions of the cell and of each molecular bond between the cell and surface are recorded as the cell rolls over or moves relative to the other surface in shear flow. In the near-contact region between the sphere and plane, adhesion molecule pairs are stochastically tested for bond formation according to their deviation length-dependent binding kinetics. Other surface interactions, such as electrostatic repulsion, and body forces such as gravity are included in the model. A summation of all external forces and torques enables a mobility calculation to determine the translational and rotational velocities of the sphere.

A one-step dissociation model of single biomolecular bonds is generally used (Bell, 1978),

$$k_r = k_r^0 \exp\left[\frac{r_0 F}{k_b T}\right], \tag{12.1}$$

which relates the rate of dissociation k_r to the magnitude of the force on a single bond F. The unstressed off-rate k_r^0 and reactive compliance r_0 have been experimentally determined for the selectins interacting with their respective ligands (Smith et al., 1999). The rate of formation k_f directly follows from the Boltzmann distribution for affinity. The expression for the binding rate must also incorporate the effect of the relative motion of the two surfaces (Chang and Hammer, 1999). The solution algorithm for AD is as follows:

- All unbound molecules in the contact area are tested for bond formation using the probability

$$P_f = 1 - \exp(-k_f \Delta t). \tag{12.2}$$

- All of the currently bound molecules are tested for dissociation using the probability

$$P_r = 1 - \exp(-k_r \Delta t). \tag{12.3}$$

- The external forces and torques on each cell are summed.
- The mobility calculation is performed to determine the rigid body motions of the cells.
- Cell and bond positions are updated according to the kinetics of particle motion.

Unless firmly adhered to a surface, leukocytes can be modeled effectively as rigid spheres. Typical values of physical parameters yield Reynolds numbers, Re $= \dot{\gamma}a^2/v = O(10^{-3})$, where shear rate $\dot{\gamma} \approx 100$ s^{-1}, cell radius $a = 4$ μm, and the kinematic viscosity of the suspending fluid $v = 1$ cS. Thus, inertia can be neglected and fluid motion is governed by the Stokes equation

$$\mu \nabla^2 u = \nabla p, \quad \nabla \cdot u = 0, \tag{12.4}$$

where u is the velocity, μ is the fluid viscosity, and p is the local pressure. No-slip boundary conditions hold at the cell surfaces and at $z = 0$, the position of the planar wall.

We significantly advanced the AD algorithm to include the effects of multiple particles with the resulting simulation termed MAD (King and Hammer, 2001a; Bhatia et al., 2003) (see Fig. 12.1).

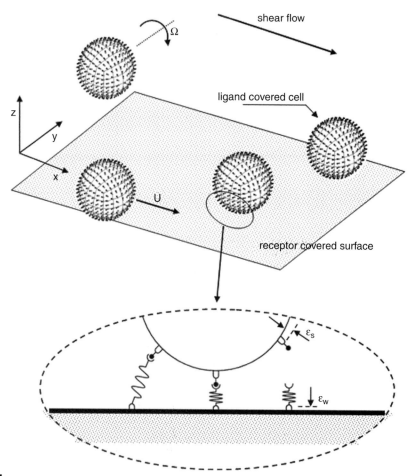

FIGURE 12.1 Schematic diagram of multiparticle adhesive dynamics. Linear shear flow is applied, and the motion of cells is calculated using a boundary element method. Cells (or beads) and the reactive plane have their own surface roughness of dimensions ϵ_s and ϵ_w. The receptor–ligand bonds are represented by linear springs whose endpoints remain fixed with respect to either surface.

The central technique in MAD is the completed double-layer boundary integral equation method (CDL-BIEM), which refers to a boundary element method for calculation of the hydrodynamic mobilities of a suspension of small particles in a viscous fluid (Kim and Karrila, 1991). Applying the standard boundary element method to the Stokes flow problem produces a Fredholm integral equation of the first kind, which is generally ill-conditioned. By posing the mobility problem in terms of a compact double-layer operator and completing the range with the velocity field resulting from a known distribution of point forces and torques placed inside each particle, one can derive a fixed point iteration scheme for solving an integral representation of the creeping flow equation,

$$u(x) - u^{\infty}(x) = u^{RC}(x) + \oint_S K(x - \zeta) \cdot \phi(\zeta) dS(\zeta), \qquad (12.5)$$

where u^{∞} is the ambient fluid velocity (e.g., simple shear flow), u^{RC} is a "range completing" velocity generated by point forces and torques that accounts for the fact that the ill-behaved single-layer integral has been discarded, K is the double-layer operator, and ϕ is the unknown double-layer distribution. After the spectral radius of the corresponding discretization is reduced, the solution is found to converge rapidly. The presence of a single wall is treated by incorporating the singularity solutions corresponding to a point force near a plane. To speed the calculation, a coarse discretization is used that does not resolve the cell–cell and cell–plane lubrication forces, which are added from known solutions as additional "external forces." As a model of the roughness of the spherical and planar surfaces, it is normally assumed that both surfaces are covered with small bumps of sufficient coverage to support the cell, yet of a dilution that permits the flow disturbance caused by the bumps to be neglected (for instance, $\varepsilon_s = 175$ nm on the spheres and $\varepsilon_w = 50$ nm on the wall).

III. HYDRODYNAMIC RECRUITMENT: CELL COLLISIONS RESULTING IN NEW ATTACHMENTS

The recruitment of leukocytes to inflammatory sites can occur through several mechanisms (see Fig. 12.2).

Primary capture refers to the initial tethering of a cell far from other cells, by binding receptors on a cell with ligands on the surface. In contrast, *multicellular tethering,* that is, the recruitment of a new cell in the vicinity of other adherent cells on a reactive surface, is caused by hydrodynamic interactions as well as the biochemical interaction of leukocyte receptors with leukocyte ligands. Transient tethering between a rolling leukocyte and a freely flowing leukocyte, mediated by L-selectin–PSGL-1 binding, has been suggested as an initial step of multicellular tethering (Fig. 12.2B) (Walcheck et al., 1996). In addition, we have reported that a distinct hydrodynamic interaction between leukocytes can cause the recruitment of a flowing cell (King et al., 2001; King and Hammer, 2001b). First, a free-stream cell that approaches a previously bound cell can be pulled down to a reactive distance from the wall. Binding then can occur with the surface without physical contact between leukocytes, which is referred to as "upstream hydrodynamic recruitment" (see Fig. 12.2C). During events of "downstream hydrodynamic recruitment," as depicted in Fig. 12.2D, the free-stream cell is irreversibly shifted off of its initial streamline due to contact repulsion, moving either around the side of the rolling cell or up over the top, depending on its initial position. After such a collision between a rolling cell and a free-stream cell occurring near the wall, the cell is pulled back down to the wall due to far-field hydrodynamic interactions, allowing an adhesive tether to form between the free-stream cell and the wall.

Figure 12.3 illustrates data collected from a representative simulation and different experiments comparing the effects of hydrodynamic interactions between cells (or microspheres) recruited to a reactive surface. Figure 12.3A is the height trajectory of a free-stream cell during hydrodynamic interaction with an adherent cell obtained from MAD simulations. First, the cell is pulled down toward the wall, where it can bind. Then, as the free-stream cell closely approaches the bound cell, it is lifted slightly off the wall, decreasing the

A. Primary recruitment

B. Secondary recruitment

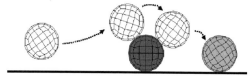

C. Upstream hydrodynamic recruitment

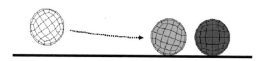

D. Downstream hydrodynamic recruitment

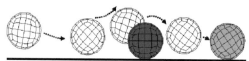

FIGURE 12.2 Classification of the type of recruitment of a free-stream cell (or microsphere) on a reactive surface. (A) Primary capture is the isolated tethering of a free-stream cell on a reactive surface with no bound cells nearby. (B) Secondary tethering occurs when the receptor on a free-stream cell binds a ligand on a prebound cell (or vice versa), resulting in physical contact and adhesion. The two attached cells rotate as a doublet down toward the surface due to the surrounding flow, and the second cell can later tether with the surface. In contrast, hydrodynamic recruitment is defined as the tethering of a free-stream cell in the vicinity of an adherent cell not by physical binding but due to far-field hydrodynamic interactions. According to the relative position of a newly tethering cell relative to the previously bound cell, these events are classified as either upstream attachment (C) or downstream attachment (D).

probability of bond formation with the wall. After this collision, the free-stream cell is pulled back down toward the wall due to the hydrodynamic disturbance created by the two cells, and the free-stream cell can again bind with the wall. This vertically directed motion occurs even in the absence of simple gravitational sedimentation, as shown by King and Hammer (2001b).

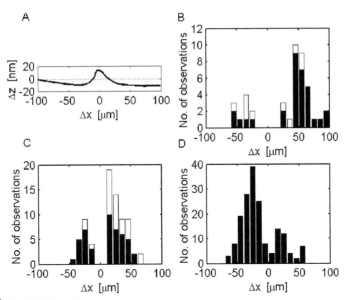

FIGURE 12.3 (A) Height trajectory of a simulated free-stream cell above the reactive surface during hydrodynamic collision with an adherent cell in shear flow, obtained from the MAD simulation (side view). (B, C) Relative locations of hydrodynamically recruited microspheres (B) and human neutrophils (C) during *in vitro* flow chamber experiments with P-selectin-coated surfaces. (D) Similar experiments observing leukocytes within hamster venule walls *in vivo*. $\Delta x = 0$ indicates the axial position of a previously bound cell (or bead). The black bars represent shear rates of $80 \ s^{-1}$ (B), $100 \ s^{-1}$ (C), $108 \ s^{-1}$ (and 94) s^{-1} (D), and the open bars represent shear rates of $160 \ s^{-1}$ (B) and $200 \ s^{-1}$ (C), respectively.

Figures 12.3B, C, and D show where attachment was found to occur experimentally due to hydrodynamic recruitment. The details of each set of experimental conditions are as follows:

- sLe^{X}-coated microspheres (radius: $5.4 \ \mu m$) were perfused into a parallel-plate flow chamber containing a lower P-selectin-coated surface with a surface density of 180 molecules/μm^2 (Fig. 12.3B). The wall shear stresses were 0.8 and 1.6 dyn/cm^2, and the total number of observations of cell recruitment was 42. The data are adapted from King and Hammer (2001b).
- Isolated human neutrophils in Hanks' balanced salt solution containing Ca^{2+} and human serum albumin were perfused into a parallel-plate flow chamber, the surface of which was coated with sLe^{X} at a density of 10^5 to 10^6 molecules/μm^2 (Fig. 12.3C). The wall shear stresses used were 1.0 and 2.0 dyn/cm^2, and the total number of observations of cell recruitment was 74.
- Hydrodynamic recruitment was also observed in two postcapillary venules of diameter 24 and 30 μm in hamsters. The wall shear rate in these two vessels was 108 and 94 s^{-1}, and the data are reproduced from King et al. (2003) (Fig. 12.3D).

As can be seen in Figs. 12.3.B, C, and D, all of the leukocyte capture events occurring in the vicinity of a previously adherent leukocyte obtained from different experimental systems exhibit the same trend, that an adherent leukocyte promotes the capture of a free-stream cell in either the upstream or downstream region due to hydrodynamic interactions. During hydrodynamic recruitment events, no leukocyte–leukocyte adhesion was evident, that is, behavior indicative of secondary tethering occurring due to leukocyte–leukocyte transient adhesion. Very good agreement was found between *in vitro* and *in vivo* experiments, specifically that the hydrodynamic recruitment events at a shear stress around 1 dyn/cm^2 are strongly concentrated around two positions, upstream and downstream of a previously adherent cell. This has been also demonstrated in the results obtained from MAD simulations (King and Hammer, 2001b). Note that because of the geometrical differences between a cylindrical microvessel and a parallel-plate flow chamber, upstream hydrodynamic recruitment events may be much more common in the venular microcirculation than in *in vitro* systems.

Figure 12.4 shows that the ratio of primary capture events to total adhesion events in flow chamber experiments with human neutrophils decreases with increasing wall shear stress. In other words, the frequency of multicellular tethering events increases with increasing shear. This may be due to an increase in the number of collisions between cells as the flow rate is increased. This experimental observation also agrees with model data collected by another laboratory (Zhang and Neelamegham, 2002). In contrast, multicellular tethering events are dominant in postcapillary vessels at lower shear because the smaller geometry of the microvessel compared with that of parallel-plate flow chambers may induce more frequent collisions between cells (King et al., 2003).

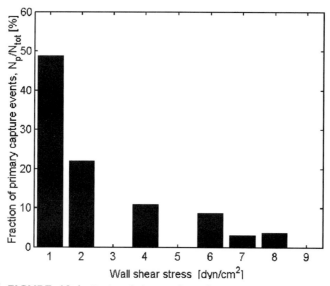

FIGURE 12.4 Ratio of the number of primary capture events to the number of total recruitment events at different wall shear stresses. Isolated human neutrophils were perfused into a parallel-plate flow chamber with a sLe^X coated surface, and both isolated captures and multicellular recruitment events on the reactive surface were monitored through the entire viewing window (0.14 mm^2).

IV. MULTIPARTICLE ADHESIVE DYNAMICS: FAR-FIELD INTERACTIONS BETWEEN ROLLING CELLS

We have presented *in vitro* and *in vivo* experiments and a simulation technique capable of explaining the hydrodynamic interactions between adhesive cells reversibly interacting with an adhesive surface under laminar shear flow (King and Hammer, 2001a, b; King et al., 2005b). To determine the effect of spatial separation on average rolling velocity between nearby cellular pairs, carbohydrate microspheres were perfused into a parallel-plate flow chamber at a wall shear stress of 0.8 dyn/cm². Figure 12.5 illustrates the geometrical definition used to quantify the pair velocity. Pairs of sLe^X-coated beads rolling on a P-selectin–coated surface were selected for measurement only when they were at least 20 radii away from any other adherent bead, to isolate binary interactions. As shown in Fig. 12.6A, the average rolling velocity of a pair of beads decreases as the separation distance between the two beads decreases. This result was also consistent with the increasing cell–cell (force-free) drag as illustrated by King and Hammer (2001b). To focus on binary leukocyte–leukocyte hydrodynamic interactions *in vivo*, leukocytes undergoing P-selectin–mediated rolling on stimulated endothelial cells were directly observed in the postcapillary venules of mouse cremaster muscle at various wall shear stresses (King et al., 2005b) (Fig. 12.6B). Note that these leukocyte–leukocyte hydrodynamic interactions are found to occur even in the presence of physiological microcirculatory concentrations of red blood cells, estimated to be 20–30% by volume. Figure 12.6D was collected from *in vivo* observations of hamster cheek pouch at a shear rate of 62 s^{-1}. Because the vessel cross section was viewed from the side in these experiments, data are displayed as function of x only.

MAD simulations of pairs of rolling cells were performed under the same conditions used in previous microsphere experiments for direct comparison. As can be seen in Fig. 12.6.C, the good level of agreement between both *in vivo* and *in vitro* experiments and simulations demonstrates that rolling cells slow each other down as they approach contact. Therefore, the simulation allows us to examine the behavior of a rolling pair of cells at a higher spatial and temporal resolution than can be readily accessed in experiments. The velocities predicted by the MAD simulation lie somewhat below the experimental velocities as cells approach each other (in Figs. 12.6A, B); this is

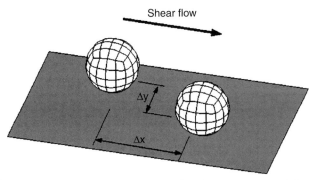

FIGURE 12.5 Coordinate system for the measurement of the rolling velocity of two neighboring rolling leukocytes (or microspheres) in close proximity on a reactive surface.

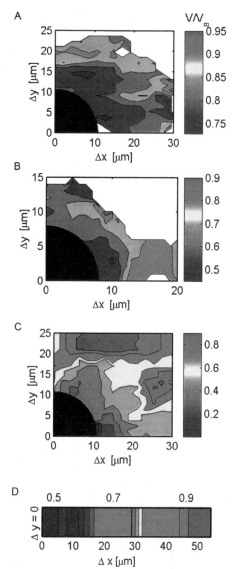

FIGURE 12.6   (A) Rolling velocity of pairs of sLe^x-coated microspheres (diameter = 10.9 μm) flowing over a P-selectin–presenting surface in a parallel-plate flow chamber. (B) Rolling velocity of pairs of leukocytes in a mouse venule (diameter = 37 μm). (C) MAD simulations for the same conditions as in (A). (D) Rolling velocity of pairs of leukocytes in a hamster venule (diameter = 22 μm). V/V_∞ represents the dimensionless rolling velocity calculated from the rolling velocity of an isolated bead (or leukocyte), V_∞. (See Color Plate 11)

most likely because, in experiments, the binary system of cells is a necessary approximation and because the distribution of surface ligand is not perfectly uniform.

V. CONCLUSIONS

We have presented numerical and experimental results that aid in understanding hydrodynamic interactions between circulating cells in blood and the tissues that they contact. Specifically, this work elucidates a mechanism for the recruitment of leukocytes in the vicinity of an adherent leukocyte and the behavior of pairs of leukocytes interacting with an adhesive surface under flow. We conclude that hydrodynamic interactions in blood flow play a significant role in the attachment and rolling of leukocytes in microvessels.

One of the ultimate goals of numerical models such as adhesive dynamics is a complete, predictive simulation of blood flow that includes all relevant cellular interactions. Although the dynamics of adhesion is controlled primarily by the chemistry of adhesion molecules and hydrodynamic interactions in blood flow, cellular features such as cell deformability and distribution of receptors or ligands can affect the tethering and rolling behavior (King et al., 2005a). Indeed, recent studies have indicated that neutrophil rolling is significantly slower and relatively smoother than that of microspheres, and that receptors on leukocytes can become rapidly redistributed (Green et al., 2004; King et al., 2005a,c). Furthermore, the distribution of P-selectin on the vascular wall is not uniform (Kim and Sarelius, 2004; King, 2004). Recent work with adhesive dynamics has succeeded in simulating more realistic cellular behavior, including the effects of capping of L-selectin on leukocytes or the effect of nonuniform distributions of adhesion molecules on the endothelial surface. The relation between blood flow-induced hydrodynamic shear forces and chemical adhesive interactions and their roles in the microcirculatory system is an active area of current research, one that is very relevant to the understanding of inflammation and cardiovascular disease. This work investigating the effects of hydrodynamic interactions between blood cells has recently been extended to *branched* postcapillary venules, where most inflammation processes of leukocytes actually occur, and to arterial vessels, specifically the sites of atherosclerosis where pulsatile flow is dominant.

ACKNOWLEDGMENTS

The authors gratefully acknowledge Dr. Phil Knauf for critical reading of the manuscript, Kyle Matthews for data analysis, and the National Institutes of Health for funding (Grant HL10828).

REFERENCES

Alon, R., Fuhlbrigge, R. C., Finger, E. B., and Springer, T. A. (1996) Interactions through L-selectin between leukocytes and adherent leukocytes nucleate rolling adhesions on selectins and VCAM-1 in shear flow. *J. Cell Biol.* **135**: 849–865.

Bell, G. I. (1978). Models for the specific adhesion of cells to cells. *Science* **200**: 618–627.

Bhatia, S. K., King, M. R., and Hammer, D. A. (2003) The state diagram for cell adhesion mediated by two receptors. *Biophys. J.* **84**: 2671–2690.

Burns, M. P., and DePaola, N. (2005) Flow-conditioned HUVECs support clustered leukocyte adhesion by coexpressing ICAM-1 and E-selectin. *Am. J. Physiol. Heart. Circ. Physiol.* **288**: H194–H204.

Chang, K. C., and Hammer, D. A. (1999) The forward rate of binding of surface-tethered reactants: Effect of relative motion between two surfaces. *Biophys. J.* **76**: 1280–1292.

Chang, K. C., and Hammer, D. A. (2000) Adhesive dynamics simulations of sialyl-Lewis(x)/E-selectin-mediated rolling in a cell-free system. *Biophys. J.* **79**: 1891–1902.

Chang, K. C., Tees, D. F., and Hammer, D. A. (2000) The state diagram for cell adhesion under flow: Leukocyte rolling and firm adhesion. *Proc. Natl. Acad. Sci. USA* **97**: 11262–11267.

Coughlin, M. F., and Schmid-Schonbein, G. W. (2004). Pseudopod projection and cell spreading of passive leukocytes in response to fluid shear stress. *Biophys J* **87**: 2035–2042.

Dwir, O., Solomon, A., Mangan, S., Kansas, G. S., Schwarz, U. S., and Alon, R. (2003) Avidity enhancement of L-selectin bonds by flow: shear-promoted rotation of leukocytes turns labile bonds into functional tethers. *J. Cell Biol.* **163**: 649–659.

Finger, E. B., Puri, K. D., Alon, R., Lawrence, M. B., von Andrian, U. H., and Springer, T. A. (1996) Adhesion through L-selectin requires a threshold hydrodynamic shear. *Nature* **379**: 266–269.

Green, C. E., Pearson, D. N., Camphausen, R. T., Staunton, D. E., and Simon, S. I. (2004) Shear-dependent capping of L-selectin and P-selectin glycoprotein ligand 1 by E-selectin signals activation of high-avidity beta2-integrin on neutrophils. *J. Immunol.* **172**: 7780–7790.

Hammer, D. A., and Apte, S. M. (1992) Simulation of cell rolling and adhesion on surfaces in shear flow: General results and analysis of selectin-mediated neutrophil adhesion. *Biophys. J.* **63**: 35–57.

Kim, M. B., and Sarelius, I. H. (2004) Role of shear forces and adhesion molecule distribution on P-selectin-mediated leukocyte rolling in postcapillary venules. *Am. J. Physiol. Heart. Circ. Physiol.* **287**: H2705–H2711.

Kim, S., and Karrila, S. J. (1991) *Microhydrodynamics: Principles and Selected Applications.* Butterworth–Heinemann, Stoneham, MA.

King, M. R. (2004) Scale invariance in selectin-mediated leukocyte rolling. *Fractals* **12**: 235–241.

King, M. R., and Hammer, D. A. (2001a) Multiparticle adhesive dynamics: Interactions between stably rolling cells. *Biophys. J.* **81**: 799–813.

King, M. R., and Hammer, D. A. (2001b) Multiparticle adhesive dynamics: Hydrodynamic recruitment of rolling leukocytes. *Proc. Natl. Acad. Sci. USA* **98**: 14919–14924.

King, M. R., Heinrich, V., Evans, E., and Hammer, D. A. (2005a) Nano-to-micro scale dynamics of P-selectin detachment from leukocyte interfaces: III. Numerical simulation of tethering under flow. *Biophys. J.* **88**: 1676–1683.

King, M. R., Kim, M. B., Sarelius, I. H., and Hammer, D. A. (2003) Hydrodynamic interactions between rolling leukocytes in vivo. *Microcirculation* **10**: 401–409

King, M. R., Rodgers, S. D., and Hammer, D. A. (2001) Hydrodynamic collisions suppress fluctuations in the rolling velocity of adhesive blood cells. *Langmuir* **17**: 4139–4143.

King, M. R., Ruscio, A. D., Kim, M. B., and Sarelius, I. H. (2005b) Interactions between stably rolling leukocytes in vivo. *Phys. Fluids* **17**: Art. No. 031501.

King, M. R., Sumagin, R., Green, C. E., and Simon, S. I. (2005c) Rolling dynamics of a neutrophil with redistributed L-selectin. *Math. Biosci.* **194**: 71–79.

Lawrence, M. B., and Springer, T. A. (1991) Leukocytes roll on a selectin at physiologic flow rates: Distinction from and prerequisite for adhesion through integrins. *Cell* **65**: 859–873.

McEver, R. P. (2002) Selectins: Lectins that initiate cell adhesion under flow. *Curr. Opin. Cell Biol.* **14**: 581–586.

Moazzam, F., DeLano, F. A., Zweifach, B. W., and Schmid-Schonbein, G. W. (1997) The leukocyte response to fluid stress. *Proc Natl Acad Sci USA* **94**: 5338–5343.

Moore, K. L., Patel, K. D., Bruehl, R. E., Li, F., Johnson, D. A., Lichenstein, H. S., Cummings, R. D., Bainton, D. F., and McEver, R. P. (1995) P-selectin glycoprotein ligand-1 mediates rolling of human neutrophils on P-selectin. *J. Cell Biol.* **128**: 661–671.

Sheikh, S., Rainger, G. E., Gale, Z., Rahman, M., and Nash, G. B. (2003) Exposure to fluid shear stress modulates the ability of endothelial cells to recruit neutrophils in response to tumor necrosis factor-alpha: A basis for local variations in vascular sensitivity to inflammation. *Blood* **102**: 2828–2834.

Smith, M. J., Berg, E. L., and Lawrence, M. B. (1999) A direct comparison of selectin-mediated transient, adhesive events using high temporal resolution. *Biophys. J.* **77**: 3371–3383.

Springer, T. A. (1994) Traffic signals for lymphocyte recirculation and leukocyte emigration: The multistep paradigm. *Cell* **76**: 301–314.

St. Hill, C. A., Alexander, S. R., and Walcheck, B. (2003) Indirect capture augments leukocyte accumulation on P-selectin in flowing whole blood. *J. Leukoc. Biol.* **73**: 464–471.

Vestweber, D., and Blanks, J. E. (1999) Mechanisms that regulate the function of the selectins and their ligands. *Physiol. Rev.* **79**: 181–213.

Walcheck, B., Moore, K. L., McEver, R. P., and Kishimoto, T. K. (1996) Neutrophil–neutrophil interactions under hydrodynamic shear stress involve L-selectin and PSGL-1: A mechanism that amplifies initial leukocyte accumulation of P-selectin in vivo. *J. Clin. Invest.* **98**: 1081–1087.

Zhang, Y., and Neelamegham, S. (2002) Estimating the efficiency of cell capture and arrest in flow chambers: Study of neutrophil binding via E-selectin and ICAM-1. *Biophys. J.* **83**: 1934–1952.

13

DYNAMICS OF PLATELET AGGREGATION AND ADHESION TO REACTIVE SURFACES UNDER FLOW

NIPA A. MODY and **MICHAEL R. KING**

Department of Biomedical Engineering, University of Rochester, Rochester, New York, USA

Platelets are the key participants in thrombotic events that take place following vascular injury. These tiny ellipsoid blood cells are recruited to the exposed subendothelial region at the vascular lesion and aggregate to form a hemostatic plug that blocks further blood loss and seals the wound. The shear effects of hemodynamic flow play an important role in governing the ability of platelets to (1) contact, tether to, and adhere at the subendothelial surface, (2) become activated, and (3) form spontaneous platelet aggregates in the blood. Hemodynamic forces influence the binding kinetics of platelet–surface molecular bonds and regulate platelet function in different regions of the vasculature such as arterioles and veins, which are subject to very different flow conditions. Any abnormalities in the flow pattern and shear rate can cause irregularities in platelet behavior giving rise to a variety of pathological consequences ranging from minor to life-threatening. Examples of cardiovascular or inflammatory diseases resulting from pathological thrombosis are venous stasis, atherosclerosis, heart attacks, and stroke. The purpose of this chapter is to familiarize the reader with the physiological process of platelet adhesion, the key receptors and ligands involved, and the effects of shear flow on platelet motion, adhesion, and aggregation. Current understanding of the biophysical aspects of platelet adhesive dynamics, pathological consequences arising from abnormalities either in participating receptor–ligand binding properties or in flow behavior, and recent theoretical studies that probe further into the effects of shear flow on platelet adhesive phenomena are discussed here.

I. INTRODUCTION

Platelet tethering and deposition at sites of vascular injury is the critical primary step in the hemostatic cascade — the natural mechanism used by the body to seal wounds and curtail further blood loss. *Hemostasis* describes the physiological process of platelet adhesion to the injured vascular endothelium, platelet accumulation at this site, and activation of the coagulation cascade to convert fibrinogen to fibrin for solidification and strengthening of the blood clot. Platelets display an abundance of adhesion receptors on their surface to mediate platelet deposition and aggregation at the site of lesion. They also harbor various chemical agonists and important signaling molecules in intracellular storage granules, which they release at the opportune moment to create a strong pro-inflammatory environment at the region of injury. While the key physiological role of platelets is widely known, promotion of healthy blood clotting, the role of platelets in the origin and progression of pathological inflammation is now being uncovered. Platelets have been implicated in a number of chronic and acute inflammatory and cardiovascular diseases such as atherosclerosis, deep vein thrombosis, heart disease, and stroke. To understand the critical factors that promote thrombotic events, both normal and pathological, it is important to determine and quantify both the binding kinetics of the participating receptors and their ligands and the influence of the flow environment, such as flow rate, shear stress, hydrodynamic forces, and flow fields, on the initiation and progression of platelet tethering, activation, and aggregation events. The shear flow acting on platelets plays an important role in determining (1) the frequency with which platelets contact the vascular wall and (2) the hydrodynamic forces acting on the platelet and on associating or dissociating bonds. The magnitude of shear forces also influences the force-dependent binding kinetics of the receptor–ligand bonds that participate in the platelet tethering process. Abnormally high shear rates can also cause platelet activation and thrombus formation, even in the absence of a physiological thrombotic signal. Much research has focused on decoding the signaling mechanisms in the platelet activation process, with an emphasis on understanding integrin activation, granule secretion, and the factors that promote the cytoskeletal rearrangements that accompany platelet shape change during activation. By gaining more knowledge about the physical and biochemical factors that induce thrombotic events in the vasculature, we can be better positioned to control or prevent pathological thrombosis, and improve the condition of patients with hemorrhagic disorders or chronic inflammation.

The aim of this chapter is to familiarize the reader with current concepts in the field of platelet tethering under flow and to discuss recent investigations into the effects of shear flow on the dynamics of platelet flow and adhesion. Any in-depth study of the effects of hydrodynamic shear flow on the motion and tethering patterns of platelets requires both experimental characterization of the distinct behavior of individual platelets in a shear flow field and development of a mathematical flow model to quantify many of the hemodynamic and adhesion parameters. Such theoretical models consist of a mathematical description of the physical flow and biochemical kinetic phenomena, and the results obtained have great utility in further interpreting and defining experimental observations. Experimentally verified computational models can be used to predict platelet adhesive behavior for different scenarios such as different types of flow fields, different cell sizes, and different binding kinetics, for example, for a mutant receptor versus a normal receptor. In theory, by being able to quantify the various factors affecting platelet motion and adhesive behavior and by having a clear understanding of the concepts behind this physiological process, one can design the biological tools necessary to control any abnormality that disrupts this system.

The integrative study of the adhesive behavior of platelets near the vessel wall is a multidisciplinary problem and requires knowledge of the molecular properties of participating receptors and their ligands, solutions to the relevant hydrodynamic flow problem, normal and pathological behaviors of platelets and participating molecules during inflammation, effects of hydrodynamic forces on the kinetics of bond formation and breakage, effects of size and shape of the platelet on its flow properties, and effects of other flowing cells on platelet motion and ability to contact the surface. It is also important to take into account the conditions under which platelets are activated and the mechanisms of signal transduction that ultimately control the fate of the resting (unactivated) platelet.

II. A CLOSER LOOK AT PRIMARY HEMOSTASIS: PLATELET ADHESION TO THE INJURED VESSEL WALL

A. Overview of Primary and Secondary Hemostasis

The primary inflammatory response of the body to an injury of the vessel wall involves the recruitment of circulating platelets from the bloodstream to the injured endothelial surface, and this is followed by platelet activation and clot formation. The mechanisms involved in platelet binding to the damaged endothelium, subsequent platelet aggregation, and thrombus formation are the same whether the inflammatory stimulus is physiological or pathological (Fig. 13.1).

The platelet initiates contact with the injured region via reversible tether bonds that allow the platelet to attach to, and detach from, the broken vessel wall. This tethering process slows the platelet down considerably and enables the platelet to translate or "translocate" at a velocity much lower than that of the ambient blood flowing past it. As the platelet binds and detaches from the surface, it is also able to sample other molecules present on the subendothelium. The platelet becomes activated during this process, so that it can now perform two functions. The first involves secretion of various chemicals from α and dense granules housed within the cell. Release of the granules' contents to the platelet exterior is called *degranulation,* and the stored molecules in these granules include fibrinogen, coagulation factors such as factors V and VIII, ADP, growth factors such as PGDF and transforming growth factor β (TGF-β), histamine, Ca^{2+}, and certain neurotransmitters such as serotonin and epinephrine (Cotran et al., 1994). ADP binds to appropriate platelet G-protein–coupled receptors (GPCRs), which initiate a signaling

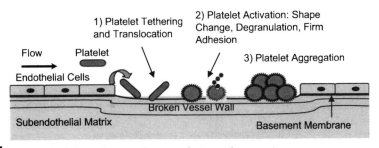

FIGURE 13.1 Schematic diagram of primary hemostasis.

cascade resulting in cytoskeletal reorganization and enhanced platelet secretion. This signaling results in full-blown platelet activation, allowing platelets to undergo platelet–platelet aggregation via their $\alpha_{IIb}\beta_3$-integrin receptors (Oury et al., 2004). This receptor is indispensable for platelet aggregation, and any congenital defect preventing normal functioning of this receptor precludes clot formation. The second function of the activated platelet is to form firmly adherent bonds with the subendothelial surface that lead to strong platelet attachment at the wound site, and platelet–platelet bonds that enable platelets to aggregate. The activation of platelets brings about a change in shape of the platelet from flat ellipsoid to globular, with many filopodium-like protrusions. Once a single layer of platelets has formed an adherent layer on the lesioned vessel wall, a subsequent stack of platelets aggregate onto the previously deposited layer. Such events of the hemostatic cascade, such as the platelet release reaction, formation of an enlarging platelet aggregate, subsequent platelet contraction, activation of the coagulation cascade, and finally thrombin production and fibrin deposition, take place during the secondary or tertiary stages of hemostasis, and are not the prime focus of this chapter.

B. Platelet Tethering via GPIbα–Von Willebrand Factor AI Binding

The dynamics of platelet tethering to the vessel wall, a key primary event during inflammation or vascular injury, can be envisioned as a "tug-of-war" between (1) fluid forces pushing adherent cells toward the direction of flow, and (2) adhesive forces generated by molecular bond tethers preventing cell detachment from the surface (Fig. 13.2).

There are many factors that govern the outcome of this force balance enabling one of the two to prevail. These influencing factors include, among others, fluid shear; kinetic properties of the receptor bonds; size, shape, and orientation of the cell; and dynamic interactions of the bound cell with other bound and unbound flowing cells. The flow environment is critical in determining not only the fate of these platelet–surface tether bonds, but also the frequency, manner, and duration of platelet contact with the surface and, hence, the probablility of tether formations. The role of tether bonds is to provide the grip by which these ellipsoid cells can adhere temporarily to the subendothelial lining of the blood vessel, even under rapid blood flows. The transient pausing enables other slower molecular platelet–surface interactions to occur.

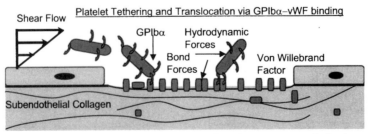

FIGURE 13.2 Schematic diagram of the flow-driven process of platelet tethering to, and translocation on, the subendothelial surface at the site of vascular injury.

Initial platelet contact with the exposed subendothelial components at physiological shear rates in arterioles (500–2000 s^{-1}), and even at pathological shear rates in stenosed regions (2000–10,000 s^{-1}) of the vasculature, is mediated by the platelet glycoprotein GPIbα surface receptor, which exists as a subunit of the GPIb–IX–V complex. The GPIbα receptor binds the A1 domain of collagen-bound von Willebrand factor (VWF) (Fig. 13.2). The latter is a large multimeric glycoprotein, with repeating identical subunits (2–100), and is secreted by platelets at the time of injury and also by endothelial cells. The latter cells have both a constitutive pathway of secretion and a regulated pathway involving storage of VWF molecules in granules and subsequent release triggered by endothelial activation. Secretion of VWF is directed either into the circulating blood or into the extravascular space (Ruggeri, 1993; Medolicchio and Ruggeri, 2005). Thus, VWF is found both in blood plasma and in certain regions of the vascular subendothelium. This multimer binds collagen types I and III through its A3 domain and collagen type VI through its A1 domain, and can also bind the platelet integrin receptor α$_{IIb}$β$_3$ (also called the GPIIb–IIIa complex) via an RGD (arginine–glycine–aspartic acid) amino acid sequence in its C1 domain (Ruggeri, 1993). GPIbα–VWF-A1 tether bond formation is critically important for enabling platelets to initiate binding to the injured exposed subendothelial surface, albeit reversibly.

The cytoplasmic domain of the transmembraneous GPIbα receptor interacts extensively with the platelet cytoskeleton. GPIbα, on association with bound VWF, can signal this binding event to the cell to activate the platelet and initiate thrombosis. This outside-in signaling is particularly important when no exogenous agonists are available at the start of hemostasis or when other signaling integrin receptors such as α$_2$β$_1$ have not had an opportunity to bind their ligands. This receptor is notably sensitive to shear as described in later parts of this section and is capable of sensing high or low shear and, accordingly, modulating cell behavior based on the flow characteristics of blood in its immediate vicinity. Therefore the GPIbα receptor is viewed as a biomechanical transducer that has the unique and intriguing property of linking shear forces in the surrounding fluid to intracellular signaling. The mechanisms by which GPIbα converts shear stress signals into biochemical signals are not yet fully known.

1. Platelet Translocation on Surface-Bound VWF

When extracellular matrix components are suddenly exposed to flowing blood in the event of an arteriolar injury, plasma VWF multimers quickly bind to collagen fibrils, and platelets flowing near the vessel wall begin making first contacts with the bound VWF molecules. The GPIbα–VWF-A1 bonds form rapidly because of the bond's intrinsic fast association rate and high tensile strength. These bonds have a short half-life and allow the platelet to quickly detach from the surface. Unbound platelets in shear flow rotate as they translate in the direction of flow. Platelets attached to the surface by reversible tether bonds can be observed to translocate or roll over the surface in the presence of a shear flow field in both *in vivo* and *in vitro* settings. This motion can be likened to the well-known selectin-mediated rolling of leukocytes observed at the start of inflammation, prior to integrin-mediated firm adhesion and transmigration of the spherical cells into the extravascular tissue. Platelets are flattened ellipsoid cells and, therefore, do not translocate by "rolling" but instead by flipping, such as, by analogy, the flipping of a coin. This flipping motion is a consequence of fluid shear action on the platelet, the formation of new tether bonds at the downstream edge of the platelet, and dissociation of old tether bonds at the upstream edge. The force mechanics involved in bond formation and breakage in the case of platelets are distinctly different from those experienced by leukocyte bonds, because of the great dissimilarity in shapes of both cells despite similarities in kinetic attributes.

2. Bond Kinetics

The GPIbα–VWF-A1 tether bond has been shown to exhibit selectin-like binding kinetics. These include both fast association and dissociation rates, strong dependence of off-rate on the magnitude of force acting on the bond, and requirement of a critical level of shear flow for adhesion to occur (Doggett et al., 2002). The kinetics of bond dissociation of GPIbα and VWF-A1 follows the Bell model characteristics. The Bell model is an expression for the force dependence of the dissociation rate of weak noncovalent bonds and is given by

$$k_{\text{off}}(F) = k_{\text{off}}^{\text{o}} \exp\left[\frac{\gamma F}{k_{\text{B}} T}\right], \tag{13.1}$$

where $k_{\text{off}}(F)$ is the bond dissociation (off) rate, $k_{\text{off}}^{\text{o}}$ is the unstressed off-rate, γ is the reactive compliance, F is the applied force on the bond, and $k_{\text{B}} T$ is the product of the Boltzmann constant and temperature (Bell, 1978). The reactive compliance represents a transition state position and describes how strongly the dissociation rate changes with force (Tees et al., 2001). The physiological importance of the characteristic binding kinetics of the GPIbα–VWF-A1 tether bond is at least twofold. First, the fast rate of association and high tensile strength of the bond enable fast-flowing platelets to attach to the surface, even at moderate to high shear rates (1000–2000 s^{-1}) as commonly experienced in arteriolar flow. Second, the fast dissociation rate effectively lowers the translational velocities of platelets to well below the free-stream velocity, thus providing more time for other platelet receptors with slower bond formation rates to interact with specific subendothelial surface molecules. The reduction in platelet velocities is essential to allow sufficient time for bond formation of activated $\alpha_{\text{IIb}}\beta_3$ with surface-bound VWF. Another useful feature of the high dissociation rate of the bond is that it may prevent flowing platelets from clustering with circulating VWF in the blood, thus ensuring that these binding partners are available for interacting at the lesioned region only when "injury calls."

The physiological significance of GPIbα–VWF-A1 bond kinetics is best illustrated by the genetic bleeding disorder, von Willebrand disease (VWD). This disease is characterized by prolonged bleeding time, and is the most common hereditary hemorrhagic disorder in humans. VWD is the result of a gain-of-function mutation either in the A1 domain of VWF (known as 2B-type VWD) or in the GPIbα receptor (known as platelet-type VWD). Reduced concentrations of plasma VWF may also cause VWD. Doggett et al. (2002, 2003) investigated the binding kinetics of GPIbα–VWF-A1 binding in which either the A1 domain mutation I546V or the GPIbα mutation

Gly233Val was present. They found that in both cases, the mutant complex exhibited a five to six times reduction in the dissociation rate constant as compared with the wild-type complex and loss of the shear threshold phenomenon.[1] The reactive compliance of the mutant variety was only slightly affected and was found to be twice that of the native receptor–ligand pair. The mutant bond is therefore more sensitive than the native bond pair to force-associated bond breakage at high shear stresses than at lower shear stress. These increased stabilizing interactions between the mutant GPIbα receptor and VWF permit spontaneous binding of flowing platelets with soluble VWF, which depletes the availability of VWF, especially large VWF multimers, for binding to subendothelial components during injury. Thus, patients with VWD develop minor to moderate bleeding problems as a result of platelet–VWF association in the blood rather than at the injured surface.

Platelets have an abundance of GPIbα receptors on their surface, approximately 25,000 copies per cell. The multimeric nature of VWF provides numerous A1 binding sites to enable efficient bond formation with several GPIbα surface receptors on a single platelet, and this is most pronounced in the case of large VWF multimers. Ultralarge VWF multimers have enhanced thrombogenic potential, because compared with smaller VWF multimers, they form much stronger bonds with platelets. The formation of multiple bonds helps strengthen the weak GPIbα–VWF interaction (Cauwenberghs, et al., 2000). This receptor–ligand binding dynamic is akin to the Velcro concept: one weak or low-affinity bond provides poor adhesive properties, but when many weak bonds are formed simultaneously, the binding stability is significantly increased. Cruz et al. (2000) have shown that despite the weak character of the bond, the binding of only one GPIbα platelet receptor with a single VWF-A1 binding site can be sufficient to enable stable platelet–surface tethering in a flow environment.

C. Platelet Arrest, Activation, and Aggregation and the Role of $\alpha_{IIb}\beta_3$

Irreversible or firm adhesion of platelets onto the injured surface, as well as aggregation of platelets, is mediated by the $\alpha_{IIb}\beta_3$ (GPIIb–IIIa complex) integrin receptor. With 50,000 copies present per platelet, this receptor is the most abundant integrin receptor on the surface of this microscopic blood cell. The $\alpha_{IIb}\beta_3$ receptor binds fibrinogen (a plasma protein), VWF, fibronectin, and vitronectin, through recognition of an RGD sequence on the ligand (Weiss et al., 1989). VWF and fibrinogen are the key ligands that bind to $\alpha_{IIb}\beta_3$ and mediate firm platelet–surface adhesion; fibronectin and vitronectin are believed to play a small role, if any. Both VWF and fibrinogen are also the sole multimeric ligands that participate in platelet–platelet binding.

It was previously believed that fibrinogen was the single necessary ligand that bound $\alpha_{IIb}\beta_3$ and mediated platelet aggregation. However, Ruggeri and

---

[1]The platelet receptor GPIbα or leukocyte receptor L-selectin abstains from forming bonds with its respective surface ligands when subjected to a shear stress less than a certain "threshold" value. Above this threshold value, these blood cells tether to the surface by forming the appropriate bonds. This phenomenon whereby cell tethering occurs only at or above a critical value of shear is called the *shear threshold phenomenon*.

co–workers have shown that *both* VWF and fibrinogen are required for stable clot formation. These two molecules act in a synergistic fashion over the physiological range of shear rates encountered in the arteries, with VWF playing an increasingly important role at higher shear rates (Ruggeri et al., 1999). At low shear rates (600–900 s^{-1}) fibrinogen effectively mediates platelet adhesion or aggregation, but at high shear rates of flow in the arterioles, VWF is the sole ligand that participates in $\alpha_{IIb}\beta_3$ receptor-mediated irreversible adhesion and platelet aggregation (Weiss et al., 1989). Under such conditions of flow, there is no involvement of fibrinogen in platelet adhesion or clot formation (Weiss et al., 1989). It has been shown that there may also be positive interactions between VWF and fibrin that strengthen the adhesion of platelets to fibrin and stabilize the clot at the wound site, as VWF appears to bind fibrin and undergo crosslinking in the fibrin chains (Ruggeri et al., 1983).

In the resting platelet, the $\alpha_{IIb}\beta_3$ receptor maintains an inactive conformation and cannot bind surface-bound VWF and firmly attach to the surface via $\alpha_{IIb}\beta_3$-vWF permanent bonds. Also, the unstimulated $\alpha_{IIb}\beta_3$ receptor does not interact with soluble (circulating or unbound) VWF or fibrinogen and, therefore, is incapable of mediating interplatelet bond formation. Stimulation of this receptor is achieved either by platelet binding to extracellular mediators such as ADP, thrombin, and thromboxane A_2, which are secreted by activated platelets, by platelet binding to collagen via its $\alpha_2\beta_1$ receptor, or by intracellular signaling promoted by GPIbα–VWF binding. On the other hand, $\alpha_{IIb}\beta_3$ can bind insoluble or immobilized fibrinogen to form irreversible adhesive bonds without prior activation. This interaction subsequently converts all unbound $\alpha_{IIb}\beta_3$ into the activated state, rendering it capable of binding soluble ligands such as fibrinogen and VWF, allowing the platelet to aggregate with other activated platelets (Savage et al., 1992).

A common route of platelet activation involves binding of GPIbα to VWF. This induces phospholipase C activation, which results in elevated cytosolic Ca^{2+} concentrations (Ikeda et al, 1993) and protein kinase C activation (Kroll et al., 1993). The signaling mechanisms that lead to platelet activation and aggregation are not fully known. However, it has been shown that with the generation of a Ca^{2+} peak, the $\alpha_{IIb}\beta_3$ integrin receptor becomes capable of binding surface-bound VWF firmly to arrest the translocating platelets onto the surface. Moreover, the partial degranulation of the platelet contents accompanying the Ca^{2+} peak results in secretion of ADP, which then binds to specific receptors on the platelet surface, initiating a second and more intense Ca^{2+} peak. This fully activates $\alpha_{IIb}\beta_3$, allowing the unoccupied receptors to bind either soluble fibrinogen or plasma VWF. At elevated shear rates, binding of soluble VWF to activated $\alpha_{IIb}\beta_3$ on adherent platelets is usually preceded by VWF–GPIbα interactions. Other platelets flowing nearby or *translocating* on the surface undergo $\alpha_{IIb}\beta_3$-mediated attachment to the adherent activated platelets by interacting with platelet-bound fibrinogen or VWF (Fig. 13.3).

In this manner, incoming platelets stack up on the previously deposited platelet layer(s), leading to formation of platelet aggregates. Often, flowing platelets first bind to adherent platelets containing bound VWF via their GPIbα receptors, and then subsequently bind via their integrin receptors when they have sufficiently decelerated. The importance of the $\alpha_{IIb}\beta_3$-integrin receptor is demonstrated in patients with the congenital hemorrhagic disease Glanzmann's thrombasthenia; such patients are deficient in the $\alpha_{IIb}\beta_3$-integrin

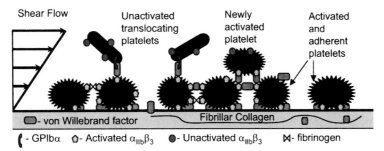

FIGURE 13.3 Schematic diagram of the step-by-step process of the capture of flowing platelets into an enlarging platelet aggregate. This figure depicts the various platelet–platelet and platelet–surface bonds that are formed during the process.

cell surface receptor, and because of this their platelets only weakly participate in physiological thrombus formation.

D. Other Platelet Surface Receptors Involved in Platelet Adhesion

Platelets express a variety of integrins on their surface, the most important of which is the $\alpha_{IIb}\beta_3$-integrin, the role of which in thrombosis was discussed in Section II.C. These various platelet surface integrins also participate in platelet binding to the subendothelial surface and include at least three types of β_1-integrin and two types of β_3 integrin surface receptors. The $\alpha_2\beta_1$ (GPIa–IIa) surface integrin receptor mediates binding with collagen types I–VIII and has approximately 1000 copies present on the platelet (Konstantopoulos et al., 1998). $\alpha_5\beta_1$ binds fibronectin and laminin, and $\alpha_6\beta_1$ interacts with laminin in the extracellular matrix. The β_3-integrins present on the platelet surface are $\alpha_V\beta_3$, which can bind vitronectin and $\alpha_{IIb}\beta_3$. Although $\alpha_5\beta_1$, $\alpha_6\beta_1$, and $\alpha_V\beta_3$ have been implicated in a number of platelet adhesive events, their essentiality in, as well as their level of contribution to, the hemostatic process is yet to be determined. However, these receptors may play increasingly important roles when $\alpha_{IIb}\beta_3$ and GPIbα contributions to platelet surface deposition are considerably decreased due to genetic disorders that result in the diminished functionality of these molecules (Konstantopoulos et al., 1998).

The specific receptor–ligand combinations that mediate platelet adhesion, activation, and aggregation are determined by both the local hydrodynamic forces and the composition of the subendothelium exposed at the injured site. To understand the complementary roles of various platelet receptors and their ligands in promoting thrombosis, it is useful to look at platelet binding characteristics at both low and high shear rates. More often than not, different receptor proteins or ligands mediate *either* low or high shear thrombus formation, but not both.

1. Thrombus Formation at Low Shear Rates (20–500 s^{-1})

Low shear rates are common in large arteries and veins. Both fibronectin and laminin, which are present in the subendothelium, support firm platelet adhesion at low shear rates by binding to corresponding β_1-integrin receptors on the platelet surface. Platelets adhere and aggregate on type VI collagen at low shear rates (100 s^{-1}) via VWF bridging (Ross et al., 1995). This collagen type, on the other hand, does not support platelet adhesion at arterial shear stresses (1000 s^{-1}). Fibronectin binding can induce platelet activation but is less efficient in promoting adhesion than VWF or fibrinogen (Grunkemeier et al., 2000). GPIbα– or $\alpha_2\beta_1$-mediated adhesion is still common, even at flow rates of $\sim$100 s^{-1}. However, tether bond formation via GPIbα–vWF bonds is not a requirement for adhesion to occur. Conversely, $\alpha_{IIb}\beta_3$ easily binds fibrinogen under these flow conditions, and clots formed in these regions of the vasculature are rich in fibrin.

2. Thrombus Formation at Moderate to High Shear Rates ($\geq$1000 s^{-1})

Shear rates of >500 s^{-1} are typically observed in arterioles. Here again, the β_1- and β_3-integrins support firm platelet adhesion on the surface, but these bonds can be formed only after the platelet has tethered to the surface via GPIbα–VWF interactions. It has been observed that at high shear rates, blocking GPIbα rather than $\alpha_{IIb}\beta_3$ interactions is much more effective in inhibiting platelet accumulation on collagenous surfaces. This observation demonstrates that, first, GPIbα–mediated tethering of platelets to the surface is an indispensable step preceding platelet arrest onto the surface, and second, other platelet surface receptors besides $\alpha_{IIb}\beta_3$ actively participate in firmly attaching the cell to the surface (e.g., by binding collagen; examples of such receptors are $\alpha_2\beta_1$ and GPVI, a nonintegrin glycoprotein receptor on the platelet surface). However, blocking $\alpha_{IIb}\beta_3$ functioning does completely abolish platelet aggregation and, thus, shows the criticality of this receptor for thrombus growth (Savage et al., 1998). $\alpha_2\beta_1$ contributes to irreversible platelet adhesion and outside-in platelet signaling at elevated shear rates, but again requires the initial GPIbα-mediated platelet contact with the surface to act as a brake on the fast platelet flow velocities, allowing $\alpha_2\beta_1$ sufficient time to form bonds with collagen. The contribution of $\alpha_2\beta_1$ to stable platelet attachment to the surface becomes increasingly important under conditions of low surface density of collagen on which $\alpha_{IIb}\beta_3$, by itself, is insufficient to anchor the platelet firmly to withstand the high fluid shear stresses (Savage et al., 1998).

E. Platelet Rolling on Stimulated Endothelium

Platelets can transiently tether and also roll on activated venular endothelium *in vivo* as a result of coordinated formation and breakage of tether bonds between endothelial P-selectin and an as yet undefined platelet ligand. Platelets constitutively express P-selectin ligand on their surface and do not require activation to exhibit this rolling behavior. Their rolling velocities on stimulated endothelium were found to be roughly an order of magnitude greater than that of leukocytes (Frenette et al., 1995). Activated platelets, unlike resting platelets, have P-selectin redistributed from the intracellular α granules to the platelet surface and are capable of binding with leukocytes and rolling together, as observed *in vivo* in mice venules (Frenette et al., 1995). The corresponding pathological consequence of platelet–leukocyte binding is demonstrated by the chronic inflammatory disease venous stasis, in which proper functioning of the veins in the legs is gradually lost. Those with venous stasis have been shown to possess significant concentrations of circulating monocyte–platelet aggregates in their bloodstream, where the aggregation of monocytes and platelets is dependent on platelet activation.

III. IMPORTANCE OF SHEAR IN PLATELET ADHESIVE MECHANISMS

The mechanical fluid forces caused by the shear flow of blood affect platelet function *in vivo*, and by determining the link between platelet function and shear stress, one can better comprehend the various biophysical mechanisms governing thrombosis under conditions of blood flow. The physical and mathematical definitions of shear rate and shear stress and the differences between the two deserve clarification. *Shear rate* is, by definition, the rate of increase or decrease in velocity of a fluid flowing in the x direction per unit distance in the y direction, and has units of reciprocal time. In the simplest scenario, laminar flow over an infinite planar surface is linear shear, and Fig. 13.4A shows that for flow in the x direction (parallel to the wall), the velocity changes linearly with distance in the z direction (normal to the wall). The shear rate is defined as dv_x/dz, which is equal to a constant because v_x is a linear function of distance from the wall. There are no velocity variations in the y direction. As another example, laminar flow in a circular tube gives rise to a parabolic velocity profile such that the velocity v at any point is only in the z direction (axial) and is equal to

$$v = v_{max}\left(1 - \left(\frac{r}{R}\right)^2\right),$$

where v_{max} is the velocity of the fluid at the centerline of the tube and r is the radial coordinate. Figure 13.4B describes the geometry of this system. The velocity is a function of the radial position only. The shear rate in this geometry is calculated as

$$\frac{dv_z}{dr} = v_{max}\left(\frac{-2r}{R^2}\right).$$

The negative sign exists because of the definition of the radial coordinate, set as zero at the centerline. Shear stress is given by the product of fluid viscosity and shear rate, and can be written mathematically as

$$\tau_{yx} = \mu\frac{dv_x}{dy},$$

and has units of force per unit area such as Newtons per square meter (Pascals) or, as commonly used in biomedical flow experiments, dynes per square centimeter. For particulate flow such as blood flow, shear rate

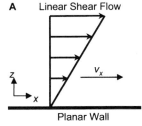
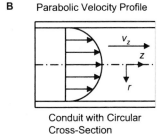

A Linear Shear Flow

v_x

Planar Wall

B Parabolic Velocity Profile

v_z

z

r

Conduit with Circular Cross-Section

FIGURE 13.4 Schematic diagrams of simple flow scenarios used as examples to calculate shear rate. (A) Linear shear flow of a viscous fluid over an infinite planar surface. (B) Fully developed laminar (parabolic) flow in a pipe with circular cross section.

physically determines the rate at which cells collide with each other and the duration of the collisions, whereas shear stress dictates the magnitude of hydrodynamic normal and shear forces that act on the cells, cell surfaces, or cell aggregates. Shear stress increases proportionately with shear rate and therefore in arterioles, where velocities are faster, the drag forces — the sum total of shear and normal forces — on blood cells are much greater in this region of the vasculature than in veins or large arteries. It should be noted that close to the wall in large arteries, where cell radius is much smaller than vessel radius and distance from the wall is much smaller than vessel radius, curvature of the vessel wall can be neglected and the tube geometry reverts back to linear shear flow near a planar wall.

As platelets must encounter a wide range of shear stresses in the vasculature, these cells have developed numerous pathways that depend on the level of shear to regulate their adhesive behavior and therefore either propagate or limit thrombosis in the body. The key is to determine the effects of shear flow on platelet adhesive mechanisms at both slow and fast flow rates and then, by comparing and contrasting this behavior, distinguish the role played by shear forces in promoting thrombus formation from the intrinsic properties of the platelet that are critical to normal platelet function. For example, GPIbα cannot bind VWF under static conditions (no flow), and shear flow is found to be a requirement for this association to occur. Any abnormalities in either the platelet properties or the flow behavior can have pathological consequences ranging in severity from minor to life-threatening.

A. Shear Effects on Platelet Translocation

Platelet translocation velocities on surface-bound VWF depend on several factors such as binding kinetics, shear rate, and platelet size and shape. *In vitro* platelet rolling velocities on immobilized VWF were found to be an order of magnitude lower than leukocyte rolling velocities on selectins at comparable shear rates (Savage et al., 1996). One of the reasons for this observation is the smaller size of the platelet (2 μm in diameter) compared with a leukocyte (8 μm), which causes the platelet and the bonds formed with the surface to experience much smaller normal and shear forces in the direction of flow. If shear stress is reduced to below 0.73 dyn/cm^2, platelets cease to interact with bound VWF (Doggett et al., 2002). This fluid shear-dependent interaction has been shown to be rapidly reversible. That is, platelet tethering commences immediately when the shear stress is raised above the critical value, and disappears when it is lowered below the threshold value. Moreover, GPIbα–VWF binding does not undergo any kind of lasting "memory" effect as a result of changes in shear stress (Doggett et al., 2002; Shankaran et al., 2003).

B. Shear-Induced Thrombosis and Its Pathological Consequences

Abnormally high shear rates in the range of 5000–20,000 s^{-1} are typical of stenosed (constricted or narrowed) regions of the arteries. Such high shear flows have been found to induce platelet activation and aggregation both *in vivo* and *in vitro,* even in the absence of any chemical agonists or vascular lesions, resulting in the formation of platelet thrombi that may occlude blood flow through the stenosed region. It is the large fluid shear stresses, rather than high shear rates, that are responsible for inducing thrombosis, because it is the magnitude of hydrodynamic forces that controls platelet activation and not the rate and duration of cell collisions (Shankaran et al., 2003). Very high levels of shear stress (>80 dyn/cm^2) (Ikeda et al., 1991; Alevriadou et al., 1993; Shankaran et al., 2003) have been shown to promote initial GPIbα–VWF binding (with no involvement of a surface (Konstantopoulos et al., 1997)), which leads to shear-induced activation of platelets and subsequent platelet aggregation via $\alpha_{IIb}\beta_3$–VWF binding. Participation of both GPIbα and $\alpha_{IIb}\beta_3$ is necessary for the formation of stable shear-induced platelet aggregates (Goto et al., 1995; Konstantopoulos et al., 1997). VWF is the only ligand that has been found capable of supporting *in vitro* platelet aggregation at high shear levels (Ikeda et al., 1991; Goto et al., 1995).

Shear-induced thrombosis has great pathological significance in that it is responsible for the thrombotic occlusion of stenosed atherosclerotic arteries. Atherosclerosis is a type of arteriosclerosis that involves deposition of low-density lipoprotein (LDL) cholesterol, triglycerides, plasma proteins such as fibrinogen and albumin, cellular products, calcium, and others, in arterial lesions or on dysfunctional endothelium, leading to the thickening and hardening of the arteriolar wall. This buildup of fatty substances, called *plaque,* leads to the progressive narrowing of the arterial lumen. High shear–induced platelet activation and formation of large platelet thrombi can critically block blood flow through the narrowed opening of the lumen, and such occurrences in coronary or cerebral arteries may result in ischemic heart disease and unstable angina (due to low oxygen supply to the heart) or stroke (due to reduced blood flow to the brain), respectively. Atherosclerosis is itself a chronic inflammatory disease and involves substantial monocyte and platelet activity at the site of the lesion. Atherosclerotic lesion-prone areas within the coronary and cerebral vasculature are high-risk sites for pathological thrombosis, and rupture of plaque or secretion of inflammatory stimuli can result in recruitment of platelets to the ruptured site and thrombus formation at the arterial wall that further obstructs blood flow. Thrombosis in advanced lesions can lead to unstable coronary syndromes or myocardial infarction. Cardiovascular diseases caused by atherosclerosis are the leading cause of illness and death in the United States (Ross, 1999). Coronary angioplasty, a well-established procedure used to widen atherosclerosed arteries, can cause injuries to the arterial wall, instigating platelet deposition and thrombus formation and resulting in restenosis of the treated vessel. Drug-coated stents that supply antithrombotic agents at a controlled rate have recently been developed for use in angioplasty treatment. Interestingly, the signaling pathway used by GPIbα to activate the platelet in response to elevated shear stress is observed to be different from the signaling mechanisms used by agonist-bound platelet receptors. This may provide a path to development of antithrombotic drugs that prevent shear-induced pathological thrombosis at atherosclerosed regions, while enabling platelets to engage in their hemostatic role (Kroll et al., 1993).

C. The Connection between Shear Rate and GPIbα–VWF-A1 Binding

The GPIbα–VWF-A1 bond exhibits a remarkable property similar to that of L-selectins — the shear threshold effect. Platelets can transiently tether or translocate on surface-bound VWF only above a critical level of shear stress equivalent to 0.73 dyn/cm^2 (Doggett et al., 2002). Most importantly, platelets do not bind to circulating vWF in the blood except at very high shear rates, at which platelets become activated, resulting in the formation of thrombi. Thus, it appears that one or both molecules have certain physical attributes such that the binding properties of the molecular pair can be modulated based on the shear level in the fluid. To explain this interesting link between GPIbα–VWF bond formation and shear stress, some researchers have hypothesized that the shear force can sufficiently perturb the conformational structure of the platelet receptor and/or VWF molecules to expose the appropriate binding site(s), and when this happens, bond formation is greatly favored. Surface-bound VWF has been shown to undergo protein unfolding as a result of shear stress. At shear stresses greater than 35 dyn/cm^2, transiently attached VWF transforms from a globular coil into an extended chain conformation (Siedlecki et al., 1996), however, this occurs when VWF is bound to a surface via nonspecific adhesion and not necessarily for the case of collagen-bound VWF, a scenario that is physiologically relevant (Novák et al., 2002).

Some researchers have also postulated that the binding of VWF to collagen may induce conformational changes that can also expose GPIbα binding sites for interactions with platelets, and that this may be sufficient to explain why platelets do not bind circulating VWF. Recent experimental evidence contests this line of argument. First, experiments in which platelets could bind to the A1 domain of a mutant VWF molecule that could not bind to collagen, but could noncovalently associate with A1 domain-negative VWF molecules bound to collagen type I fibrils, demonstrated the ability of VWF molecules to reversibly self-associate (Savage et al., 2002). Thus, platelets were shown to be capable of binding VWF not directly associated with collagen. Second, platelet translocation mediated by GPIbα on either VWF immobilized on glass or plasma VWF bound to collagen was shown to occur at the same velocities, regardless of the manner in which VWF is bound (Doggett et al., 2002; Savage et al., 1998). Thus, the binding kinetics of GPIbα–VWF is not governed by conformational changes of the VWF-A1 domain on binding to collagen.

Circulating VWF multimers can undergo homotypic association in a noncovalent, shear-dependent manner to form large VWF-sized multimers, with higher shear rates correlating with a larger average multimer size (Shankaran et al., 2003). One may speculate that this VWF self-association process, mediated by shear, provides a mechanism to regulate thrombosis. A critical observation in this regard is that circulating large and ultralarge VWF multimers can bind to flowing platelets and induce platelet activation and subsequent aggregation at shear stresses somewhat lower than in the case of smaller VWF multimers (Li et al., 2004). Several important inferences may be made from this behavior. First, the instantaneous availability of a larger number of sites for VWF binding to platelets, that is, the higher strength of adhesive bonds, may positively support platelet binding to VWF and eventual activation. Second, the larger size of these molecules may result in larger forces acting on the platelet GPIbα receptors once these VWF molecules are bound to them. Because it is the magnitude of the hydrodynamic forces acting on the GPIbα receptors that governs the GPIbα mechanotransduction activity that is responsible for cell activation (Shankaran et al., 2003), larger VWF multimers at high shear levels translate into greater platelet activity. It is interesting to note that the self-associated VWF molecules do not undergo any noticeable shape change to a more extended form as a function of shear (Shankaran et al., 2003).

Although ultralarge VWF multimers possess maximum binding activity, they are not usually detectable in the blood, except at the time of secretion from endothelial cells. This is due to an inherent mechanism of the body that regulates VWF multimer size to protect against unwanted shear-induced platelet aggregation (Mendolicchio and Ruggeri, 2005). Unusually large VWF multimers are secreted from Wiebel–Palade bodies of endothelial cells into the blood during injury and are cleaved into smaller fragments by a proteolytic mechanism involving a metalloproteinase called ADAMTS13 located at the endothelial surface, which cleaves a Tyr–Met bond within the VWF-A2 domain (Mendolicchio and Ruggeri, 2005). Secreted thrombospondin-1, originally present in platelet α granules, can also depolymerize VWF multimers by reducing the disulfide bridges between any two VWF subunits in the multimer (Ruggeri, 2003). Both thrombospondin and ADAMTS13 serve to regulate VWF size and therefore reduce or prevent thrombotic mishaps. Deficiency in ADAMTS13 activity can otherwise result in formation of abnormally large thrombi at the site of vascular injury or shear-induced platelet activation, which can create major problems especially in diseased arteries.

Crystal structure studies of the GPIbα N-terminal domain and the A1 domain of VWF have revealed some key insights into the physical mechanisms of shear-dependent activation (Uff et al., 2002). First, the activation appears to be triggered by conformational changes in the N-terminal region of the GPIbα receptor and not by modifications in the platelet cytoskeleton or as a result of clustering of the GPIbα receptors. More importantly, the GPIbα appears to exist in either an inactive or active state that is determined by the positioning of a certain R-loop. In the inactive state, this loop shields a high-affinity binding site for VWF-A1. An open conformation of the R-loop may result from the application of some critical magnitude of force that can convert this receptor to an active state, thereby enabling efficient binding with VWF to commence. More insight into the GPIbα–VWF binding mechanism may result from a comparison of the magnitude of forces that act on these bonds when they are formed at the wall as opposed to in the blood. This may explain why spontaneous binding of soluble VWF and platelets is not observed except at very high shear stresses. Shear forces thus appear to play the role of a key that unlocks a physical or energetic barrier to allow spontaneous GPIbα-VWF binding to occur. A full explanation of the shear threshold phenomenon would shed great light on the elusive biomechanical properties of this receptor–ligand pair.

D. Perfusion Chambers to Study Shear-Induced Platelet Interactions

To study and quantify the effects of flow and shear forces on cellular behavior, a number of flow devices have been designed to recreate *in vivo* flow conditions. Shear stress and flow rate can be accurately controlled and varied in these *in vitro* flow systems. Parallel-plate flow chambers are commonly used to simulate shear flow conditions in the arterioles and to observe their effects on platelet adhesive behavior. This experimental system provides a rectangular chamber for flow with a well-defined, fully developed parabolic flow profile, and the ability to functionalize the bottom surface with appropriate adhesion molecules. In a common configuration, the channel length, height, and width are determined by a flexible gasket that is attached to a circular block containing inlet and outlet connections, and this block is placed within a snug-fitting culture dish. The block and dish are brought together into airtight contact using vacuum. A computer-controlled syringe pump can be used to control flow through the chamber.

The instrument of choice for studying the effects of shear stress on cellular function in the bulk flow region, such as measuring shear-induced thrombus formation, is the rotational viscometer. This device allows researchers to investigate the direct effects of bulk shear stress on platelets by minimizing platelet–surface interactions, to mimic the environment in a stenosed artery with intact endothelium. Platelet–surface interactions are reduced to a minimum by coating the shear stress–generating surfaces with a nonthrombogenic lubricant such as silicone. A rotational viscometer can be of two types: the cone-and-plate viscometer and the Couette viscometer. The former consists of a stationary flat plate on which an inverted cone is placed such that the apex just contacts the plate. The angle between the conical and flat surfaces is kept very small, typically less than a degree. The platelet-rich plasma or platelet solution is sandwiched between these two surfaces, and the cone is rotated at some constant angular velocity to produce uniform shear stress in the fluid. Cone-and-plate viscometers can also be used to analyze platelet–surface interactions by coating the flat plate with appropriate molecules. The Couette viscometer comprises two coaxial cylinders, the outer of which can be rotated at a constant angular velocity. Flows in these rotational systems are steady-state laminar.

IV. COMPUTATIONAL MODELING OF PLATELET ADHESION AND AGGREGATION UNDER SHEAR FLOW

Modeling of spherical particles flowing near a flat wall or in a tube, and collisions of spheres in an unbounded or bounded fluid, has been a topic of great research interest over the past several decades. These fluid dynamic models have been adapted by biomedical researchers to study blood cell adhesive phenomena under flow by incorporating into the model the binding kinetics of the appropriate receptor–ligand pairs taking part in the cell–cell interactions. Because fluid mechanical solutions of spheres in a variety of bounding conditions and flow patterns are abundant, much computational modeling of cellular flows and adhesion has focused on single leukocyte (spherical cell) tethering to the endothelium (Hammer and Apte, 1992; Jadhav et al., 2005), cell–cell hydrodynamic effects on leukocyte tethering to the vascular wall (King and Hammer, 2001a, b; King et al., 2001), and spherical blood cell collisions and subsequent aggregation in shear flow (Tandon and Diamond, 1998; Long et al., 1999). Erythrocyte and platelet aggregation has also been modeled by approximating these cells as perfect spheres (Helmke et al., 1998; Tandon and Diamond, 1997). Activated platelets are somewhat spherical, although they have a rough surface with many filopodia extending in different directions. However, inactivated platelets are far from spherical, instead appearing as flattened ellipsoids. Such a vast difference in shape is expected to play a large influencing factor on the manner in which the cells collide, their frequency of collision, the duration of contact, the collision contact area, and the magnitude of shear and normal forces acting on the platelet(s)

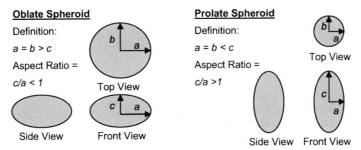

FIGURE 13.5 Diagram defining an oblate spheroid and a prolate spheroid and the differences between the two.

and on interplatelet bonds formed between two or more cells. Solutions for nonspherical particulate flows are much more difficult to obtain compared with that for suspended spheres, however, because of the difficulties in determining the hydrodynamic interactions between nonspherical particles or between a nonspherical particle and a bounding surface. Computational models of platelet adhesion to a surface under flow have been lacking until very recently because of the dearth of theoretical studies on nonspherical particle–wall interactions under flow. A few studies have looked at collisions between oblate spheroids (Yoon and Kim, 1990). However, none of these have been incorporated into platelet aggregation studies to date. Figure 13.5 visually depicts a spheroid showing convenient coordinate definitions. Theoretical platelet aggregation studies that have been conducted to date do not include the presence of a bounding wall of any sort. The first study to look at cell–cell (leukocyte) collisions and their effects on cell rolling on a flat bounding surface was done by King and Hammer (2001b). However, such studies for nonspherical particles are lacking in the literature.

A. Computational Modeling of Platelet Aggregation in Unbounded Shear

Theoretical studies of platelet aggregate formation are aimed at quantifying certain physical aspects of the system that are otherwise difficult to determine experimentally such as the hydrodynamic efficiency of contact, which includes collision frequency and contact time, forces of collision, and hydrodynamic forces acting on the bonds formed between the cells. These hydrodynamic variables are unknowns in the fluid dynamical equations that, along with the mathematical expressions describing the intrinsic properties of the particles, such as the reactivity of the surfaces, form the composite theoretical model. Certain theoretical predictions such as the overall aggregation efficiency are compared with experimental observations to validate the model. The mathematical model can then be used to draw insights into cellular behavior and related physical concepts for various scenarios of flow or varying properties of the cells and receptor–ligand pairs. This section gives an overview of the progress to date on the theoretical modeling of the formation of aggregates of activated platelets. In all of the previous studies, the fluid is unbounded and the platelets are assumed to be perfectly spherical. A number of mathematical models of platelet deposition or aggregation have been developed that either do not include the hydrodynamics of the flow and hydrodynamic interactions between particles (Kuharsky and Fogelson,

2001; Laurenzi and Diamond, 2002) or approximate platelets as infinitesimal points rather than finite-sized particles (Bluestein et al., 1999). These models are not discussed here because they neglect the key aspect of treating platelets as finite force–bearing surfaces.

Tandon and Diamond (1997) studied $\alpha_{IIb}\beta_3$-mediated aggregation of platelets crosslinked with fibrinogen. They studied the collisions of unequally sized smooth rigid spheres in linear shear flow, with each sphere representing either an individual platelet or a platelet aggregate, and included the effects of the hydrodynamic interactions between the particles on the trajectories of the spheres. The colliding particles were assumed to possess 100% activation of their $\alpha_{IIb}\beta_3$ receptors, and formed bonds if sufficient contact time was available for fibrinogen, bound to one $\alpha_{IIb}\beta_3$ receptor, to bind to the opposing cell–surface receptor on the second platelet. They observed that at low shear rates, a head-on collision provided sufficient time to allow bonds to form, but at higher shear rates $>750\ s^{-1}$, only collisions that involved one sphere rolling over the other (termed *secondary collisions*) were successful in providing the time required for aggregate formation. For a range of shear rates, they calculated the overall efficiency of aggregate formation, which includes the sum total effects of hydrodynamically predicted collision frequencies and the fraction of collisions that result in bond formation between the colliding cells.

Using the Stokes flow (Reynolds number $\ll 1$) equations as a starting point, Shankaran and Neelamegham (2004) computed the hydrodynamic forces that act on platelet doublets (two platelets joined by one or more tethers), a neutrophil–platelet doublet, a platelet GPIbα receptor, and VWF dimers, under conditions of constant linear shear flow. The doublets consisted of two unequally sized spheres separated by a distance equal to the tether length. This model produced some interesting results. They found that (1) the shear flow exerted much greater extensional and compressive forces (normal forces) as compared with shear forces on VWF molecules, and (2) the magnitude and nature of the hydrodynamic forces depended on, among other factors, the ratio of the radii of the two spheres constituting the doublet and the separation distance between the two spheres. For a radius ratio greater than 0.3, normal forces predominated, but when the ratio was less than 0.3, shear forces became more important. The authors speculate that in the latter case, pertaining to cells with globular surface receptors such as GPIbα, the shear forces could be responsible for triggering a conformation change in the surface receptors. They also hypothesized that in the case of cell–surface receptors that act as mechanotransducers, the longer their length, the greater the receptivity to shear forces and accompanying changes in functionality.

B. Computational Models of Platelet Tethering to Surface-Bound VWF

Mathematical models of one or more platelets flowing near a surface under linear shear flow, and undergoing reversible binding to a reactive surface, have been virtually nonexistent until recently. Very little has been done to date to model the adhesive behavior of an ellipsoidal-shaped cell near a surface because of the limited availability of fluid mechanical equations for flow of an ellipsoid in a bounded fluid. Such dynamic simulations of platelet adhesion would have immense utility in understanding and modeling healthy and pathological thrombotic phenomena. The few fluid mechanical solutions that do exist for ellipsoids are numerical in nature, as opposed to analytical, and become increasingly complex, because they involve several additional variables such as the shape (i.e., aspect ratio) of the particle and the angular orientation of the cell in all three directions of the Cartesian coordinate system at each point in time. The majority of the hydrodynamic solutions that are available are not adaptable to multiparticle flows, or only calculate forces and torques acting on the ellipsoid or spheroid but not the trajectories of the particle. In this section, we discuss our recent efforts that have overcome this technical barrier; these new models take into account the important geometrical aspects of the system, such as wall effects on flow, and the characteristics of platelet shape, to generate insights into platelet motion and tethering.

1. Two-Dimensional Studies of Platelet Adhesive Behavior under Flow

We developed a two-dimensional analytical model obtained from the solution of the creeping flow equations to characterize the flipping motion of a tethered platelet on a surface under linear shear flow (Mody et al., 2005) (Fig. 13.6).

This theoretical model accurately predicts the flipping trajectories of platelets as observed in *in vitro* flow experiments. The microscale motions of platelets on surface-coated VWF, as observed in flow chamber experiments, were precisely tracked using an image processing algorithm that models the platelets as rigid ellipsoids. These experimental trajectories compare very well with the theoretical predictions. The two-dimensional model can also be used to estimate the hydrodynamic forces acting on the tethered platelet, and therefore on the bonds formed between the platelet and the surface, as a function of orientation angle α of the platelet. The radial forces acting on the platelet are predicted to be compressive during the first half of the flip as defined by orientation angles $180° > \alpha > 90°$, and tensile during the second half, that is, for angles $90° > \alpha > 0°$ (Fig. 13.7).

Studies on spherical doublets (similar in shape to prolate spheroids) rotating and translating in linear shear have shown that doublets experience both normal (radial) and shear forces, with the normal forces also alternating between compressive and tensile in each successive quadrant of the orbit of the particle's major axis of revolution (Tees et al., 1993; Neelamegham et al., 1997). The compressive force promotes surface contact between the cells constituting the doublet and strengthens the bonds through additional bond formation, whereas the extensional forces favor bond rupture. Thus, the two-dimensional platelet flipping model demonstrates that platelets experience alternating compressive and tensile forces during successive flips over a reactive surface.

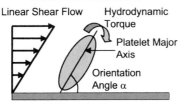

FIGURE 13.6 Schematic diagram illustrating the two-dimensional platelet flipping model for which theoretical equations based on Stokes equations were developed.

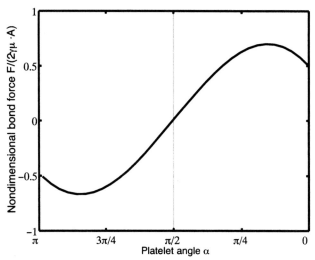

FIGURE 13.7 Plot of the nondimensionalized radial force acting on the platelet–surface bond as a function of platelet orientation angle α, as estimated by a two-dimensional flipping model. As can be observed, the force on the bond is compressive during the first half of the platelet's flip on a reactive surface (orientation angles: $\pi/2$–π), and is tensile during the second half of the flip (orientation angles: 0–$\pi/2$). γ is the shear rate, μ is the viscosity of the fluid, and A is the cross-sectional area of the platelet perpendicular to the plane of flow.

We found that the angular orientations at which flowing platelets attach to a surface in flow experiments are consistently >90°, while the detachment angles are in all cases <90°. The experimental observations of bond formation and dissociation show a striking correlation with the normal forces acting on the platelet bonds. These results suggest that the radial forces experienced by the platelet regulate the patterns of platelet binding to, and release from, the surface. Thus, our findings imply that compressive hydrodynamic forces play an important role in initiating platelet tether bond formation with the surface, by promoting intimate contact of the platelet surface with the bound VWF on the subendothelial wall and/or by providing a critical threshold compressive force that may be required for the bond association process (see also Chapter 2 for additional discussion of compressive force effects on bond formation). According to the model, the radial force acting on the platelet increases in a linear fashion with shear rate. Thus, the threshold shear rate of 73 s^{-1} is equivalent to a critical normal force of 1 pN. Whether a critical force is required for binding to occur remains to be determined.

2. Three-Dimensional Studies of Platelet Adhesive Behavior under Flow

Previous hydrodynamic studies of spheroidal particle motion near a wall in shear flow did not consider particle contact with the surface or the effect of surface contact on subsequent flow behavior. This prompted us to develop a fully three-dimensional computational model to study platelet motion very close to a planar surface, and to include the influence of the proximity of the wall and the consequences of surface contact on the cell's translational and rotational trajectories. This model was developed along lines similar to those along which the Multiparticle Adhesive Dynamics method was developed by King and Hammer (2001b), to study the effects of hydrodynamic interactions between flowing and bound leukocytes on the adhesive behavior of these cells. The three-dimensional platelet flow model approximates the platelet as an oblate spheroid 2 μm in diameter, with an aspect ratio (ratio of minor axis to major axis) of 0.25 (Frojmovic et al., 1990). The platelet surface is discretized into 96 elements, with each element comprising one node at the center and three nodes per edge, totaling 9 nodes per element and 386 nodes per surface. The simulation results derived from the model provide a number of insights into platelet flow behavior (Mody and King, 2005). It was found that spheroid motion in such a geometry is dramatically different from sphere motion. Platelets, unlike spherical cells, experience significant lateral motion toward or away from the surface, especially when flowing close to the surface, and also demonstrate three distinct regimes of motion based on the platelet's distance from the surface. These distinct regimes can be found to exist if the platelet axis of revolution has minimal tilt about the *x* axis (Fig. 13.8) that is, the flow is symmetric about the major axis of the platelet that lies in the plane of flow; for significant out-of-plane motions the behavior is more complex.

Platelet flow simulations were initiated with the platelet's major axis lying parallel to the surface. In flow regime 1, a platelet too far from the surface to make direct contact rotates periodically while translating in the direction of flow. Regime 2 flow exists when the platelet centroid starts at a distance within 1.1 and 0.75 platelet radii from the wall. The platelet makes contact with the surface at an oblique angle, flips over the surface, and then flows away. After an irreversible displacement away from the wall, the platelet lightly contacts the wall at regular intervals. During the flipping motion, the platelet contacts the surface at an angle α > 90° and remains in close contact until its angular orientation α becomes <90°. At this point, the platelet edge lifts off from the surface (Fig. 13.9).

A platelet flowing very close to the surface, with a centroid height less than 0.75 times the platelet major radius, exhibits regime 3 flow. Such a platelet does not undergo any form of periodic rotation and, in fact, "glides" over the surface with its angular orientation marginally oscillating about α = π. During regime 3 motion the platelet does not contact the surface. Even with small initial angular tilts about the *x* or *y* axis, the platelet continues to display regime 3 flow behavior with no rotation or surface contact. However, there is a distinct critical angle of tilt about the *x* axis and the *y* axis, above which the platelet will contact and flip over the surface, leading

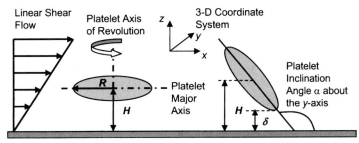

FIGURE 13.8 Schematic diagram of a platelet in two different orientations over a surface and subject to linear shear flow. The figure defines important variables in the three-dimensional platelet hydrodynamic flow calculations and the coordinate system used in this geometry. *H* is the distance of the platelet centroid from the surface, and δ is the closest distance of approach of the platelet to the surface.

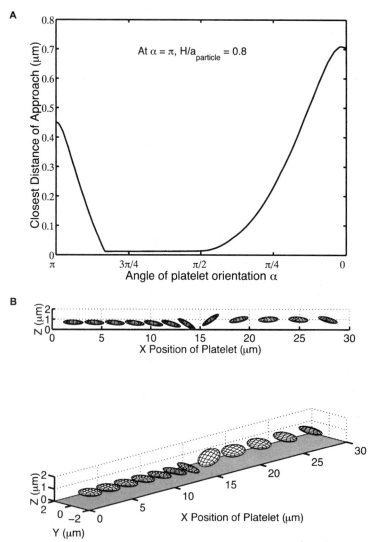

FIGURE 13.9   Platelet motion as predicted by the three-dimensional platelet flow simulations we conducted. (A) Plot of the closest distance of approach δ as a function of the orientation angle of a platelet ($a = b = 1$ μm, $c = 0.25$ μm) that begins flow in a horizontal orientation ($\alpha = \pi$) at a distance 0.8 μm from the surface. The platelet makes contact with the surface when oriented at $\alpha = 5\pi/6$ to the surface and ceases to make contact when $\alpha < \pi/2$. (B) Two- and three-dimensional plots of the motion of a platelet in linear shear flow initially located 0.8 μm from the surface and with an initial orientation of $\alpha = \pi$.

to a repeated contact–flip–flow–contact motion (i.e., regime 2 behavior). The magnitude of this critical angle of tilt depends on the initial platelet centroid height from the surface.

By combining the hydrodynamic calculations described by this three-dimensional flow model with equations governing the stochastic binding kinetics of the GPIbα–VWF-A1 bond, we can predict the platelet tethering behavior on surface-bound VWF. The adhesion model used in our simulations is based on the AD algorithm developed by Hammer and Apte (1992) and refined by King and Hammer (2001a). The 25,000 GPIbα surface receptors are equally distributed among the 386 nodes on the platelet surface, resulting in 65 receptors at each node. When the platelet nears the surface and is within binding range, bond formation between GPIbα and surface-bound VWF is tested. Bonds that form are represented as linear springs with fixed endpoints on either surface. The bond forces and torques acting on the cell are determined by the length and orientation of each of the bond springs that bind the cell to the surface. Both bond formation and dissociation rates are strongly dependent on the deviation bond length from equilibrium, i.e., on the physical state of compression and extension. The Bell model parameters for single-bond dissociation kinetics were obtained by Doggett et al. through flow experiments in which VWF-coated beads were perfused over surface-coated platelets and pause-time distributions were measured for varying shear stresses. An expression for the rate of formation is derived from the Boltzmann distribution for affinity (Bell et al., 1984).

Figure 13.10 depicts, for a single flipping event, the platelet adhesive behavior and hydrodynamic motion resulting from platelet–surface contact and subsequent formation and breakage of GPIbα–VWF bonds. The platelet simulation begins with the platelet close to the surface and at an angular orientation of $\alpha = 154°$ with respect to the y axis. As the platelet contacts and flips over the surface, tether bonds form and break. The platelet–surface bonds pull on the platelet as it rotates over the surface, slowing down its flipping motion. For the flipping event shown in Fig. 13.10, the time taken for the platelet to flip from angular orientation $\alpha = 154°$ to $\alpha = 30°$ is increased by more than 370% as compared with that for a corresponding platelet flipping event that is free of adhesive interactions with the wall. The tethered platelet then proceeds to one of two possible fates: as it rotates toward the surface on the downstream side ($\alpha < 90°$) it can either (1) detach from the surface or (2) flip over completely so that it rests parallel to the surface. Platelet detachment then occurs from the surface after some finite time at $\alpha \sim 0°$. Figure 13.10 shows the three-dimensional simulation results of a representative binding event, in which the platelet attains a more or less horizontal position over the surface after the flipping event and remains attached to the surface for a short duration before detaching and flowing away.

The AD simulation method is a novel technique that can be used to provide solutions to many of the unanswered questions discussed in this chapter. For example, this method can clarify (1) the effects of hydrodynamic flow on platelet–surface contacting frequency, (2) the percentage of surface contacts that result in a binding event, (3) the effects of hydrodynamic forces on the lifetime of the tether bonds and on the duration for which the platelet remains tethered to the surface, (4) the distribution of hydrodynamic forces on the platelet surface as well as on the individual tether bonds, and (5) the number of bonds formed, their respective spatial distributions on the platelet surface, and the magnitude of the bond forces and torques. The simulation can be adapted to consider multiple adhesive and nonadhesive interactions of spherical and nonspherical cells with one another and with the surface. With the inclusion of the hydrodynamic flow effects of other blood cells on the adhering cell, AD simulations can more realistically mimic the platelet tethering phenomena that occur *in vivo*. Additionally, by inclusion of cell–cell interactions in the form of hydrodynamic effects and multiple stochastic bond formations/breakages, the platelet adhesive dynamics (PAD) algorithm can be used to study platelet–platelet and platelet–leukocyte aggregation studies in

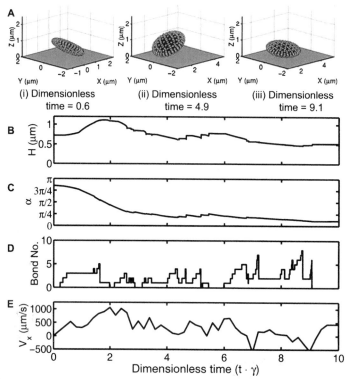

FIGURE 13.10 Simulation results for the adhesive behavior of a platelet and its hydrodynamic motion over a surface coated with VWF under linear shear flow. (A) Three-dimensional instantaneous plots of a tethered platelet flipping on a VWF-coated surface showing platelet orientation at three different times (i–iii) during a single flipping event. (B) Plot of the height H of the platelet centroid with respect to time, nondimensionalized with the shear rate. The platelet centroid height is highest when it is oriented at $\alpha = 90°$ to the surface, while maintaining surface contact. H is lowest when the platelet is oriented almost parallel to the surface, while remaining surface-bound. (C) Plot of the orientation angle α of the platelet about the y axis as a function of nondimensionalized time. As the platelet flips over the surface, the angle α decreases with time. (D) Plot of the number of GPIbα–VWF bonds between the platelet and the surface as a function of time. (E) Plot of the platelet translational velocity (V_x) in the x direction as a function of time. The negative x velocities depict instances when the platelet traverses in the y direction (transverse motion) due to the breakage of an off-center bond cluster. The unstressed bond association rate was set at 10 s^{-1}. The equilibrium bond length is 120 nm (Doggett et al., 2002). The shear rate γ was 1000 s^{-1}.

the vicinity of a boundary. In this way, PAD can be used to elucidate the role of hydrodynamic effects in platelet adhesion and aggregation phenomena.

V. CONCLUSIONS

Despite being the smallest of the blood cells, platelets represent a complex and robust system that possesses a myriad of receptor molecules and signaling pathways to ensure that successful hemostasis prevails over the wide range of physiological flow environments and varied distribution of subendothelial and plasma adhesive ligands in the vasculature. Of great biomedical interest are the mechanisms that control platelet adhesion and activation, that is, the primary events that precede the onset of fully developed thrombosis. One of the key factors in this process is fluid shear; this physical variable has overwhelming influence on the path the platelet takes to activate and aggregate when a thrombotic signal is present. The manner in which fluid shear regulates functioning of the GPIbα–vWF bond is an intriguing question that remains an active area of research. The key to resolving the underlying critical events that initiate thrombosis is understanding the nonlinear integrative effects of the flow environment, including the hydrodynamic effects of the bounding surface(s), nearby flowing cells, and type and rate of flow on the ability of platelets to tether to surfaces or form aggregates in the blood. Equally important is the quantification of these effects to understand their relative importance in different thrombotic situations. Much research is underway, first, to develop a range of predictive platelet adhesive simulations that not only corroborate experimental evidence but also provide a level of detail and understanding beyond what experiments can realistically achieve, and second and importantly, to uncover, through experiment, the remaining mysteries of the GPIbα–VWF and integrin–ligand bonds.

ACKNOWLEDGMENT

The authors gratefully acknowledge funding from The Whitaker Foundation.

SUGGESTED READING

Kroll, M. H., Hellums, J. D., McIntire, L. V., Schafer A. I., and Moake J. L. (1996) Platelets and shear stress. *Blood* **88**: 1525–1541.

This review article gives a detailed description of the basic principles of shear flow, the viscometers used to measure shear, and the specific effects of shear on platelet adhesion, aggregation, and signaling mechanisms. Also discussed are the effects of shear stress on endothelial cell behavior, including the synthesis and secretion of vasoactive substances and antithrombotic agents.

Mazzucato, M., Pradella, P., Cozzi, M. R., De Marco, L., and Ruggeri, Z. M. (2002) Sequential cytoplasmic calcium signals in a 2-stage platelet activation process induced by the glycoprotein Ibα receptor. *Blood* **100**: 2793–2800.

This article provides insight into this research group's findings on the signaling paths linking hydrodynamic stressing of the GPIba–VWF-A1 bond to activation of the $\alpha_{LIB}\beta_3$ receptor. Figure 7 of the above article provides a helpful visual depiction of the signaling mechanisms involved in this process.

Mendolicchio, G. L., and Ruggeri, Z. M. (2005) New perspectives in von Willebrand factor functions in hemostasis and thrombosis. *Semin. Hematol.* **42**: 5–14.

This excellent review article discusses the progress to date in our understanding of the structure and functions of von Willebrand factor and the role this multimeric protein plays in regulating hemostasis and arterial thrombosis.

Ross, R. (1999) Atherosclerosis: An inflammatory disease. *N. Engl. J. Med.* **340**: 115–126.

This review article provides a comprehensive overview of the role of inflammation in the origin and progression of atherosclerosis.

REFERENCES

Alevriadou, B. R., Moake, J. L.,Turner, N. A., Ruggeri, Z. M., Folie, B. J., Phillips, M. D., Schreiber, A. B., Hrinda, M. E., and McIntire, L. V. (1993) Real-time analysis of shear-dependent thrombus formation and its blockade by inhibitors of von Willebrand factor binding to platelets. *Blood* **81:** 1263–1276.

Bell, G. I. (1978) Models for the specific adhesion of cells to cells. *Science* **200:** 618–627.

Bell, G. I., Dembo, M., and Bongrand, P. (1984) Competition between non-specific repulsion and specific bonding. *Biophys. J.* **45:** 1051–1064.

Bluestein, D., Gutierrez, C., Londono, M., and Schoephoerster, R. T. (1999) Vortex shedding in steady flow through a model of an arterial stenosis and its relevance to mural platelet deposition. *Ann. Biomed. Eng.* **27:** 763–773.

Cauwenberghs, N., Vanhoorelbeke, K., Vauterin, S., and Deckmyn, H. (2000) Structural determinants within platelet glycoprotein Ibα involved in its binding to von Willebrand factor. *Platelets* **11:** 373–378.

Cotran, R. S., Kumar, V., and Robbins, S. L. (1994) *Robbins Pathologic Basis of Disease.* W. B. Saunders, Philadelphia.

Cruz, M. A., Diacovo, T. G., Emsley, J., Liddington, R., and Handin, R. I. (2000) Mapping the glycoprotein Ib-binding site in the von Willebrand factor A1 domain. *J. Biol. Chem.* **275:** 252–255.

Doggett, T. A., Girdhar, G., Lawshé, A., Schmidtke, D. W., Laurenzi, I. J., Diamond, S. L., and Diacovo, T. G. (2002) Selectin-like kinetics and biomechanics promote rapid platelet adhesion in flow: The GPIbalpha–vWF tether bond. *Biophys. J.* **83:** 94–205.

Doggett, T. A., Girdhar, G., Lawshe, A., Miller, J. L., Laurenzi, I. J., Diamond, S. L., and Diacovo, T. G. (2003) Alterations in the intrinsic properties of the GPIbalpha–VWF tether bond define the kinetics of the platelet-type von Willebrand disease mutation, Gly233Val. *Blood* **102:** 152–160.

Frenette, P. S., Johnson, R. C., Hynes, R. O., and Wagner, D. D. (1995) Platelets roll on stimulated endothelium *in vivo:* An interaction mediated by endothelial P-selectin. *Proc. Natl. Acad. Sci. USA* **92:** 7450–7454.

Frojmovic, M., Longmire, K., and Van de Ven. T. G. M. (1990) Long-range interactions in mammalian platelet aggregation: II. The role of platelet pseudopod number and length. *Biophys. J.* **58:** 309–318.

Goto, S., Salomin, D. R., Ikeda, Y., and Ruggeri, Z. M. (1995) Characterization of the unique mechanism mediating the shear-dependent binding of soluble von Willebrand factor to platelets. *J. Biol. Chem.* **270:** 2335–23361.

Grunkemeier, J. M., Tsai, W. B., McFarland, C. D., and Horbett. T. A. (2000) The effect of adsorbed fibrinogen, fibronectin, von Willebrand factor and vitronectin on the procoagulant state of adherent platelets. *Biomaterials* **21:** 2243–2252.

Hammer, D. A., and Apte, S. M. (1992) Simulation of cell rolling and adhesion on surfaces in shear flow: General results and analysis of selectin-mediated neutrophil adhesion. *Biophys. J.* **62:** 35–57.

Helmke, B. P., Sugihara-Seki, M., Skalak, R., and Schmid-Schönbein, G. W. (1998) A mechanism for erythrocyte-mediated elevation of apparent viscosity by leukocytes *in vivo* without adhesion to the endothelium. *Biorheology* **35:** 437–448.

Ikeda, Y., Handa, M., Kawano, K., Kamata, T., Murata, M., Araki, Y., Anbo, H., Kawai, Y., Watanabe, K., Itagaki, I., Sakai, K., and Ruggeri, Z. M. (1991) The role of von Willebrand factor and fibrinogen in platelet aggregation under varying shear stress. *J. Clin. Invest.* **87:** 1234–1240.

Ikeda, Y., Handa, M., Kamata, T., Kawano, K., Kawai, Y., Watanabe, K., Kawakami, K., Sakai, K., Fukuyama, M., Itagaki, I., et al. (1993) Transmembrane calcium influx associated with von Willebrand factor binding to GPIb in the initiation of shear-induced aggregation. *Thromb. Haemostas.* **69:** 496–502.

Jadhav, S., Eggleton C. D., and Konstantopoulos, K. (2005) A 3-D computational model predicts that cell deformation affects selectin-mediated leukocyte rolling. *Biophys. J.* **88:** 96–104.

King, M. R., and Hammer, D. A. (2001a) Multiparticle adhesive dynamics: Hydrodynamic recruitment of rolling leukocytes. *Proc. Natl. Acad. Sci. USA* **98:** 14919–14924.

King, M. R., and Hammer, D. A. (2001b) Multiparticle adhesive dynamics: Interactions between stably rolling cells. *Biophys. J.* **81:** 799–813.

King, M. R., Rodgers, S. D., and Hammer, D. A. (2001) Hydrodynamic collisions suppress fluctuations in the rolling velocity of adhesive blood cells. *Langmuir* **17:** 4139–4143.

Konstantopoulos, K., Chow, T. W., Turner, N. A., Hellums, J. D., and Moake, J. L. (1997) Shear stress–induced binding of von Willebrand factor to platelets. *Biorheology* **34:** 57–71.

Konstantopoulos, K., Kukreti, S., and McIntire, L. V. (1998) Biomechanics of cell interactions in shear fields. *Adv. Drug Delivery Rev.* **33:** 141–164.

Kroll, M. H., Hellums, J. D., Guo, Z., Durante, W., Razdan, K., Hrbolich, J. K., and Schafer, A. I. (1993) Protein kinase C is activated in platelets subjected to pathological shear stress. *J. Biol. Chem.* **268:** 3520–3524.

Kuharsky, A. L., and Fogelson, A. L. (2001) Surface-mediated control of blood coagulation: The role of binding site densities and platelet deposition. *Biophys. J.* **80:** 1050–1074.

Laurenzi, I. J., and Diamond, S. L. (2002) Bidisperse aggregation and gel formation via simultaneous convection and diffusion. *Ind. Eng. Chem. Res.* **41:** 413–420.

Li, Feng, Li, C. Q., Moake, J. L., López, J. A., and McIntire, L. V. (2004) Shear stress–induced binding of large and unusually large von Willebrand factor to human platelet glycoprotein Ibα. *Ann. Biomed. Eng.* **32:** 961–969.

Long, M., Goldsmith, H. L., Tees, D. F. J., and Zhu, C. (1999) Probabilistic modeling of shear-induced formation and breakage of doublets crosslinked by receptor–ligand bonds. *Biophys. J.* **76:** 1112–1128.

Mendolicchio, G. L., and Ruggeri, Z. M. (2005) New perspectives in von Willebrand factor functions in hemostasis and thrombosis. *Semin. Hematol.* **42:** 5–14.

Mody, N. A., and King, M. R. (2005). Three-dimensional simulations of a platelet-shaped spheroid near a wall in shear flow. *Phys. Fluids* (in press).

Mody, N. A., Lomakin, O., Doggett, T. A., Diacovo, T. G., and King, M. R. (2005) Mechanics of transient platelet adhesion to von Willebrand factor under flow. *Biophys. J.* **88:** 1432–1443.

Neelamegham, S., Taylor, A. D., Hellums, J. D., Dembo, M., Smith, C. W., and Simon, S. I. (1997) Modeling the reversible kinetics of neutrophil aggregation under hydrodynamic shear. *Biophys. J.* **72:** 1527–1540.

Novák, L., Deckmyn, H., Damjanovich, S., and Hársfalvi, J. (2002) Shear-dependent morphology of von Willebrand factor bound to immobilized collagen. *Blood* **99:** 2070–2076.

Oury, C., Sticker, E., Cornelissen, H., De Vos, R., Vermylen J., and Hoylaerts, M. F. (2004) ATP augments von Willebrand factor-dependent shear-induced platelet aggregation through Ca^{2+}-calmodulin and myosin light chain kinase activation. *J. Biol. Chem.* **279:** 26266–26273.

Ross, R. (1999) Atherosclerosis: An inflammatory disease. *N. Engl. J. Med.* **340:** 115–126.

Ross, J. M., McIntire, L. V., Moake, J. L., and Rand, J. H. (1995) Platelet adhesion and aggregation on human type VI collagen surfaces under physiological flow conditions. *Blood* **85:** 1826–1835.

Ruggeri, Z. M. (1993) Von Willebrand factor and fibrinogen. *Curr. Opin. Cell Biol.* **5:** 898–906.

Ruggeri, Z. M. (2003) Von Willebrand factor. *Opin. Hematol.* **10:** 142–149.

Ruggeri, Z. M., De Marco, L., Gatti, L., Bader, R., and Montgomery, R. R. (1983) Platelets have more than one binding site for von Willebrand factor. *J. Clin. Invest.* **72:** 1–12.

Ruggeri, Z. M., Dent, J. A., and Saldivar, E. (1999) Contribution of distinct adhesive interactions to platelet aggregation in flowing blood. *Blood* **94:** 172–178.

Savage, B, Almus-Jacobs, F, and Ruggeri, Z. M. (1998) Specific synergy of multiple substrate-receptor interactions in platelet thrombus formation under flow. *Cell* **94:** 657–666.

Savage, B., Saldivar, E., and Ruggeri, Z. M. (1996) Initiation of platelet adhesion by arrest onto fibrinogen or translocation on von Willebrand factor. *Cell* **84:** 289–297.

Savage, B., Shattil, S. J., and Ruggeri, Z. M. (1992) Modulation of platelet function through adhesion receptors. *J. Biol. Chem.* **267:** 11300–11306.

Savage, B., Sixma, J. J., and Ruggeri, Z. M. (2002) Functional self-association of von Willebrand factor during platelet adhesion under flow. *Proc. Natl. Acad. Sci. USA* **99:** 425–430.

Shankaran, H., Alexandridis, P., and Neelamegham, S. (2003) Aspects of hydrodynamic shear regulating shear-induced platelet activation and self-association of von Willebrand factor in suspension. *Blood* **101:** 2637–2645.

Shankaran, H., and Neelamegham, S. (2004) Hydrodynamic forces applied on intercellular bonds, soluble molecules, and cell-surface receptors. *Biophys. J.* **86:** 576–588.

Siedlecki, C. A., Lestini, B. J., Kottke-Marchant, K., Eppell, S. J., Wilson, D. L., and Marchant, R. E. (1996) Shear-dependent changes in the three-dimensional structure of human von Willebrand factor. *Blood* **88:** 2939–2950.

Tandon, P., and Diamond, S. L. (1997) Hydrodynamic effects and receptor interactions of platelets and their aggregates in linear shear flow. *Biophys. J.* **73:** 2819–2835.

Tandon, P., and Diamond, S. L. (1998) Kinetics of β_2-integrin and L-selectin bonding during neutrophil aggregation in shear flow. *Biophys. J.* **75**: 3163–3178.

Tees, D. F. J., Coenen, O., and Goldsmith, H. L. (1993) Interaction forces between red cells agglutinated by antibody: IV. Time and force dependence of breakup. *Biophys. J.* **65**: 1318–1334.

Tees, D. F. J., Waugh, R. E., and Hammer, D. A. (2001) A microcantilever device to assess the effect of force in the lifetime of selectin-carbohydrate bonds. *Biophys. J.* **80**: 668–682.

Uff, S., Clemetson, J. M., Harrison, T., Clemetson, K. J., and Emsley, J. (2002) Crystal structure of the platelet glycoprotein Iba N-terminal domain reveals an unmasking mechanism for receptor activation. *J. Biol. Chem.* **277**: 35657–35663.

Weiss, H. J., Hawiger, J., Ruggeri, Z. M., Turitto, V. T., Thiagarajan, P., and Hoffmann, T. (1989) Fibrinogen-independent platelet adhesion and thrombus formation on subendothelium mediated by glycoprotein Iib–IIIa complex at high shear rate. *J. Clin. Invest.* **83**: 288–297.

Yoon, B. J., and Kim, S. (1990) A boundary collocation method for the motion of two spheroids in stokes flow: Hydrodynamic and colloidal interactions. *Int. J. Multiphase Flow* **16**: 639–649.

INDEX